T0054695

SCIENCE *of* YOGA

UNDERSTAND THE ANATOMY AND PHYSIOLOGY TO PERFECT YOUR PRACTICE

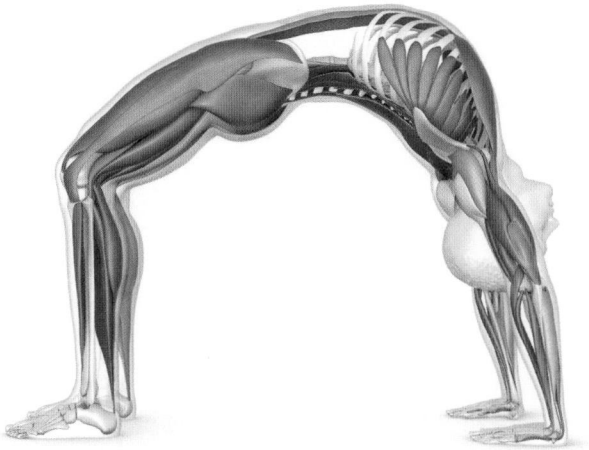

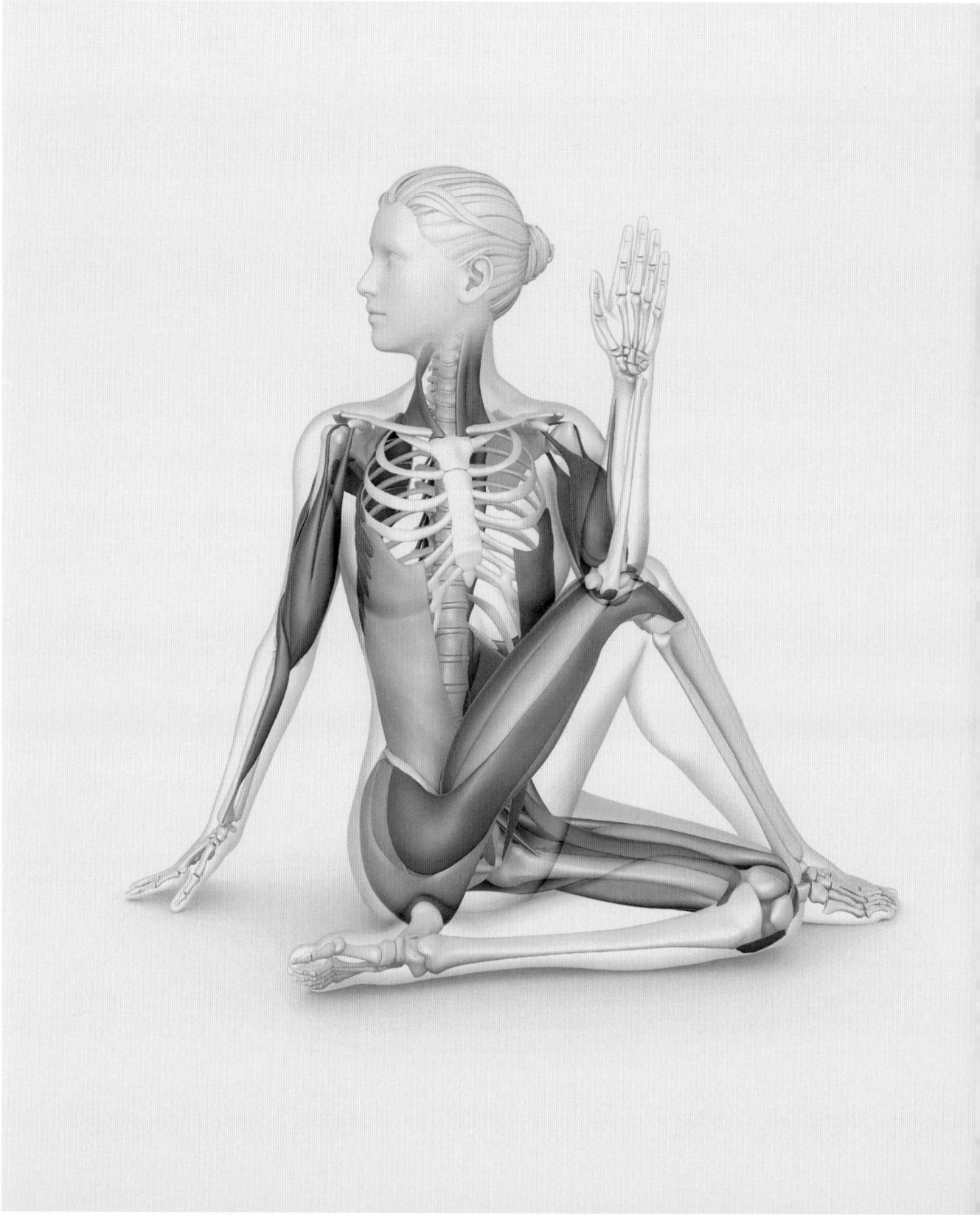

DK

SCIENCE *of* YOGA

UNDERSTAND THE ANATOMY AND PHYSIOLOGY TO PERFECT YOUR PRACTICE

Ann Swanson

Penguin Random House

Senior Editor **Ruth O'Rourke-Jones**
Senior Designer **Clare Joyce**
Editor **Alice Horne**
US Editor **Karyn Gerhard**
Senior Art Editor **Karen Constanti**
Design assistance **Philippa Nash, Louise Brigenshaw**
Editorial Assistant **Megan Lea**
Senior Jacket Creative **Nicola Powling**
Jacket Co-ordinator **Lucy Philpott**
Pre-production Producers **Heather Blagden, Tony Phipps**
Production Controller **Igrain Roberts**
Creative Technical Support **Sonia Charbonnier**
Managing Editor **Dawn Henderson**
Managing Art Editor **Marianne Markham**
Art Director **Maxine Pedliham**
Publishing Director **Mary-Clare Jerram**
Illustrations **Arran Lewis**

First American Edition, 2019
Published in the United States DK Publishing
1450 Broadway, Suite 801, New York, NY 10018

Copyright © 2019 Dorling Kindersley Limited
DK, a Division of Penguin Random House LLC
Text copyright © Ann Swanson 2019

21 22 23 16 15 14 13 12
043–310296–Jan/19

Health warning All participants in fitness activities must assume the responsibility for their own actions and safety. If you have any health problems or medical conditions, consult with your physician before undertaking any of the activities set out in this book. The information contained in this book cannot replace sound judgment and good decision making, which can help reduce risk of injury.

All rights reserved. No part of this publication may be reproduced, stored in or introduced into a retrieval system, or transmitted, in any form, or by any means (electronic, mechanical, photocopying, recording, or otherwise) without the prior written permission of the copyright owners. Published in Great Britain by Dorling Kindersley Limited

A catalog record for this book is available from the Library of Congress.
ISBN: 978-1-4654-7935-8

Printed and bound in China

For the curious
www.dk.com

MIX
Paper from responsible sources
FSC™ C018179
FSC www.fsc.org

This book was made with Forest Stewardship Council™ certified paper – one small step in DK's commitment to a sustainable future. For more information go to www.dk.com/our-green-pledge

CONTENTS

Preface **6**

HUMAN ANATOMY 8

Cell to system **10**
Skeletal system **12**
Muscular system **18**
Nervous system **22**
Endocrine system **28**
Respiratory system **30**
Cardiovascular system **34**
Lymphatic system **36**
Digestive system **38**
Urinary system **40**
Reproductive system **41**

THE ASANAS 42

SEATED ASANAS 44
Accomplished *Siddhasana* **46**
Bound Angle *Baddha Konasana* **50**
Cat *Marjaryasana* **54**
Cow *Bitilasana* **56**
Cow Face *Gomukhasana* **60**
Side Bend *Parivrtta Janu Sirsasana* **64**
Seated Twist *Ardha Matsyendrasana* **68**
Child's Pose *Balasana* **72**
Camel *Ustrasana* **76**
King Pigeon *Eka Pada Rajakapotasana* **80**

STANDING ASANAS 84
Mountain *Tadasana* **86**
Forward Fold *Uttanasana* **90**

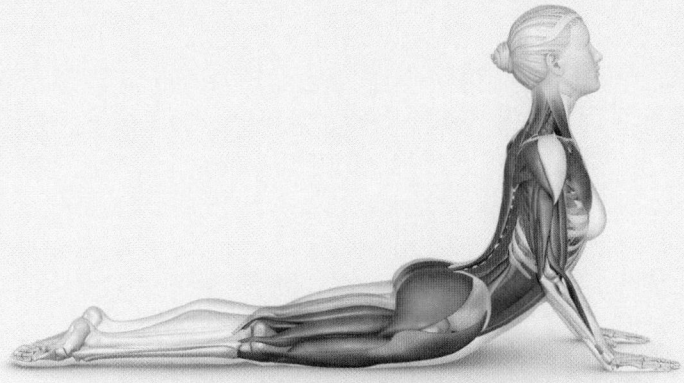

Chair *Utkatasana* **94**

Crescent Lunge *Anjaneyasana* **98**

Warrior II *Virabhadrasana II* **102**

Warrior III *Virabhadrasana III* **106**

Tree *Vrksasana* **110**

Dancer *Natarajasana* **114**

Triangle *Trikonasana* **118**

INVERSION ASANAS 122

Downward-Facing Dog *Adho Mukha Svanasana* **124**

Headstand *Sirsasana* **128**

Half Shoulderstand *Ardha Sarvangasana* **132**

Bridge *Setu Bandhasana* **136**

Wheel *Urdhva Dhanurasana* **140**

FLOOR ASANAS 144

Crow *Bakasana* **146**

Plank *Kumbhakasana* **150**

Side Plank *Vasisthasana* **154**

Cobra *Bhujangasana* **158**

Locust *Salabhasana* **162**

Supine Leg Stretch *Supta Padangusthasana* **166**

Supine Twist *Supta Matsyendrasana* **170**

Q&AS 174

Joints and flexibility **176**

Spinal care **178**

Life stages **180**

Meditation **184**

Savasana **186**

Stress **188**

The brain and mental well-being **192**

Chronic pain **194**

Yoga therapy **196**

Transformation **198**

On the frontiers of science **200**

Cautions **202**

Glossary **206**

Index **208**

Bibliography **214**

About the author and acknowledgments **216**

PREFACE

As the daughter of a NASA scientist I was raised to have an analytical mind. A part of me craves method, data, and evidence. I started journaling at the age of seven, carrying my notebooks everywhere. I filled them with charts, graphs, observations, and plans concerning everything from what I ate that day to what to rent at the video store.

I was a curious child, constantly asking, "Why?" My parents would send me to the trusty encyclopedia to look up the answer.

At the same time, I have always been artistic, creative, and interested in spirituality. My notebooks are also filled with elaborate stories, poetry, and colorful drawings.

My undergraduate studies in art led to burnout. Like many people, I came to yoga hoping to relieve stress and anxiety during a difficult time—with the added bonus of staying fit. I didn't expect that yoga would transform me in an ineffable, seemingly magical way.

When I started practicing, I aimed to make the picture-perfect poses. I slowly realized that yoga isn't about performing the pose "perfectly," but instead about being perfectly okay with my body and mind in the moment. Now I know that many of the most profound effects of poses transcend my anatomy of muscles and bones to shape my neurology, psychology, and energetic body. I vividly remember lying on my mat at the end of a yoga class with my eyes wide open, looking impatiently around when I was supposed to be relaxing. I thought "What a waste of time; I have work to do!" With practice, I started to enjoy the way relaxation and meditation practices made me feel.

Now, through reading research, I know that when I meditate, I am literally reshaping my brain. Ultimately, I am impacting every single system of my body, and optimizing function. What more important work could I possibly do?

My shifting mindset drew me to the Himalayas to study yoga, massage, and healing arts. My teacher, Yogi Sivadas, renewed my interest in science. I returned to the US and completed the pre-medicine courses, in pursuit of understanding how and why yoga works in such life-changing ways.

I will never forget the first time I held a human brain in the cadaver lab. The experience was neither antiseptic nor clinical, but deeply spiritual. That three-pound folded gray mysterious mass once both computed mathematics and felt the depths of love. Holding that brain, I knew that the mind-body connection was a key mechanism behind yoga's benefits.

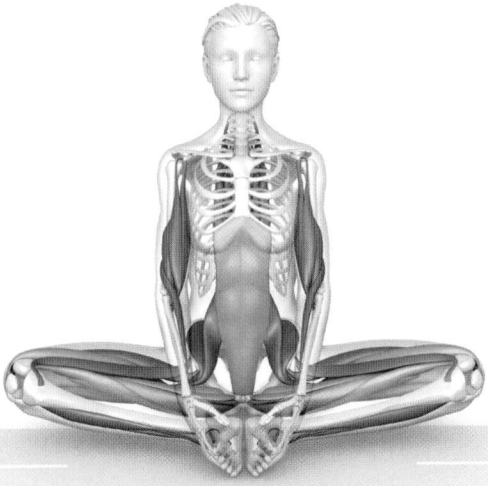

> ***Scientific** principles and evidence have **demystified** so much of the practice*

Science of Yoga is the book I wanted to read when I first started practicing yoga. In classes, teachers offer (sometimes conflicting) cues and claims—"Calm your nervous system by elongating your exhales;" "This pose will boost your immunity;" "Align your knee over your ankle"—and I constantly wondered, "Why?"

For the past decade, through workshops, reading research papers, and completing my Master of Science in yoga therapy at Maryland University of Integrative Health, I have continued to fill my notebooks with facts, figures, sketches, and stories. *Science of Yoga* summarizes the notes I found most fascinating as a yoga student and teacher. This book is intended as neither a comprehensive text on human anatomy and yoga, nor a medical reference book; it is just the beginning. My intention is for this material to spark more curiosity and discussion about the science of yoga, and lead to more inspired yoga practitioners and professionals, more rigorous research, more public policies that encourage yoga in schools and healthcare, and, ultimately, more accessibility and acceptance.

Through my research, scientific principles and evidence have demystified so much of the practice. Surprisingly, this made my transformative experiences feel even more magical. There is just so much more to discover. In the grand scheme of scientific inquiry, yoga research is in its infancy. However, now is an exciting and pivotal point in the field, with a remarkable increase in the quality and quantity of yoga research papers in the past few decades; the evidence supporting yoga's benefits continues to grow rapidly.

Science can explain the hows and whys of many things, but research studies, no matter how rigorously conducted, cannot compare to your personal, experiential evidence of healing and transformation. Only you can harness the power of yoga through practice. As with any scientific inquiry, I hope this book leaves you with more questions than answers, and brings out your inner child to playfully enquire, "Why?"

Be well,

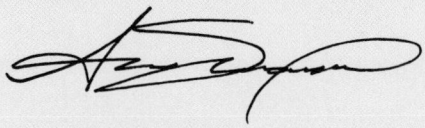

Ann Swanson, MS
Mind-body science educator and certified yoga therapist
Go to www.scienceof.yoga for valuable resources

HUMAN ANATOMY

Most yoga anatomy books and courses focus on the musculoskeletal system, but research shows that practicing yoga affects all body systems. This section breaks down the key effects and benefits for each one. Study your anatomical systems as modern biology defines them—then, challenge yourself to shift to a yogic perspective, that of unity. Experience your extraordinary body as an interconnected whole.

CELL TO SYSTEM

As in design, a key concept in biology is "form follows function"—this means that the physical structures of your body reflect their specific tasks. Anatomy is the study of these body structures and physiology is the study of their functions, or how your body works.

Telomeres

Telomeres are like caps on the tips of chromosomes. With aging, telomeres tend to shorten. Studies on the cutting edge of molecular biology have shown that a yogic lifestyle (including exercise, yoga practice with meditation, social support, and a plant-based diet) seems to increase telomere length, which may have an impact on increased longevity and health.

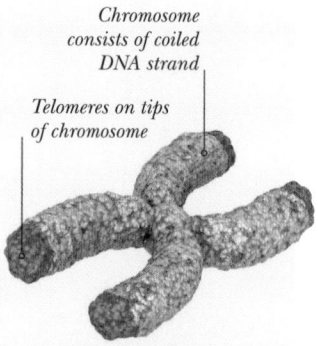

CHROMOSOME

BUILDING BLOCKS

Atoms are the building blocks of matter; cells are the building blocks of biological life. Approximately 37 trillion body cells are vibrating in your body right now. They create four basic tissue types and 11 organ systems. All of these parts and pieces create an integrated whole called the human body.

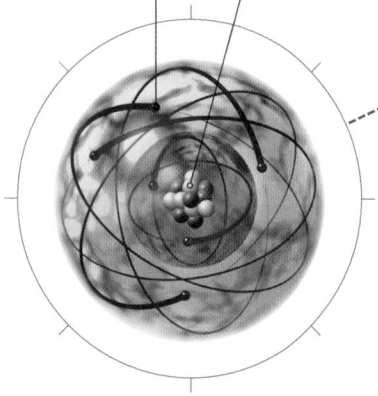

Atom

These chemical building blocks contain protons, neutrons, and electrons. They bond together to make important molecules, such as water (H_2O).

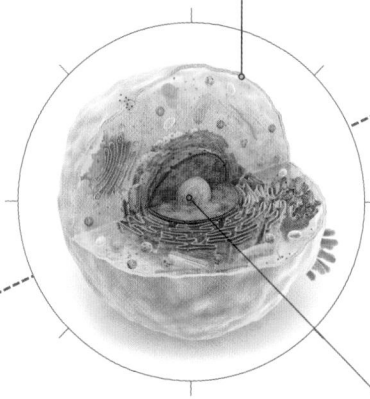

Cell

Cells are the smallest unit of life. Most cells contain a nucleus in the center, cytoplasm, and an outer layer called the cell membrane. Small functional units inside the cell are called organelles.

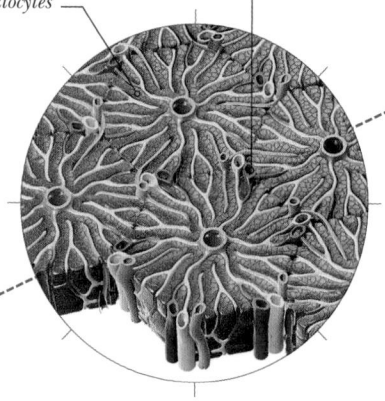

Tissue

Cells come together to form tissues, which are like unique fabrics. This specifically shaped tissue is located in the liver.

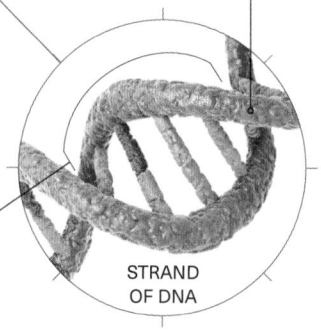

STRAND OF DNA

Organ

Tissues come together to form organs, like your liver (shown below). This large organ receives blood from all over your body for processing and purification. It also makes bile, which is used to break down fats in the digestive process.

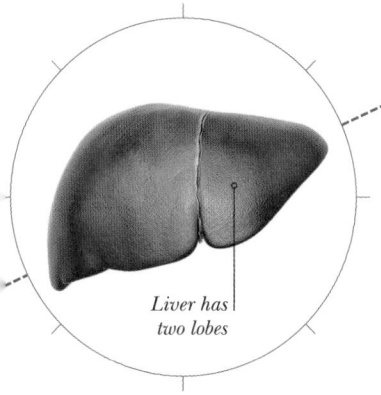

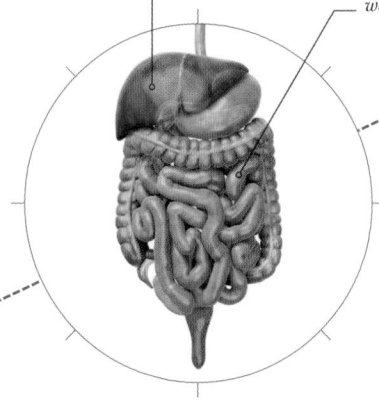

System

Organs come together to form organ systems, including: integumentary, skeletal, muscular, nervous, endocrine, respiratory, cardiovascular, lymphatic, digestive (shown above), urinary, and reproductive.

Integumentary system

The integumentary system includes hair, nails, skin and associated structures like sweat glands. Some claim that hot yoga causes you to "sweat out toxins." However, your liver is responsible for such detoxification processes. What you are actually sweating out is water, leading to dehydration. If you sweat a lot or practice hot yoga, make sure you drink plenty of water to replenish your losses.

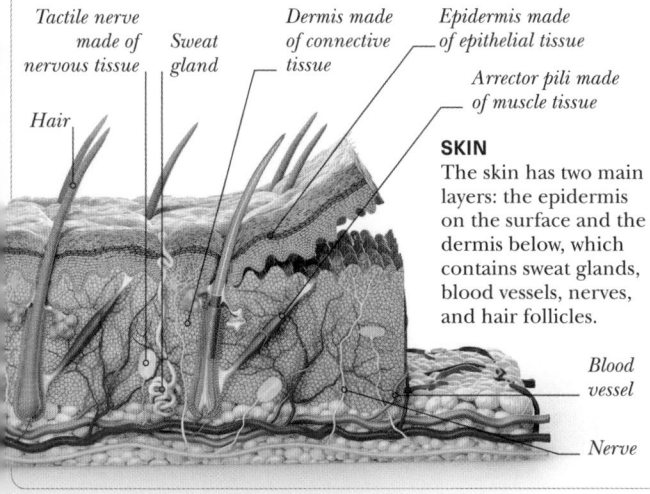

SKIN

The skin has two main layers: the epidermis on the surface and the dermis below, which contains sweat glands, blood vessels, nerves, and hair follicles.

Human body

Organ systems come together to form an organism. As a human being, you are made up of all of this, functioning as a dynamic, living whole.

SKELETAL SYSTEM

The 206 bones that make up your skeleton are dynamic, living organs. Together they form a framework for your body that provides structure and protection, and has the ability to move.

SYSTEM OVERVIEW

Your bones are made of collagen and they store calcium, a mineral that makes them strong and is vital for bodily functions. They also contain bone marrow where blood cells are produced. Bones form joints, which are supported by cartilage and structures such as ligaments. Yoga can support your bone and joint health.

Skull
These fused plates of bone protect your brain

Mandible
Lower jawbone that forms the only movable joint in your skull

Clavicle
Also called the collarbone, it connects your scapulae and sternum

Sternum
Also called the breastbone, it connects your ribs

Ribs
The 12 pairs of bones that form your ribcage

Pelvis
Two hip bones connected by your sacrum

Carpals
Eight small bones form each wrist

Metacarpals
Five long bones run through each palm

Phalanges
Each hand has 14 bones forming your fingers

Patella
Also called the kneecap, it is attached to your quadriceps tendon

Tarsals
The seven small bones that form your ankle

Metatarsals
Five long bones that run through your foot

Phalanges
The 14 bones in each foot that form your toes

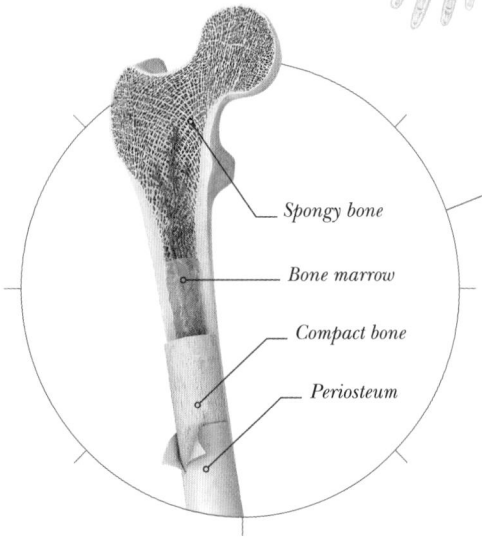

Bone structure

Bone has a smooth outer connective tissue shell called periosteum. Inside this is a strong, dense layer known as compact bone. Deeper still is spongy bone with honeycomblike spaces; this is strong yet light.

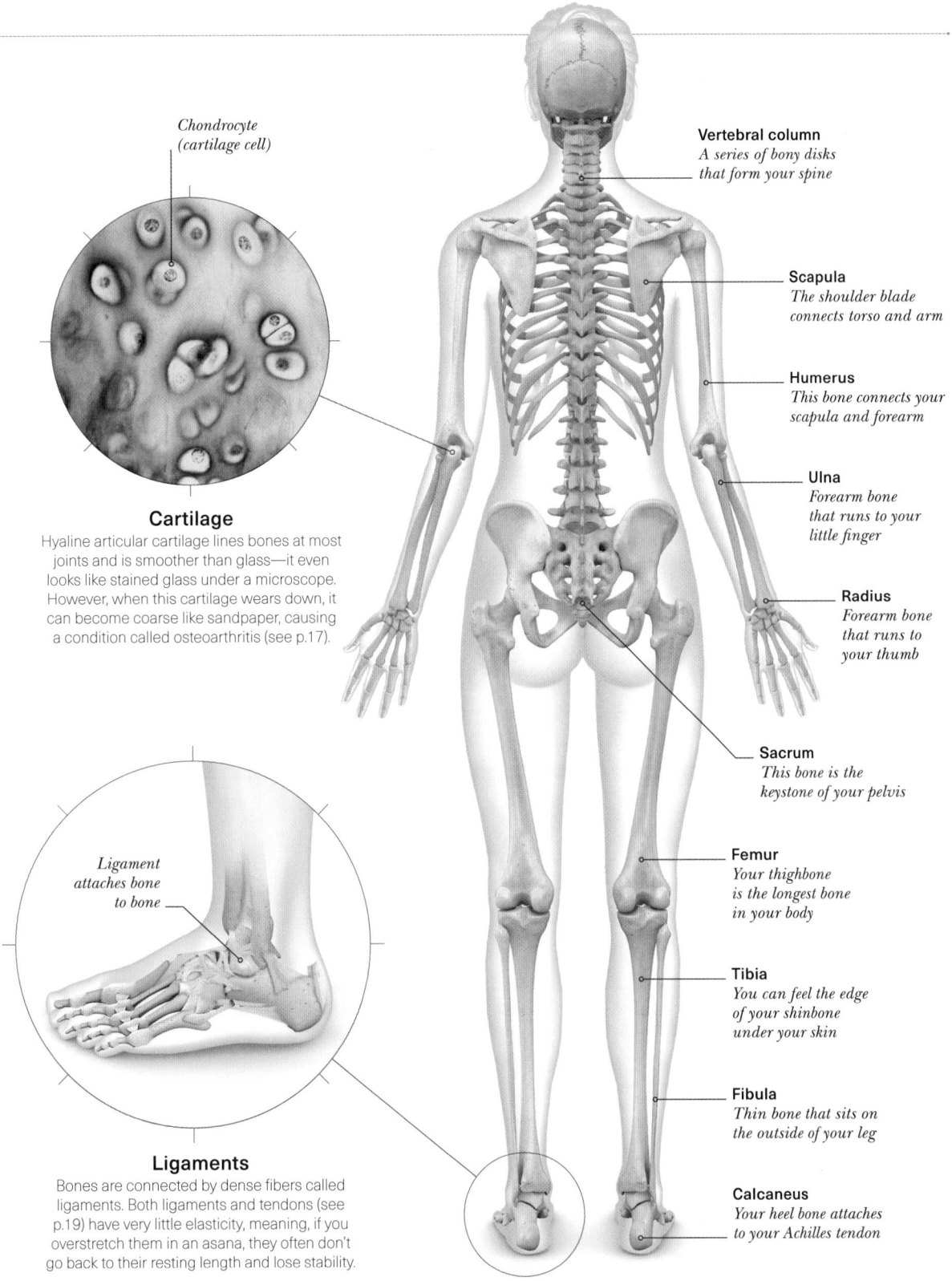

Cartilage

Hyaline articular cartilage lines bones at most joints and is smoother than glass—it even looks like stained glass under a microscope. However, when this cartilage wears down, it can become coarse like sandpaper, causing a condition called osteoarthritis (see p.17).

Ligaments

Bones are connected by dense fibers called ligaments. Both ligaments and tendons (see p.19) have very little elasticity, meaning, if you overstretch them in an asana, they often don't go back to their resting length and lose stability.

SPINE

Your vertebrae sit on top of each other to create natural curves. This is called a "neutral spine." It alternates between curving inward (lordosis) and outward (kyphosis) to absorb shock like a coiled spring. Your vertebrae are like wedges stacked to form these curves in order to bear your body weight most efficiently.

Cervical curvature
There is a natural lordosis in the neck

Thoracic curvature
There is a natural kyphosis in the mid-back

Lumbar curvature
There is a natural lordosis in the lower back

Hole for spinal cord

Space for vertebral artery

CERVICAL VERTEBRA

Articulates with rib

Body of vertebra

THORACIC VERTEBRA

Articular process

Enlarged body to support weight

LUMBAR VERTEBRA

Neutral spine

Many asanas incorporate a neutral spine, such as seated meditation poses. Poor posture and other considerations can lead to a multitude of spinal structural deviations, including common conditions like hyperlordosis and hyperkyphosis. Yoga works your spine in unique ways and enhances body awareness to improve your overall posture.

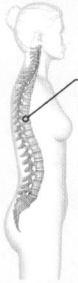

Gentle, even curves

NEUTRAL SPINE
These natural curves create the strongest, most stable alignment of the spine. In this ideal, the spine is also not twisted or leaning to either side.

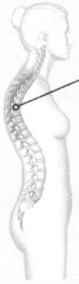

Curvature in upper spinal column

KYPHOSIS
Hyperkyphosis of the thoracic spine is often simply called a kyphosis or hunchback. This exaggerated curvature is common in osteoporosis.

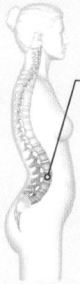

Curvature in lower spinal column

LORDOSIS
Hyperlordosis of the lumbar spine is sometimes just called a lordosis or swayback. This exaggerated curvature is natural during pregnancy.

PELVIS

Your pelvis includes two hip (coxal) bones connected by your sacrum. The sacrum, which means "sacred" in Latin, is the triangular bone with the tailbone at the lower, or inferior, end; it acts like the keystone to an arched bridge, forming a structurally sound base for your spine.

Anterior superior iliac spine
Your "hip points" can be felt under your skin

Sacroiliac joint
Commonly called the SI joint, this is slightly movable

Acetabulum
The socket of your hip joint which articulates with your femur

Greater sciatic notch
This creates a space for the sciatic nerve to pass

Coccyx
These fused bones are known as the tailbone

Ischial tuberosity
Your "sitting bones" are at the base of your pelvis

Pubic symphysis
This joint is made of fibrocartilage like your intervertebral disks

FEMALE PELVIS

Neutral pelvis

A neutral pelvis facilitates a neutral spine and vice versa. Imagine your pelvic bowl filled with water. Finding a neutral spine and pelvis means that the water wouldn't spill backward, forward, or to the side—such as when one of your hip points is lifted or your pelvis is rotated.

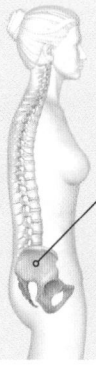

Pelvic bowl tilts backward, flattening lumbar curve

POSTERIOR TILT

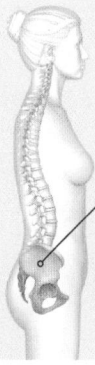

Pelvic bowl is balanced, with neutral lumbar curve

NEUTRAL

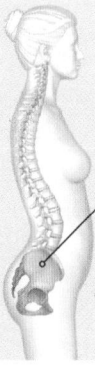

Pelvic bowl tilts forward, exaggerating lumbar curve

ANTERIOR TILT

JOINTS

Joints are where bones unite and articulate to allow movement. There are three joint types: fibrous, cartilaginous, and synovial. Fibrous joints are immobile, such as the sutures in your skull. Cartilaginous joints are slightly mobile, like your pubic symphysis. Synovial joints are most mobile and are very important for asanas.

JOINT ACTIONS

Synovial joints of your body can move in many directions. Hinge joints in your elbow and knee mainly perform flexion and extension, like the hinge of a door. Larger ball and socket joints like in your shoulder and hip can also perform abduction, adduction, rotation, and circumduction, which is a combination of all of the above movements.

TYPES OF MOVEMENT	
Flexion	Angle at joint generally gets smaller
Extension	Angle at joint generally gets larger
Abduction	A limb moves away from the body
Adduction	A limb moves closer toward the body
External rotation	A limb rotates outward
Internal rotation	A limb rotates inward
Axial rotation	The spine twists on its axis
Plantar flexion	Pointing the feet
Dorsiflexion	Flexing the feet

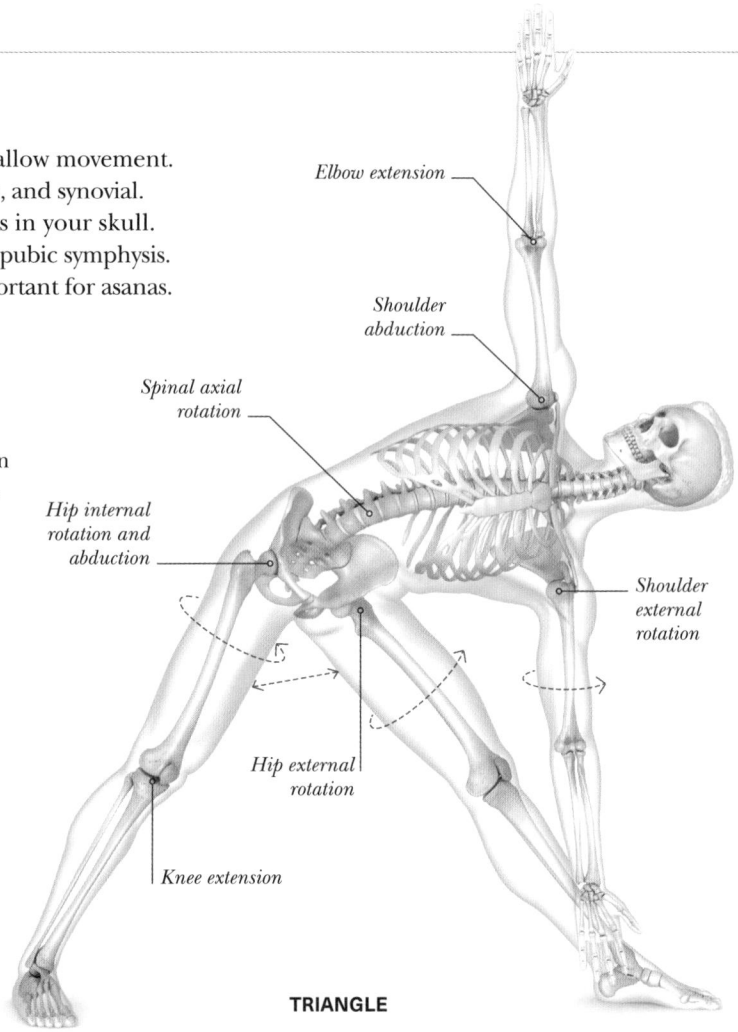

TRIANGLE

Inside a joint

Synovial fluid lubricates and cushions. It is a "non-Newtonian fluid," which means it gets more viscous or thicker in response to pressure, similar to solutions of cornstarch in water. With a sedentary lifestyle, synovial fluid may become thin and less effective. However, impact from the practice of yoga asanas causes synovial fluid to thicken, reducing pain and better protecting joint structures such as cartilage.

SYNOVIAL JOINT

Synovial joints allow movement while protecting bone ends from touching each other, which would cause damage. They are the most common type of joint in the body.

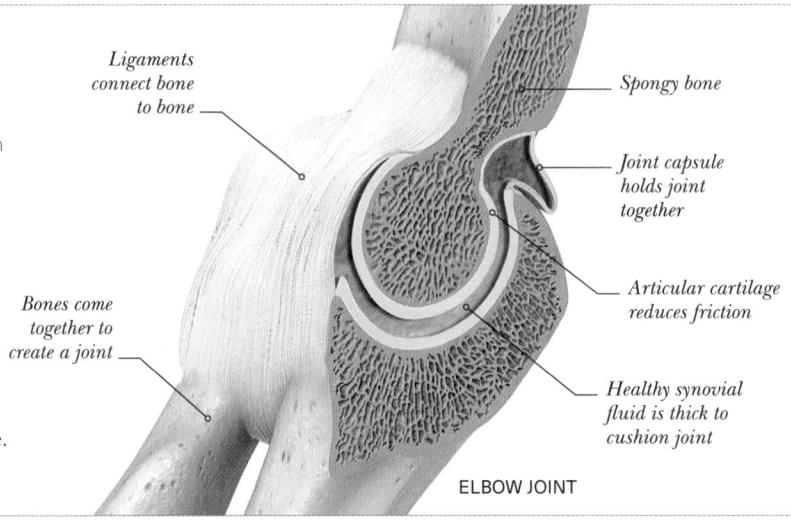

ELBOW JOINT

THE POSES
Yoga asanas move joints in all directions. Visualize or try doing these poses to experience the joint actions in your body. Imagine or feel each joint action internally.

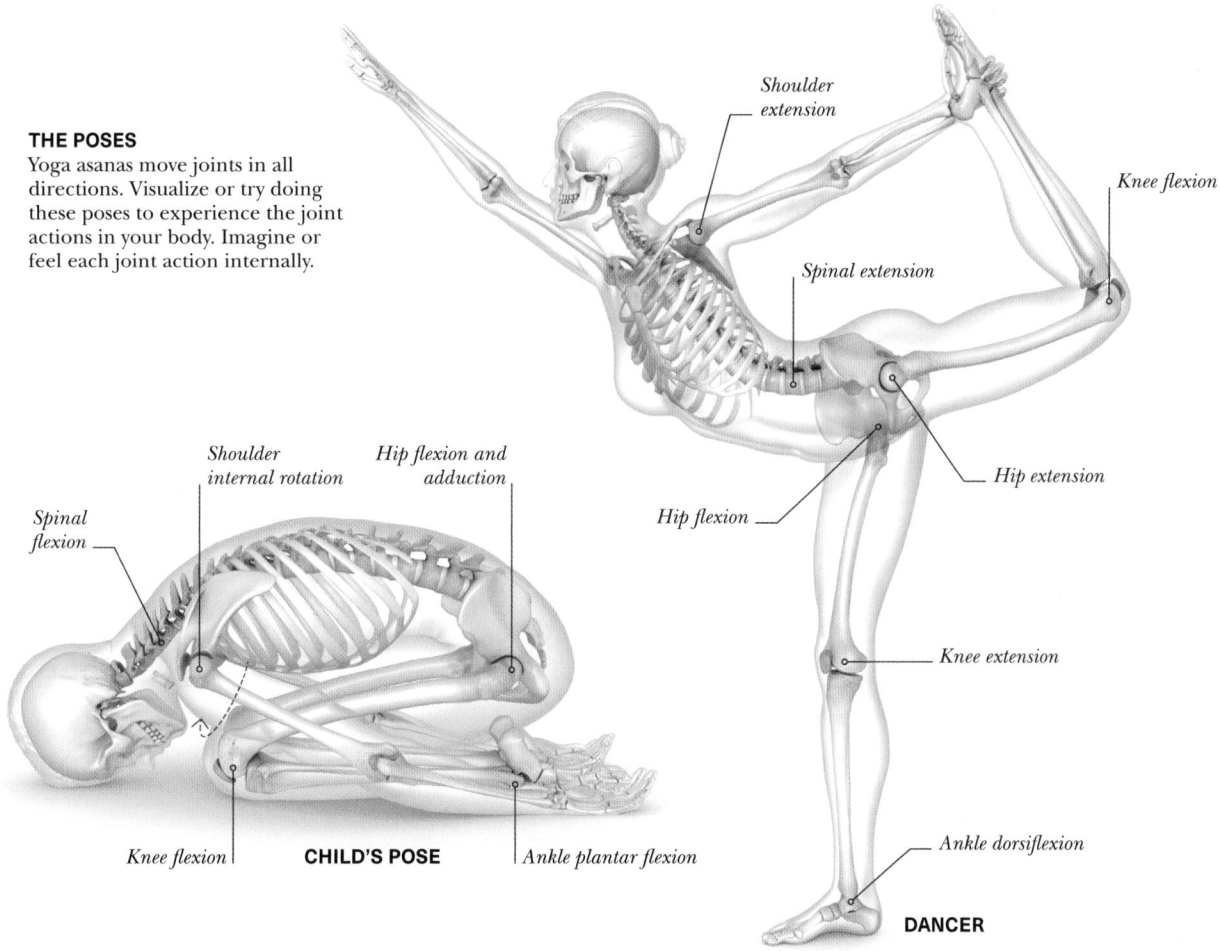

Arthritis

Wear and tear on joints can lead to osteoarthritis. In a 7-year clinical trial, researchers found that yoga is safe and effective in managing both osteoarthritis and rheumatoid arthritis (see p.37). After an 8-week yoga class, participants showed a reduction in pain by 25 percent, along with statistically significant improvements in physical fitness and quality of life.

PROGRESSION
As cartilage degrades there is less space in the joint, leading to inflammation and pain. Bone spurs or osteophytes can form as the condition progresses.

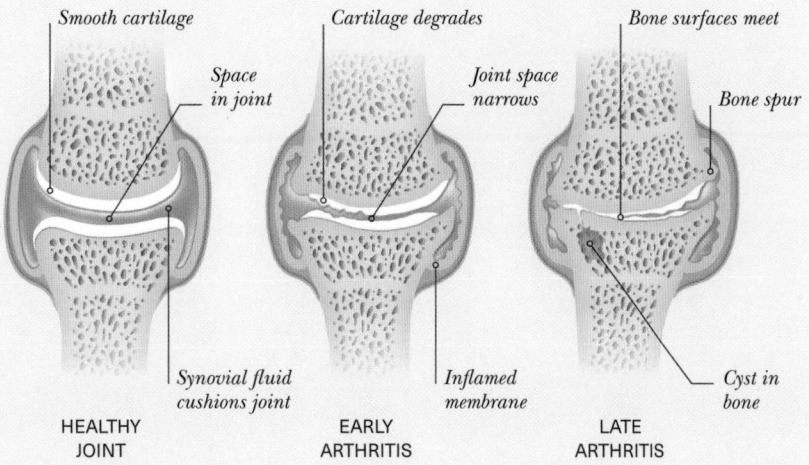

MUSCULAR SYSTEM

There are about 640 muscles in your body. Your skeletal muscles are attached to your bones, allowing you to move. Some muscles are superficial (close to the surface) and others are deep.

SYSTEM OVERVIEW

As you study each of these key chosen muscles, try to palpate or physically touch them while visualizing their internal location. This will help you learn better, while improving your mind-body connection. Most of the muscles here are categorized into groups based on their actions.

Pectorals
Pectoralis major
Pectoralis minor

Intercostal muscles

Brachialis

Abdominals
Rectus abdominis
External abdominal obliques
Internal abdominal obliques (deep, not shown)

Transversus abdominis

Hip flexors
Iliopsoas (iliacus and psoas major
Rectus femoris (see quadriceps)
Sartorius
Adductors (see below)

Elbow flexors
Biceps brachii
Brachialis (deep)
Brachioradialis

Adductors
Adductor longus
Adductor brevis
Adductor magnus
Pectineus
Gracilis

Quadriceps
Rectus femoris
Vastus medialis
Vastus lateralis
Vastus intermedius (deep, not shown)

Ankle dorsiflexors
Tibialis anterior
Extensor digitorum longus
Extensor hallucis longus

SUPERFICIAL **DEEP**

Muscle fibers are arranged in parallel orientation

Striations are visible stripes from internal structures (see p.21)

Skeletal muscle

There are three types of muscle tissue: cardiac, smooth, and skeletal. We will focus on skeletal muscle as it is responsible for the movement of joints in asana. This is what it looks like under a microscope.

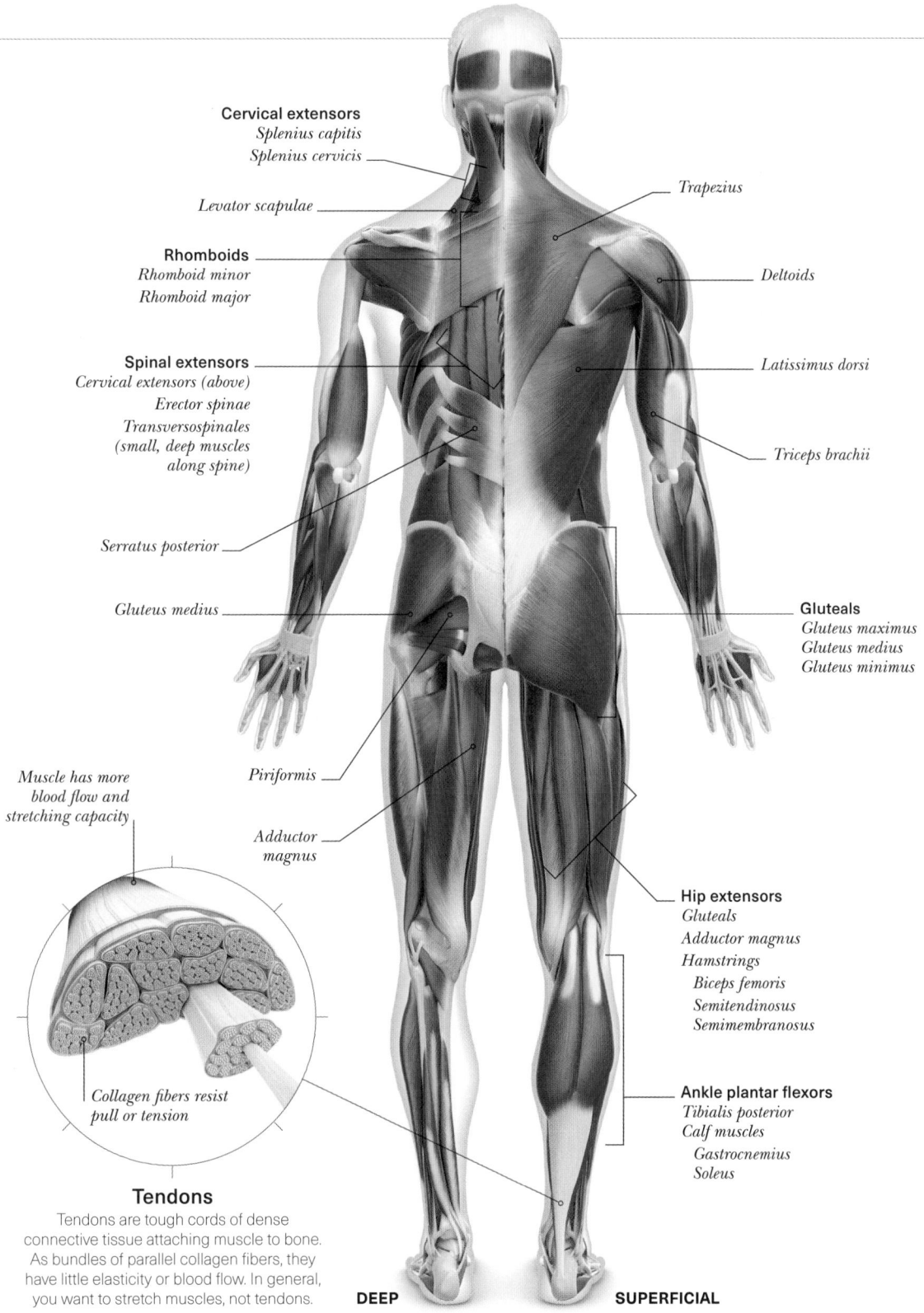

Tendons

Tendons are tough cords of dense connective tissue attaching muscle to bone. As bundles of parallel collagen fibers, they have little elasticity or blood flow. In general, you want to stretch muscles, not tendons.

MUSCLE STRUCTURE

Skeletal muscles are bundles of bundles of bundles of parallel muscle cells, blood vessels, and nerves wrapped with connective tissue, including fascia. Fascia creates a network through and around muscles and other structures of your body. Microscopic proteins in your muscles cause muscle contractions.

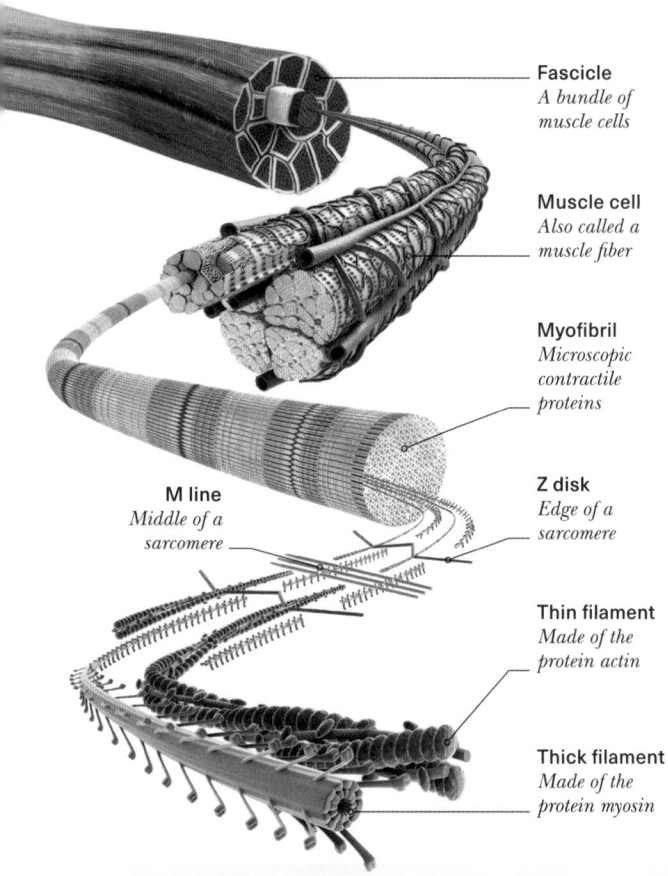

Fascia

Fascia is similar to the white pith of an orange; it separates parts yet integrates the whole. Fascia is not just found around muscles. It is also around vital organs and woven throughout your body. This body-wide network of fascia is part of the reason why a yoga pose that affects your foot can suddenly release your tight shoulders.

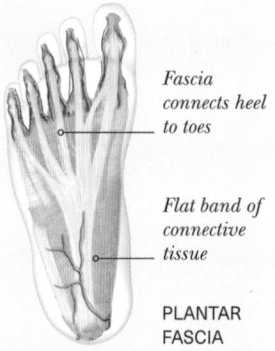

HOW MUSCLES WORK

Muscles often work in antagonistic pairs. As the agonist muscle engages, the antagonist generally releases. Synergist muscles engage around the joint to support the action.

TYPES OF CONTRACTION

Isotonic contractions involve a change in muscle length, as in the act of flexing or extending your elbow (see below) or transitioning in or out of an asana. Isometric contractions involve tension with no change in muscle length, such as when holding an asana.

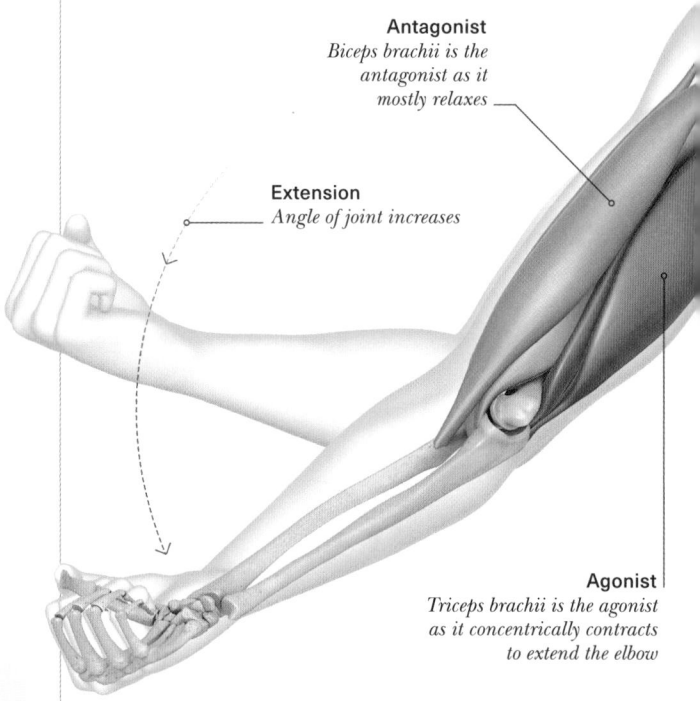

ECCENTRIC CONTRACTION

Eccentric contractions occur when muscle fibers "lengthen" to change the angle of a joint. This occurs in your biceps when extending your elbow as you lower a weight, or in your hamstrings when extending your knee as you transition from Warrior II to Triangle (see pp.118–21) pose.

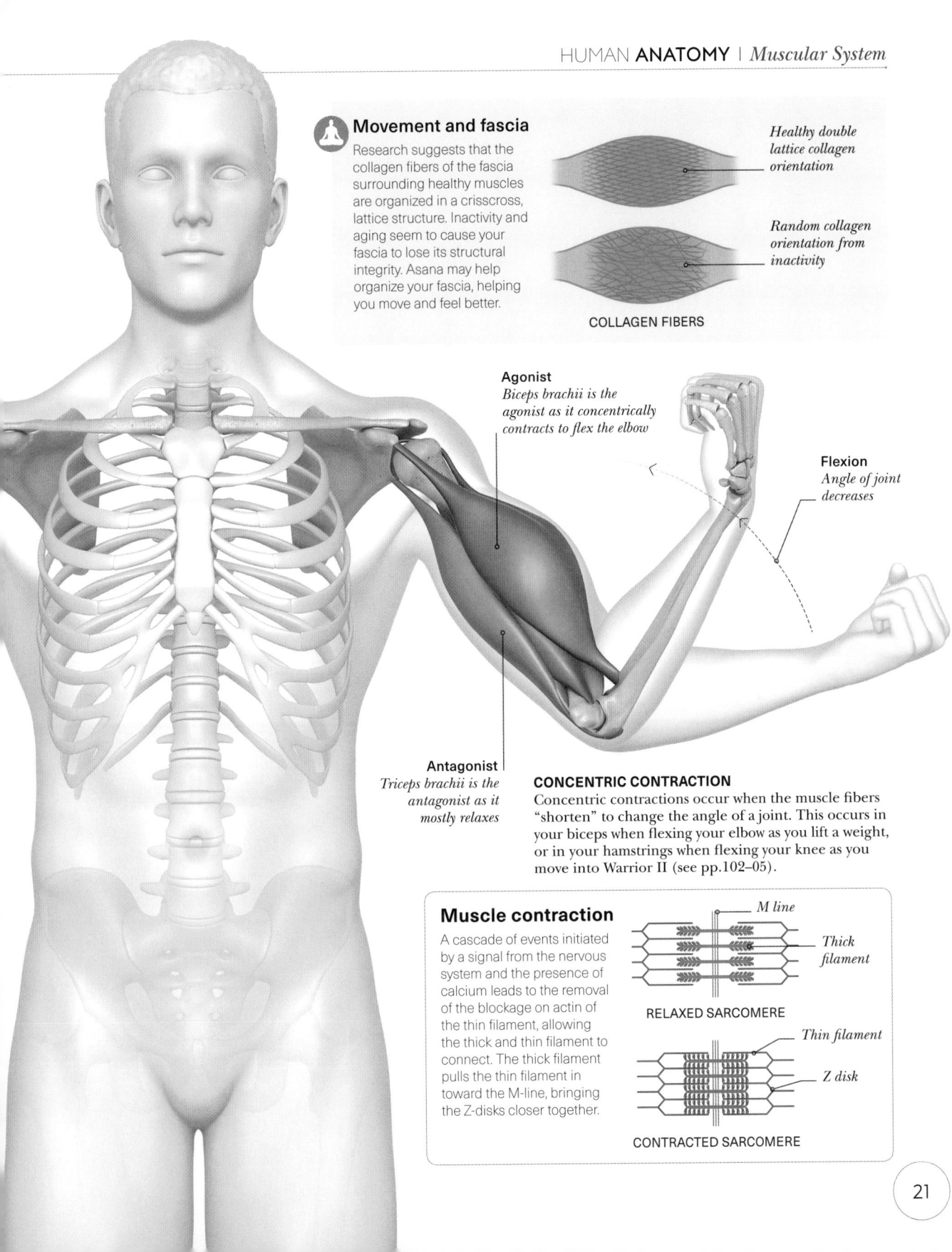

Movement and fascia

Research suggests that the collagen fibers of the fascia surrounding healthy muscles are organized in a crisscross, lattice structure. Inactivity and aging seem to cause your fascia to lose its structural integrity. Asana may help organize your fascia, helping you move and feel better.

CONCENTRIC CONTRACTION

Concentric contractions occur when the muscle fibers "shorten" to change the angle of a joint. This occurs in your biceps when flexing your elbow as you lift a weight, or in your hamstrings when flexing your knee as you move into Warrior II (see pp.102–05).

Muscle contraction

A cascade of events initiated by a signal from the nervous system and the presence of calcium leads to the removal of the blockage on actin of the thin filament, allowing the thick and thin filament to connect. The thick filament pulls the thin filament in toward the M-line, bringing the Z-disks closer together.

NERVOUS SYSTEM

The nervous system is a control network that connects all body systems. It is split into the central and peripheral nervous systems (PNS). The PNS is comprised of the somatic and autonomic nervous systems.

SYSTEM OVERVIEW

The somatic nervous system consists of nerves carrying sensory and motor signals to and from the spinal cord and brain. The autonomic nervous system (ANS) is divided into two functional systems: the sympathetic nervous system and parasympathetic nervous system, which accounts for many of yoga's benefits.

Brain
Responsible for control and cognition

Cranial nerves
12 pairs of peripheral nerves

Brachial plexus
A collection of nerves around your armpit

Spinal cord
Your body's primary communication highway

Median nerve
Can be pressed on in carpal tunnel

Lumbar plexus
A collection of nerves around lower back

Sacral plexus
A collection of nerves around your sacrum

Ulnar nerve
On the little finger side, causes tingling when hitting "the funny bone"

Femoral nerve
Supplies sensation around thigh and leg

Sciatic nerve
Largest nerve in your body

Tibial nerve
Branches off sciatic nerve

Digital branches of fibular nerve
Supply sensation to foot

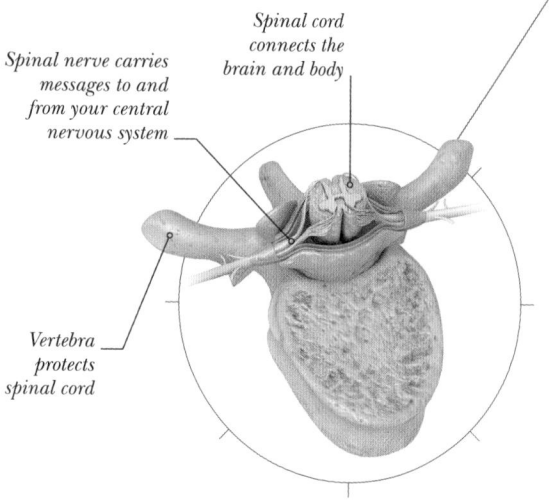

Spinal cord

In this superior, or bird's-eye, view of a vertebra, you can see how your spinal cord is protected by the bony encasement of the spinal column. Spinal nerves project off to the side, in between the vertebrae.

NERVE STRUCTURE

Neurons are the main cells of your nervous system. Axons are bundled together in your PNS to make nerves. Nerves are like highly conductive electrical wires sending signals throughout your body. Some are wrapped with a fatty substance called myelin, making their signals travel faster.

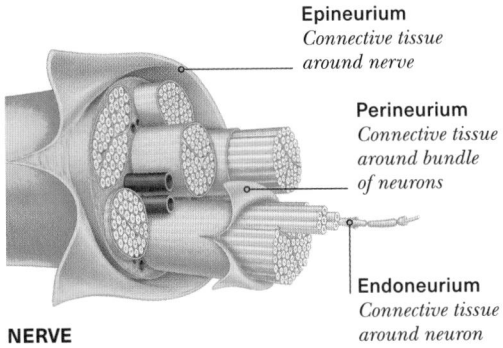

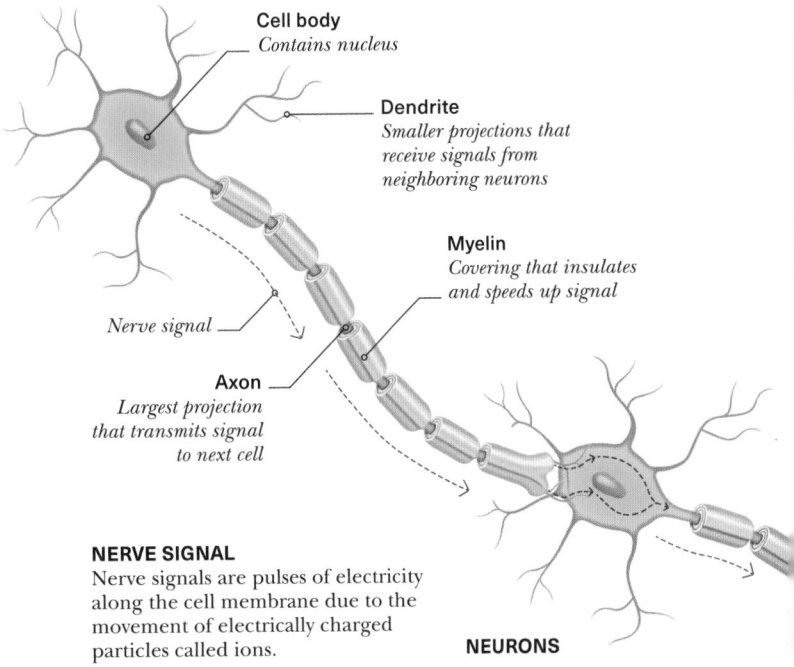

NERVE SIGNAL
Nerve signals are pulses of electricity along the cell membrane due to the movement of electrically charged particles called ions.

THE AUTONOMIC NERVOUS SYSTEM

The autonomic nervous system (ANS) can be thought of as your body's autopilot. Its functions are automatic and they include processes such as your heart rate, breathing, digestion, and excretion, which happen without you having to consciously think about them. The ANS is further divided into two systems of control that complement each other: the sympathetic nervous system (SNS) and the parasympathetic nevous system (PSNS).

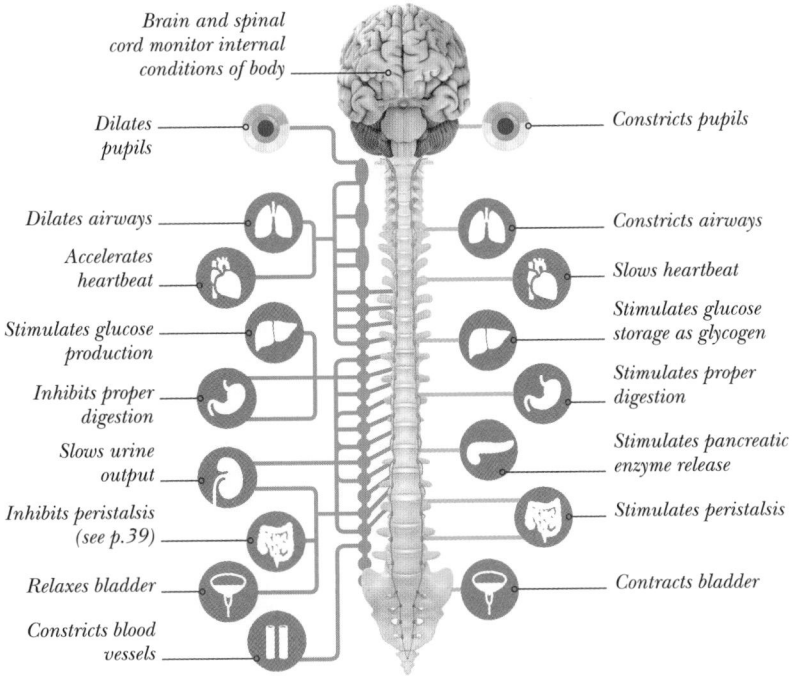

SYMPATHETIC NERVOUS SYSTEM
The SNS is known as "fight or flight" or the "stress response" because it helps you deal with stressful situations.

PARASYMPATHETIC NERVOUS SYSTEM
The PSNS is known as "rest and digest" or the "relaxation response" because it creates a restful state of optimal function.

CEREBRAL CORTEX

Compared to other mammals, our brains are massive for our bodies, with a particularly developed cerebral cortex. Most of the cortex is on the outside of the brain, except the insula. It is composed of gray matter, which is filled with synapses or connection points between neurons. Your cortex has five lobes and many functional areas.

LOBES OF THE BRAIN
The brain is separated into five main divisions, called lobes, including the insula which is inside the brain (not seen here).

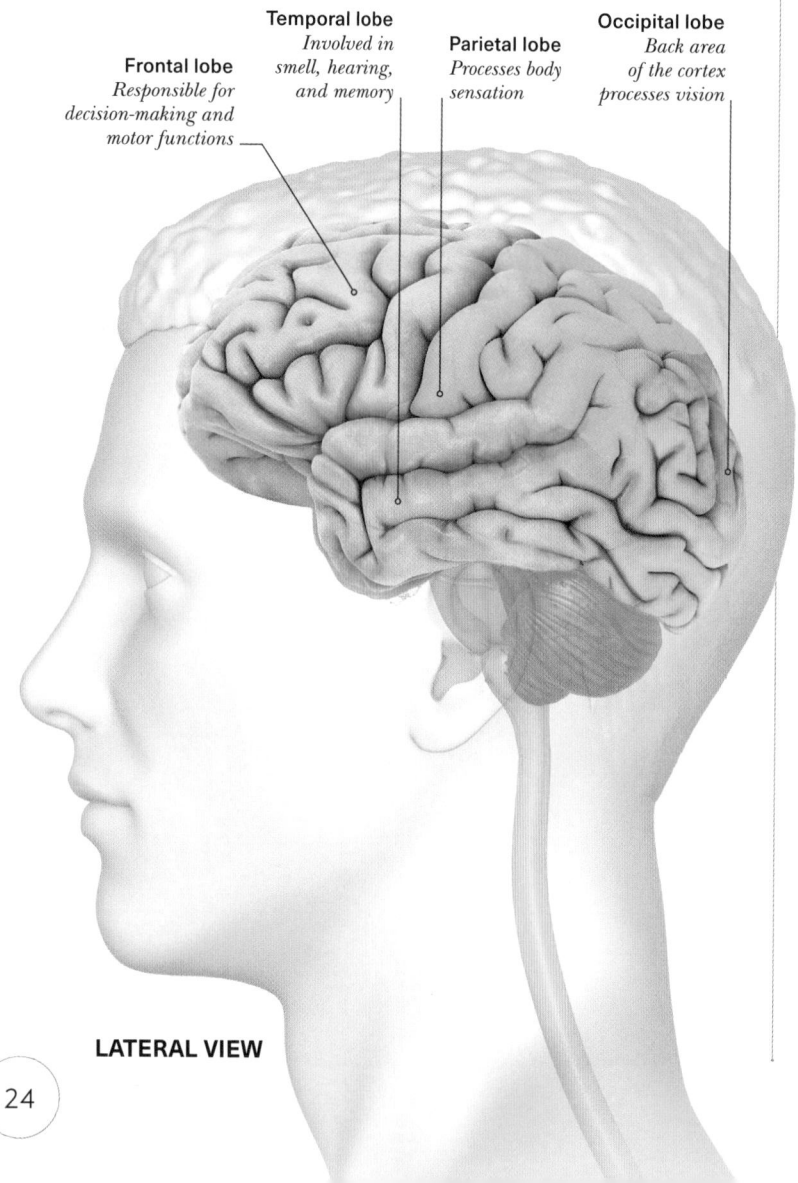

INSIDE THE BRAIN

The brain contains many different structures and scientists are still figuring out what their functions are. Some of these structures monitor conditions inside your body and relay information. The limbic system is the emotional center of your brain.

INTERNAL STRUCTURES
This image shows the brain as if it were cut in half down the middle (a midsagittal section) to reveal structures inside the cerebrum.

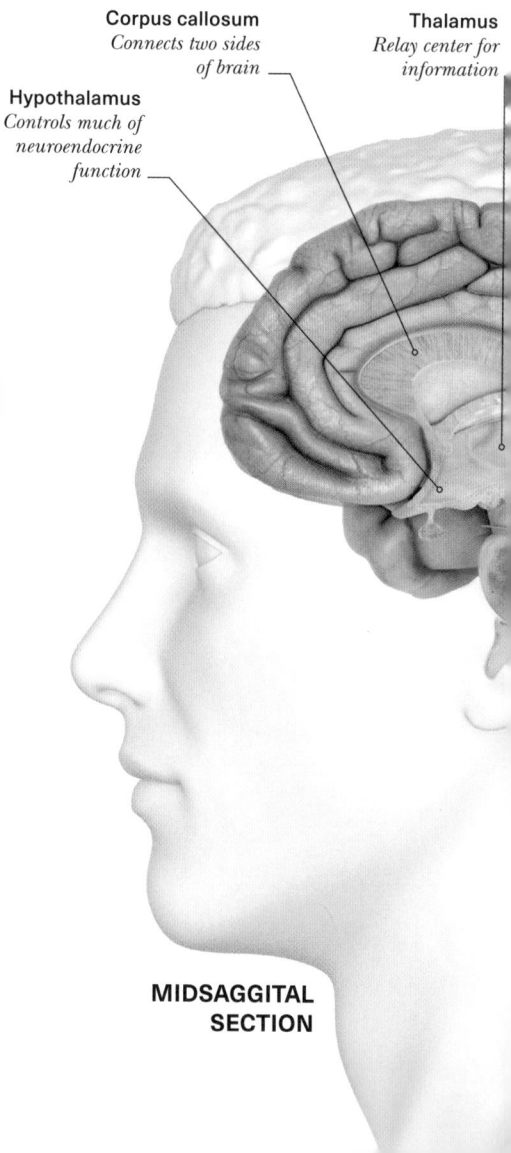

How yoga affects your brain

This chart looks at the neuroscience that may explain the vast mental and physical benefits of yoga. Modern science shows us that the brain maintains its ability to adapt across a lifetime, making it possible to break bad habits and negative patterns. It can also create the key chemicals that pharmaceutical companies synthesize in a lab. Research is uncovering the huge potential of yoga therapy to help people on a global scale. These effects stem from yoga's multidimensional approach, reflected in its 8-limb structure (see p.198), which includes guidelines on self-control and self-regulation.

↑ **Brain alpha wave activity increased** Alpha waves are associated with relaxation.

↑ **GABA increased** Gamma-aminobutyric acid counteracts anxiety and stress symptoms, leading to more relaxation.

↑ **Serotonin increased** Serotonin helps regulate your mood. Low levels of usable serotonin are associated with depression.

↑ **BDNF increased** Brain-derived neurotrophic factor is a protein responsible for neuron health and neuroplasticity. Yoga can boost levels of BDNF, which may help people with chronic pain or depression.

↻ **Dopamine regulated** Dopamine acts as your body's reward system and dysfunction is associated with addiction. Research suggests that meditation results in improved self-regulation.

↓ **Cortisol reduced** Cortisol is a stress hormone. When your baseline increases and levels are too high for too long, it can lead to inflammation and weight gain.

↓ **Norepinephrine reduced** A decrease in norepinephrine, or adrenaline, means fewer stress hormones in your system.

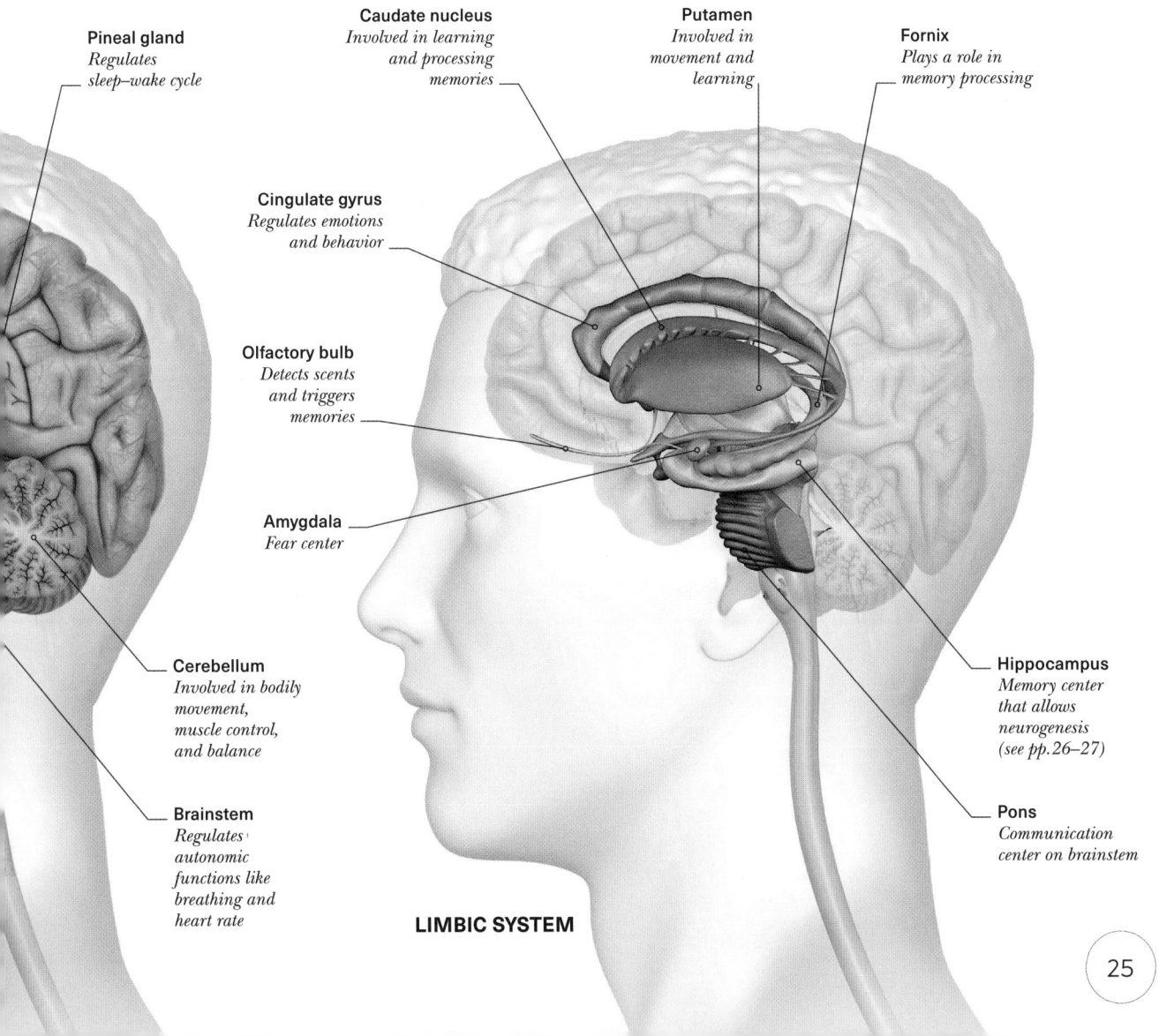

NEURAL PATHWAYS

The brain develops neural connections—and eventually becomes conditioned—based on your choices and experiences. It is said that neurons that fire together, wire together. The more you practice an activity—or a mindset—the more networks are created. With approximately 100 billion neurons, the brain's possible connections are vast. Yoga practices facilitate this process.

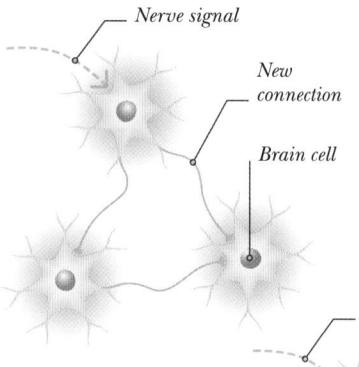

BUILDING CONNECTIONS
A new experience triggers neurons to form new connections.

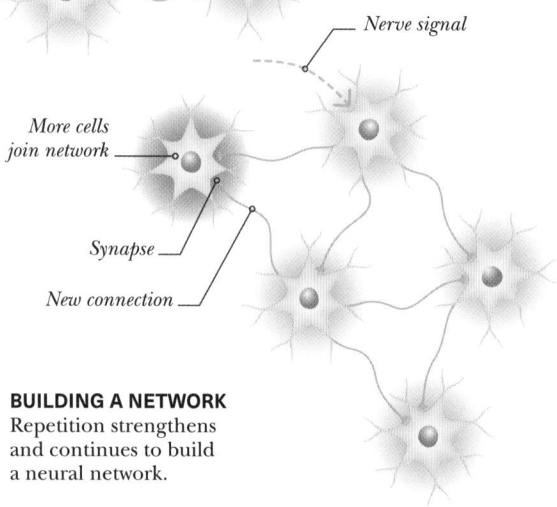

BUILDING A NETWORK
Repetition strengthens and continues to build a neural network.

CHANGING BRAIN

Neuroplasticity is the ability of your brain to be molded. Not long ago, scientists thought the brain couldn't change after childhood and degraded with age. Now we know that nervous tissue adapts. Like exercise affects your muscles, your brain tissue either develops or atrophies based on stimulation.

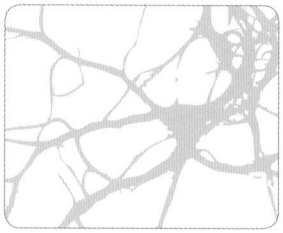

UNSTIMULATED BRAIN
Without stimulation, fewer connections are made. The brain tissue looks like a dying tree with sparse branches.

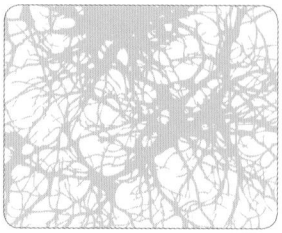

STIMULATED BRAIN
With stimulation, more connections form. The brain tissue looks like a thriving tree with dense branches.

Samskara

Yogis perhaps conceptualized neuroplasticity with *samskaras*: impressions due to past thoughts and actions. Yoga can help beat bad habits or conditioned responses by affecting neural pathways and *samskaras*. This occurs at a synaptic level each time you consciously change your thoughts and actions through awareness and practice. The more you travel that new path, the stronger the connection between the neurons gets.

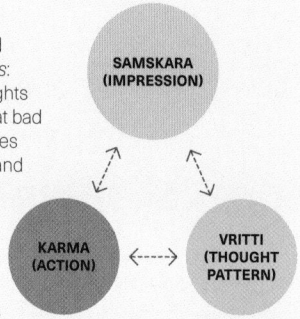

CYCLIC NATURE OF HABITS

How yoga boosts your brain

There is no neuroplasticity pill. The most effective way to shape your brain is through behavioral changes. Although any yoga practice should encourage neuroplasticity, try the tips here for improved results.

Increase the intensity
Moderate to vigorous physical activity, like from sun salutations, is one of the most effective ways of increasing brain-derived neurotrophic factor. This is a nerve growth factor, which is like a glue that helps to wire in neural connections.

Change your routine
Purposefully and consciously changing your yoga practice routine benefits your mind and your body.

Meditate
Research shows that meditation builds gray matter in your cerebral cortex.

Join a class
The act of moving with a group and following the teacher activates mirror neurons. The mirror neuron system is a recently discovered network of nerves involved in emulation of movement and developing compassion.

NEUROGENESIS

Scientists used to think that people are born with a certain number of nerve cells and that they cannot grow new ones. Research has since revealed that the growth of new neurons, or neurogenesis, can happen at any age. Neurogenesis occurs in key areas of the brain responsible for memory—the hippocampus—and smell. Neural stem cells in these regions of the brain develop new neurons.

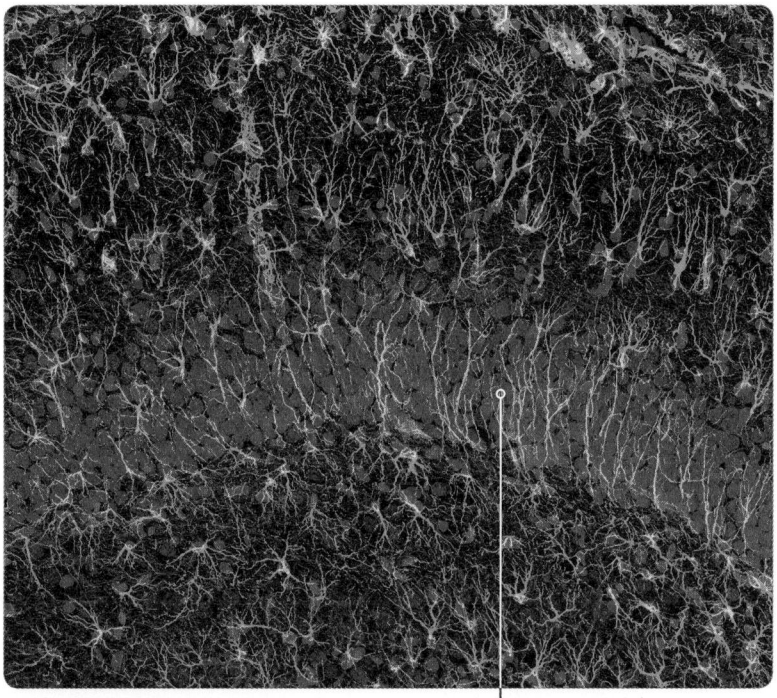

SITE FOR NEW CELLS
In this hippocampus tissue, helper cells or neuroglia are blue, axons are green, and neuron cell bodies and stem cells are pink.

Stem cells
Hippocampal stem cells can develop into new neurons, improving memory

CORTISOL LEVELS

Consistently high levels of the stress hormone cortisol are related to increased amygdala (fear center, see p.25) activity and decreased hippocampal (memory center) activity. When under these conditions, the hippocampus doesn't grow new neurons or connections well. Yoga practices are shown to reduce cortisol levels and reverse these effects, which may contribute to improving memory.

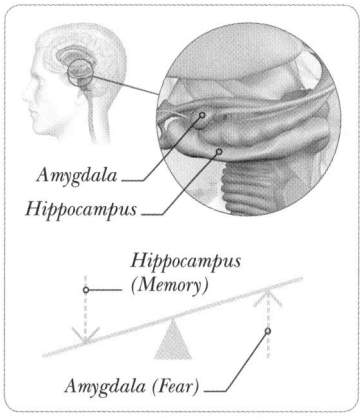

STRESS AND MEMORY
Increased activity in the amygdala is correlated with reduced activity in the hippocampus, which has an adverse effect on memory.

Practice hand mudras
Hand mudras are gestures that require concentration and awareness. Just as people who read braille have more developed hand-specific sensory areas of their brain, mudras may develop brain areas linked with sensory acuity, and fine motor skills.

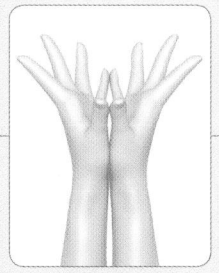

PADMA MUDRA

HAKINI MUDRA

SHUNI MUDRA

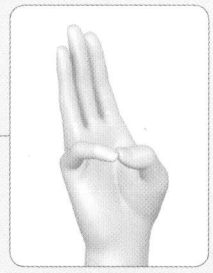

BUDDHI MUDRA

ENDOCRINE SYSTEM

The endocrine system is a slower, longer-lasting control system than the nervous system. It consists of glands that release hormones into your bloodstream to be delivered to specific cells.

SYSTEM OVERVIEW

Your brain controls the release of hormones from endocrine glands to maintain a balance inside your body, called homeostasis. Stressors—from external environmental conditions to internal or emotional factors—affect this balance, but yoga can help. For example, research suggests that yoga may prevent and improve symptoms of type 2 diabetes.

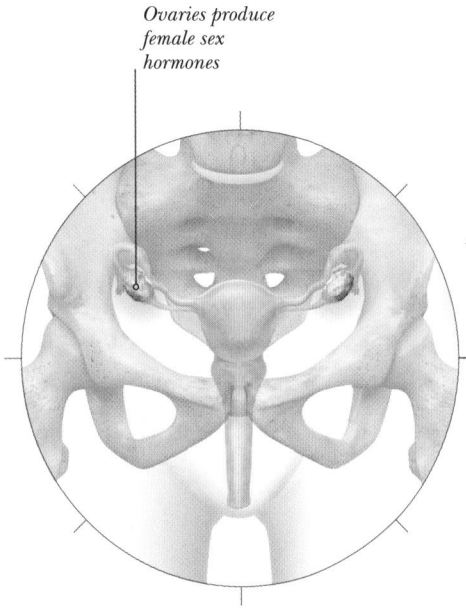

FEMALE

Pineal gland
Makes melatonin, which affects sleep

Hypothalamus
Controls other glands

Pituitary gland
Produces key hormones

Parathyroid gland
Regulates blood calcium levels

Thyroid gland
Regulates metabolism and blood calcium

Heart
Releases hormones to regulate blood pressure

Adrenal gland
Regulates salt levels and produces adrenaline in response to danger

Pancreas
Secretes insulin and glucagon to regulate blood sugar

Small intestine
Releases hormones to help with digestion

Testis
Produces male sex hormones

MALE

Homeostasis and allostasis

Homeostasis is your body's state of dynamic equilibrium. Most processes—like the control of hormone release, blood calcium and blood sugar levels, and temperature—are tightly regulated through negative feedback, which works in a similar way to a thermostat. Nature wants you to be in balance. Yogis referred to this as *samatva,* which can be translated as equilibrium or equanimity. Allostasis is a process of maintaining homeostasis amid stressors. The more intense the stress, the heavier the "allostatic load" and the more your cells have to work to maintain equilibrium. This increases the likelihood of chronic diseases. Researchers believe that yoga can reduce allostatic load.

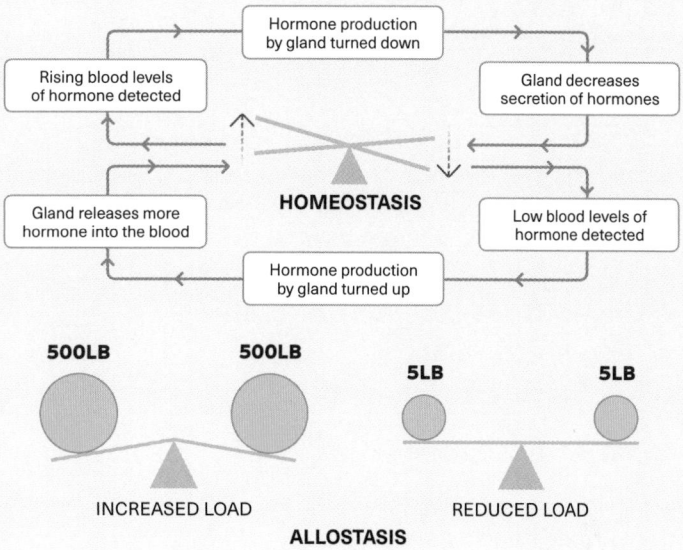

PANCREAS

Your pancreas releases insulin to help sugar get into your body cells. However, cells can become insulin-resistant, which can cause disease. A review found that yoga can improve glycemic control, lipid levels, and body composition of fat in those with metabolic syndrome and type 2 diabetes. A doctor-approved reduction in medications was also found.

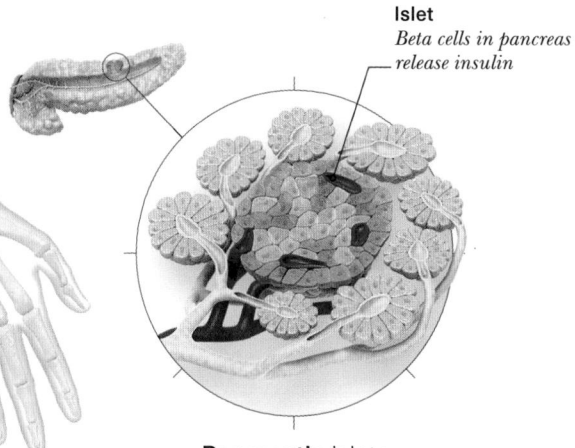

Pancreatic islets

Inside your pancreas, islets contain different types of cells. Beta cells release insulin, which allows your body cells to use glucose.

Metabolism

Most yoga practices tend to slow your metabolism, which helps your body to be more efficient with less. Although your metabolism may slightly lower from relaxation-based practices, this doesn't mean you will gain weight. A reduction in stress hormones like cortisol also prevents your body from holding onto fat.

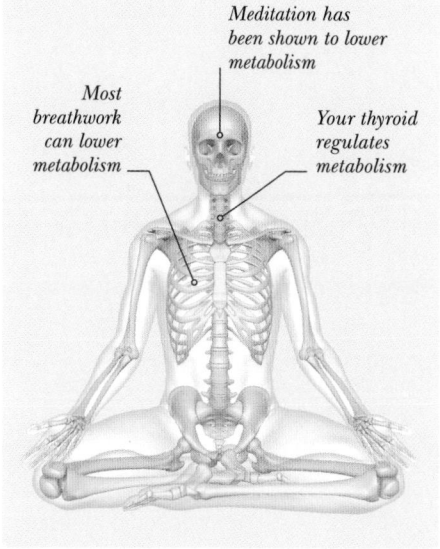

RESPIRATORY SYSTEM

You take a breath 12–20 times per minute. The purpose of your breath is to get oxygen to your cells and to get rid of waste like carbon dioxide. The respiratory system includes the nasal cavities, air passageway tubes, and lungs.

SYSTEM OVERVIEW

You don't have to think to breathe; respiration is a part of your autonomic nervous function. However, yogis claim that by controlling your breath, you can control all aspects of being. Science reveals that your breath is an access point to regulating your nervous system.

Neti pots

The neti pot is a part of traditional yogic hygiene practices. It involves pouring clean (filtered or boiled), warm salt water in one nostril to fill the sinuses and drain out the other nostril. Neti pots (or a similar sinus rinse) are recommended by many modern physicians to help allergies and respiratory illness.

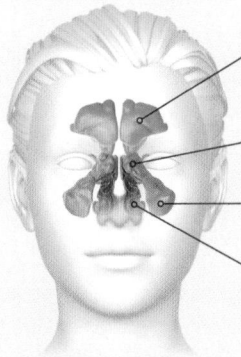

SINUSES
Your sinuses are a system of connected, air-filled cavities in your skull. They make your skull lighter, help your voice to resonate, and affect your breath.

Nasal cavity
Air entering through nose is humidified, warmed, and filtered

Nostril
Openings in the nose

Mouth
Air can enter here, though less efficient

Epiglottis
Flap that closes off trachea to keep food from entering airways

Larynx
Top part of trachea, contains vocal cords

Trachea
Also called the windpipe, carries air to lungs

Lungs
Branching airways end here; oxygen enters blood and carbon dioxide is removed

Diaphragm
Primary muscle of respiration

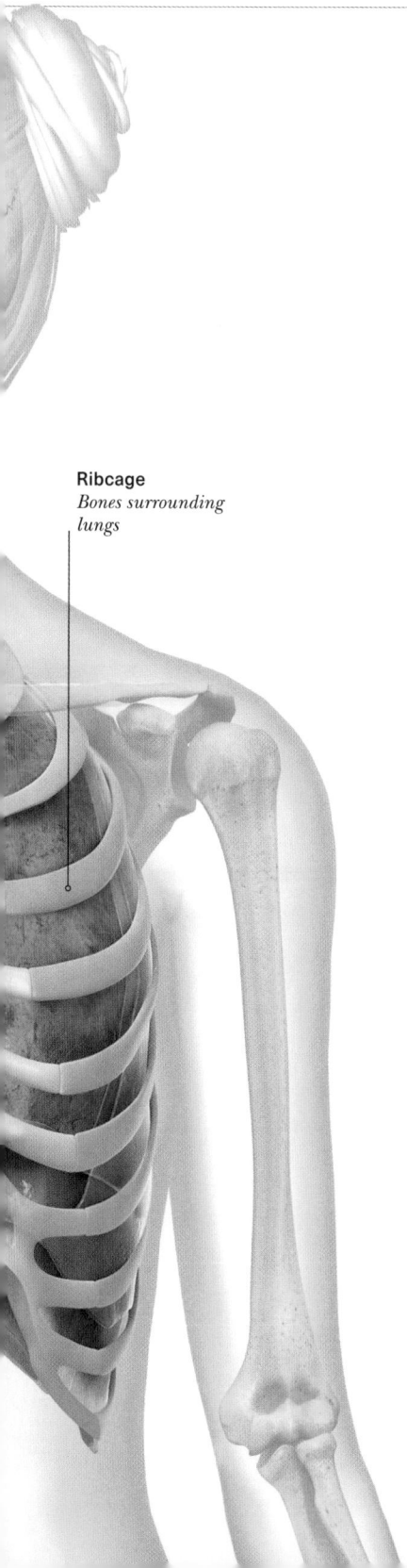

HOW WE **BREATHE**

When you inhale, the breath enters your nose, throat, and then your lungs. Your lungs and ribcage expand three-dimensionally in all directions; your diaphragm engages to flatten. When you exhale, your diaphragm relaxes to ascend, your lungs and ribcage compress, and the air releases out of your throat and then nose.

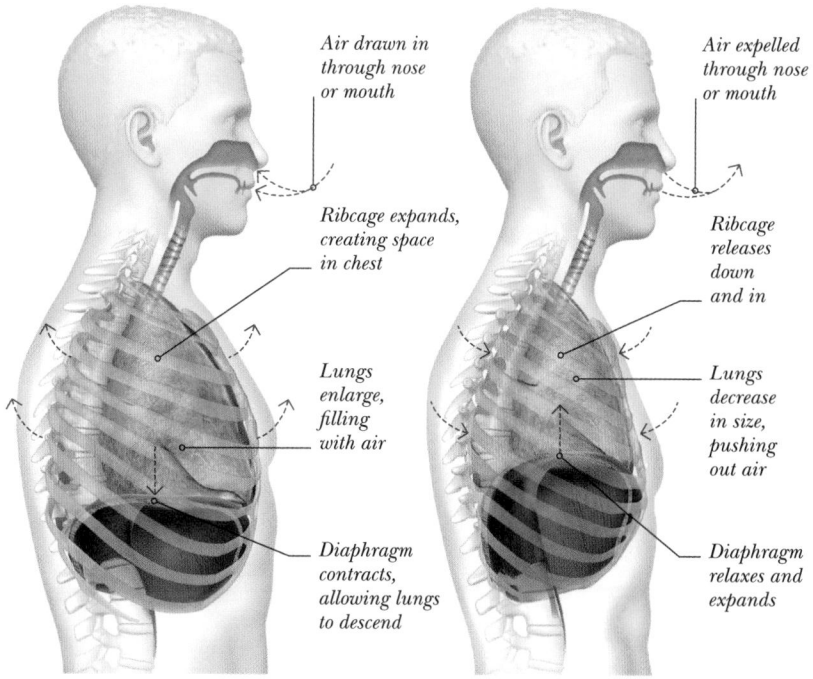

INHALE **EXHALE**

Belly breathing

"Belly breathing" doesn't mean you are actually breathing in your belly, but rather that you are allowing your belly to move freely with your breath. When your diaphragm engages with the inhale, it presses against your abdominal organs—pushing down and out, which is why this is also called diaphragmatic breathing.

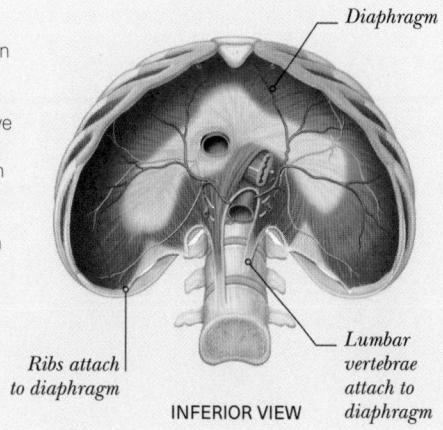

INFERIOR VIEW

PRANAYAMA

Yogis use *pranayama* or breathwork to control their *prana* and anchor to the present moment. The word *prana* in Sanskrit means vital energy or life-force energy that permeates through us and everything. Interestingly, *prana* simultaneously means breath. Yogis believe that you can change the flow and qualities of your energetics by breath control.

INHALE AND EXHALE

When you inhale, blood is shunted to your heart and lungs to help them function. Baroreceptors (see p.134) sense this increased pressure and respond by signaling to let off the brake pedal, momentarily increasing sympathetic activity. During each exhale, your heart is slightly more relaxed with increased parasympathetic activity. This explains why elongating your exhales in *pranayama* is relaxing.

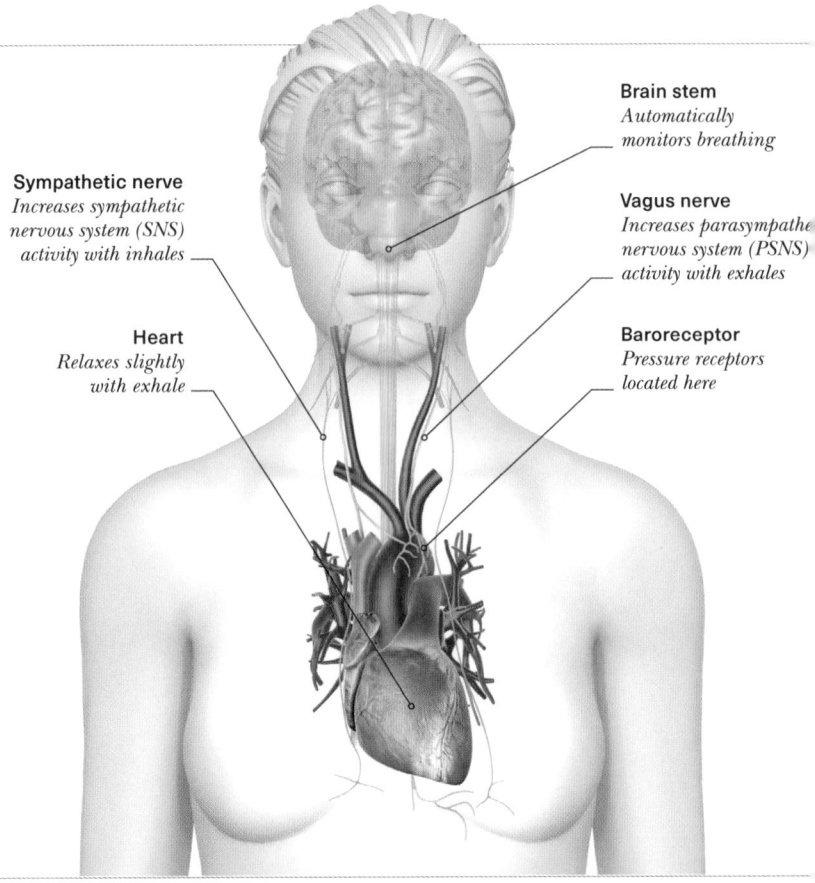

BREATHWORK PRACTICES

Modern yogis use breathwork for health benefits, including overcoming inefficient breathing patterns from a culture of poor posture and stress. Through altering your breath, you change your state of mind. For example, you may practice left nostril breathing and bee breath to calm down or right nostril breathing and *kapalabhati* for alertness.

BREATH OF FIRE (KAPALABHATI)

This is a fast breath that mimics hyperventilation, increasing your heart rate and blood pressure. It also tones your abdominals. Avoid this technique if you are pregnant or have anxiety, certain eye conditions, or high blood pressure. Similar effects and precautions apply for holding your breath (*kumbhaka*).

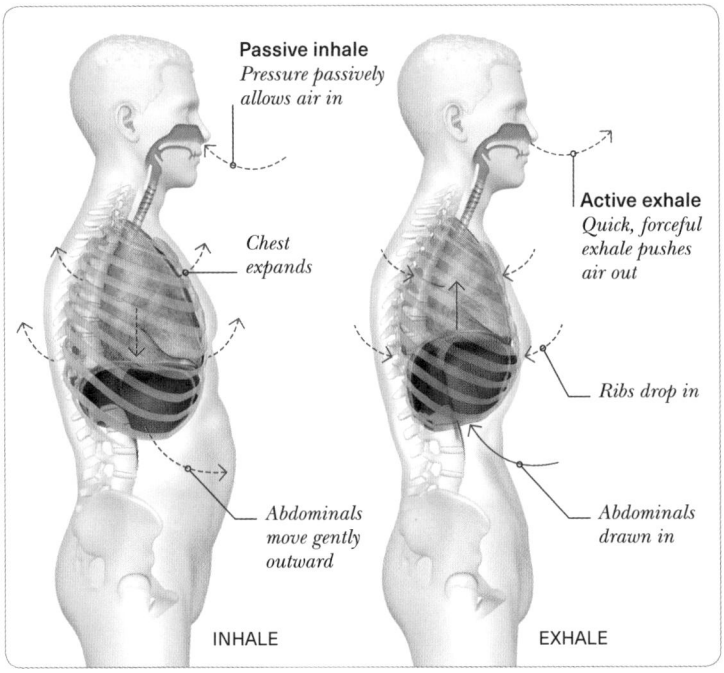

NASAL CYCLE

For many, each nostril takes turns dominating air flow (in .5- to 4-hour shifts). This is called the nasal cycle. You probably notice this more when you are congested. Openness indicates local vasoconstriction and the swollenness indicates vasodilation. Observe this cycle naturally or try purposefully covering one nostril for a desired effect (see panel right).

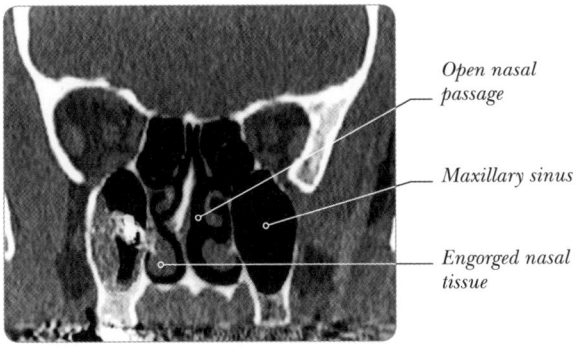

NASAL TISSUE
This image shows the right nasal passage swollen while the left is open. In this case the swelling is exacerbated by congestion.

Left brain, right brain

Each half of your body is controlled by the opposite hemisphere of your brain—meaning that your left arm is controlled by the right half of your brain. The same is true of your nostrils. This may have many implications, including a slight overall increase in SNS activity when right nostril breathing and PSNS when left, although evidence is mixed.

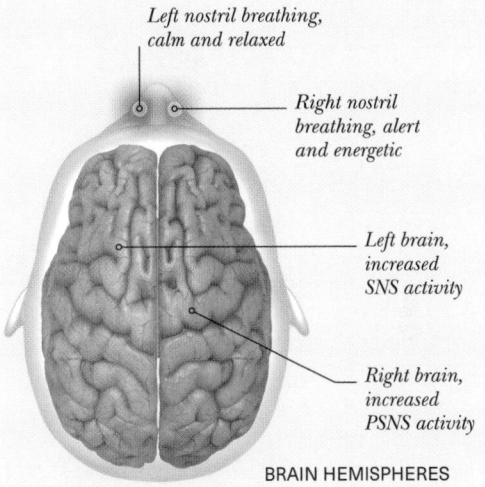

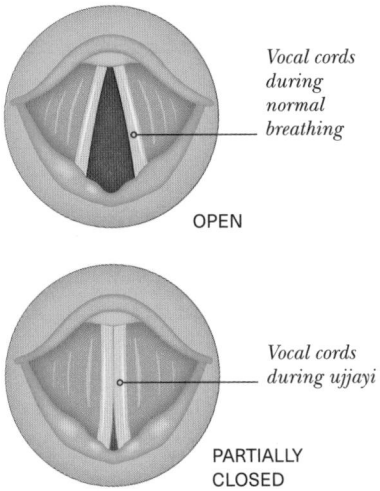

VICTORIOUS BREATH (UJJAYI)
Victorious breath involves partially constricting your vocal cords. The feeling is similar to when you whisper softly. It creates an ocean sound to give your mind a focal point.

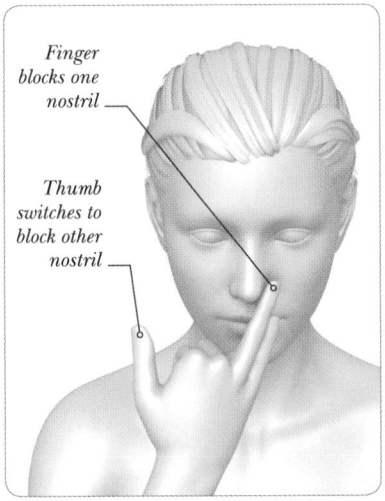

ALTERNATE NOSTRIL BREATHING
This technique may calm the mind and body. It involves focus and activation of both sides of the brain. To practice it, just remember: exhale, inhale, and switch nostrils.

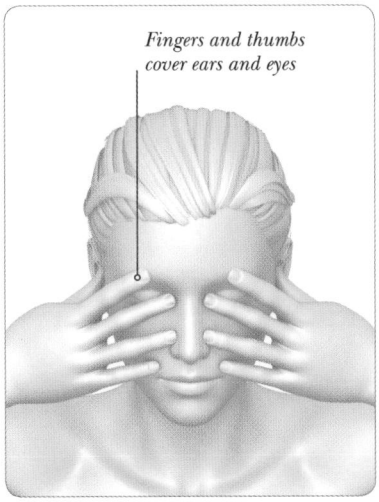

BEE BREATH (BRAHMARI)
This involves covering your eyes and ears and humming on a long exhale. Yogis used this to improve sleep. Research shows it can lower heart rate, blood pressure, and anxiety.

CARDIOVASCULAR SYSTEM

The heart, an intricate network of vessels, and the blood circulating through them make up your cardiovascular system.

SYSTEM OVERVIEW

Your heart constantly beats to pump blood around your body, removing waste and delivering vital oxygen. Research on yoga suggests profound benefits for cardiovascular health, including a reduced risk of heart disease. Yoga has been shown to clinically improve blood pressure, cholesterol levels, and cardiovascular resilience (see opposite).

Jugular vein
Returns blood from brain to heart

Caroti artery
Delive blood t brain

Subclavian artery
Carries blood to arm and hand

Superior vena cava
Returns blood back to heart

Aorta
Largest artery in your body

Heart
Muscular pump for blood

Inferior vena cava
Returns blood from lower body to heart

Abdominal aorta
Delivers blood to abdomen and lower body

Femoral vein
Delivers blood from lower limbs to heart

Femoral artery
Carries blood to thighs

Popliteal artery
Carries blood to knee and calf

Great saphenous vein
Longest vein in the body

Red blood cells
Also called erythrocytes, they deliver oxygen

White blood cells
Also called leukocytes, they fight invaders

Platelets
Also called thrombocytes, they help blood to clot

Composition of blood

Adults have about 11 pints (5 liters) of blood circulating around the body. Blood is a connective tissue composed of red blood cells, white blood cells, and platelets suspended in a liquid called plasma. It provides oxygen, nutrients, and hormones, as well as removing waste from cells.

HEART AND CIRCULATION

Circulation has two loops: pulmonary (lungs) and systemic (body). Veins carry blood to the heart, and arteries carry it away. Veins are shown in blue to represent deoxygenation, and arteries are red for oxygenation. The exceptions are pulmonary arteries (deoxygenated) and pulmonary veins (oxygenated).

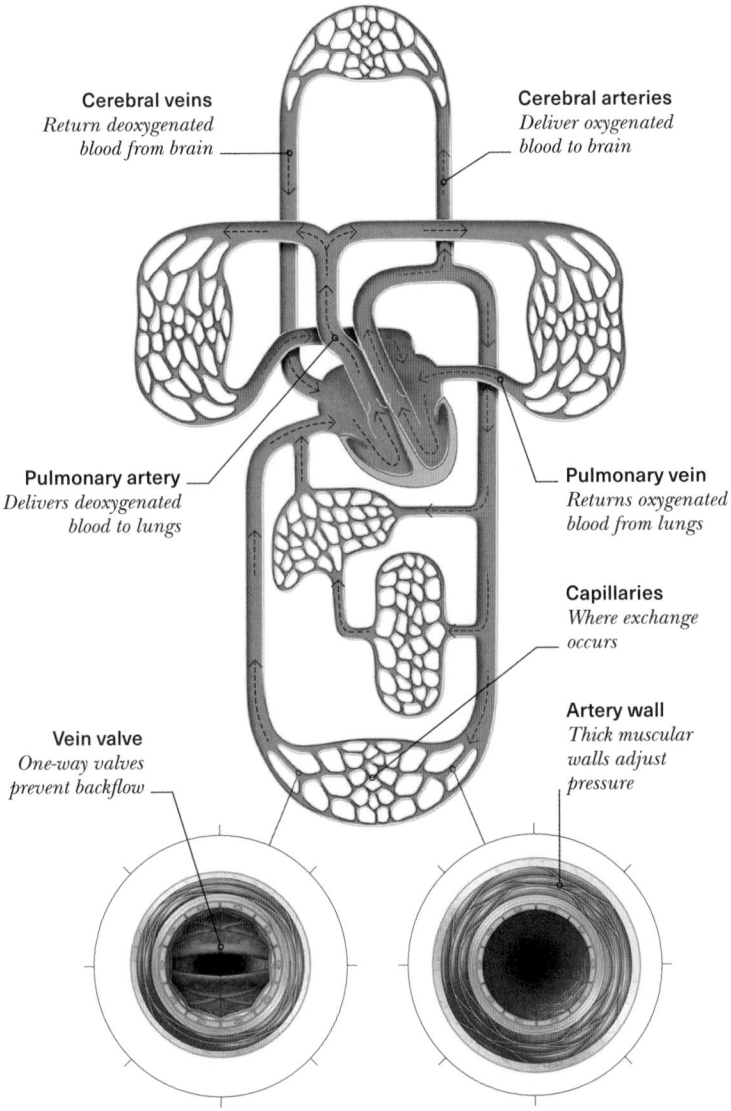

Vein

Veins return blood to your heart from cells. The blood is usually deoxygenated (oxygen-poor).

Artery

Arteries carry blood away from your heart to cells. The blood is usually oxygenated (oxygen-rich).

Heart rate variability

Heart rate variability (HRV) is the heart's ability to adapt fast. It is better for your pulse to vary rather than tick steadily. High HRV shows autonomic resilience and may lead to improved physical, emotional, and cognitive function. Yoga appears to improve HRV.

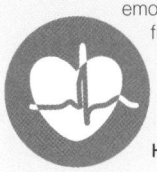

HEARTBEAT

Hypertension

Research shows that yoga can reduce blood pressure significantly. With more than 1 billion people living with hypertension, yoga offers a cost-effective adjunct to care with minimal to no side effects. Consult your doctor about any blood pressure shifts.

BLOOD PRESSURE MONITOR

Cholesterol

Reports have shown that yoga can increase "good" cholesterol (high-density lipoprotein or HDL) and decrease "bad" cholesterol (low-density lipoprotein or LDL). This reduces the risk of heart disease by preventing arterial narrowing.

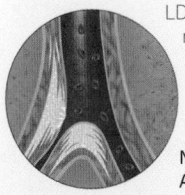

NARROWED ARTERY

Heart disease

A meta-analysis suggests that yoga reduces heart disease risk as well as or better than accepted exercise guidelines. Long-term clinical trials have shown that a yogic lifestyle— with exercise, yoga with meditation, social support, and a plant-based diet—could reverse heart disease.

DAMAGED HEART TISSUE

LYMPHATIC SYSTEM

The lymphatic and immune systems work together to fight invaders. Acute inflammation can be a helpful result of this internal war, such as when you have a cut. However, chronic inflammation is an underlying cause of many major diseases.

SYSTEM OVERVIEW

Lymph vessels collect and drain excess fluid from body tissues. They also carry immune cells around your body. Evidence suggests that yoga can help reduce chronic inflammation and it may boost immunity, helping you get sick less often and less intensely. Your body can heal itself, and yoga can help.

Tonsils
Help destroy bacteria or viruses that enter the nose or mouth

Thoracic duct
Lymph drains back into the heart through here

Axillary nodes
A concentration of lymph nodes under the arm

Spleen
Produces cells that fight infection

Cisterna chyli
Collects lymph from the lower half of the body

Inguinal nodes
A concentration of lymph nodes around the groin

Lymph node
Lymph is processed and cleaned here

Lymph vessel
Drains and transports lymph

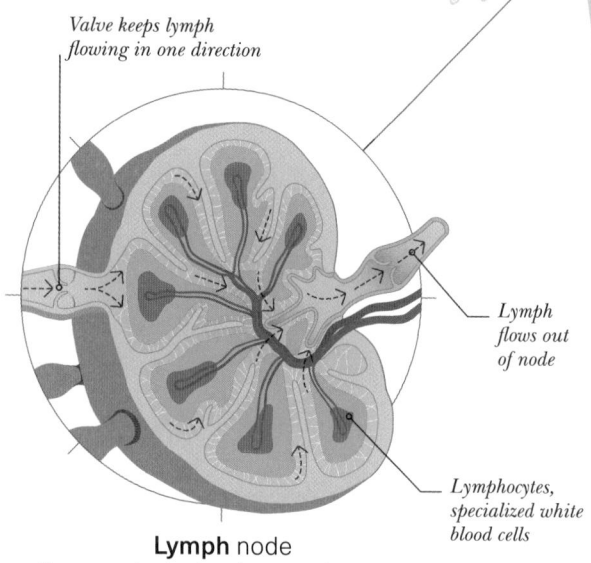

Lymph node

These are checkpoints that screen lymphatic fluid for foreign invaders. The cleaned fluid is returned to your blood. Movement in yoga asanas, particularly from sun salutations and inversions, can help facilitate lymph flow.

White blood cells

White blood cells are like warriors fighting viruses, bacteria, and cancer cells in your body. Fragments of the invaders, called antigens, are presented so the warriors can strategically fight using the right antibodies and chemical messengers, called cytokines. Communication is key—miscommunication can lead to chronic inflammation.

DENDRITIC CELL
These present antigens, which the body recognizes as a foreign invader. They activate T-cells to do their job.

MACROPHAGE
Hungry hunter cells (see phagocytosis below) that also release cytokines to induce inflammation.

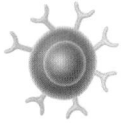

B-CELL
A type of lymphocyte that secretes antibodies, which are proteins specialized to fight specific antigens.

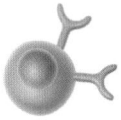

T-CELL
A type of lymphocyte that is activated to fight by the presentation of antigens. There are many specialized types.

PHAGOCYTOSIS
Macrophages (white) patrol your body on alert for invaders (red) to engulf and eat, in a process called phagocytosis.

INFLAMMATORY RESPONSE

Inflammation often involves heat, pain, redness, and swelling due to a cascade of events where white blood cells fight invaders. In an autoimmune disease, they mistakenly fight body tissue. For example, rheumatoid arthritis (see below) can flare to cause local inflammation and body-wide inflammation.

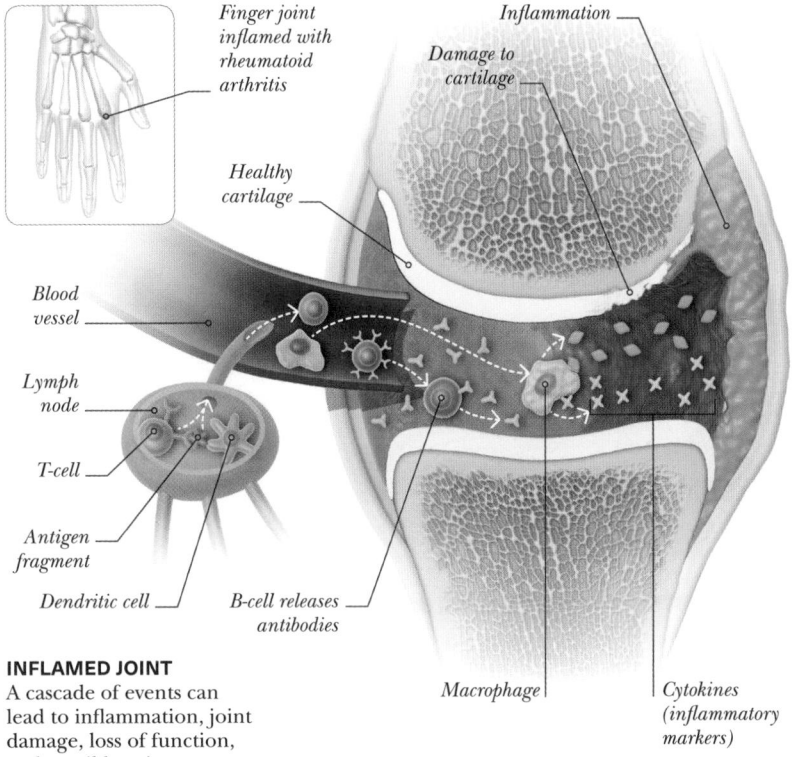

INFLAMED JOINT
A cascade of events can lead to inflammation, joint damage, loss of function, and possibly pain.

Yoga and inflammation

Yoga seems to help attenuate inflammation by reducing the stress response, which may reduce your disease risk. A review shows that yoga practice reduces cytokine count and therefore inflammation. Scientists hypothesize that a long-term, regular practice would be most effective.

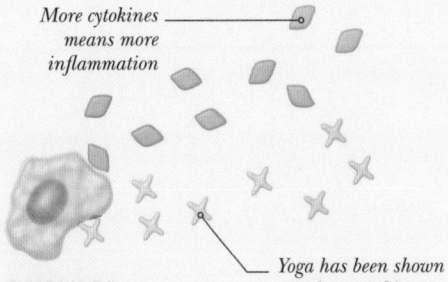

CYTOKINES
These are inflammatory markers that encourage an immune response.

DIGESTIVE SYSTEM

The digestive tract is a tube with selective membranes that control what gets into your body. Nutrients are absorbed and waste is expelled.

SYSTEM OVERVIEW

Food is broken down into absorbable units by your digestive system, from chewing in the mouth to chemical breakdown in the stomach and squeezing in the intestines. Nutrients enter the blood, and ultimately your cells. Yogis recognized that you become what you eat, equating the physical body (*anamaya*) with the "food body."

Journey of food

It is best to practice yoga asanas on an empty stomach. That may mean not eating a meal 2–4 hours before class. You may need to strategically plan a small snack, especially if you tend to have low blood sugar or other medical conditions.

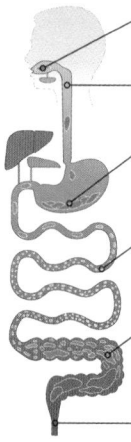

30sec
Chewing in mouth

10sec
Swallowing

2–4hr
Churning in stomach

3–5hr
Absorbing nutrients in small intestine

10–18hr
Water absorbed in large intestine

After 24–72hr
Undigested waste leaves body

Mouth
Entry point for food

Pharynx
Also called the throat

Teeth
Grind food, making it easier to swallow

Salivary glands
Release saliva to start digestion

Epiglottis
Cartilage flap that blocks entry to trachea

Esophagus
Muscular tube that pushes food down

Liver
Makes bile and processes nutrients and toxins

Stomach
Breaks down food by acid and churning

Gallbladder
Stores and releases bile

Small intestine
Digests and absorbs nutrients

Large intestine
Stores food and beneficial bacteria

Appendix
Stores extra beneficial bacteria

Rectum
Voluntary muscular release chamber

Anus
Exit point of feces

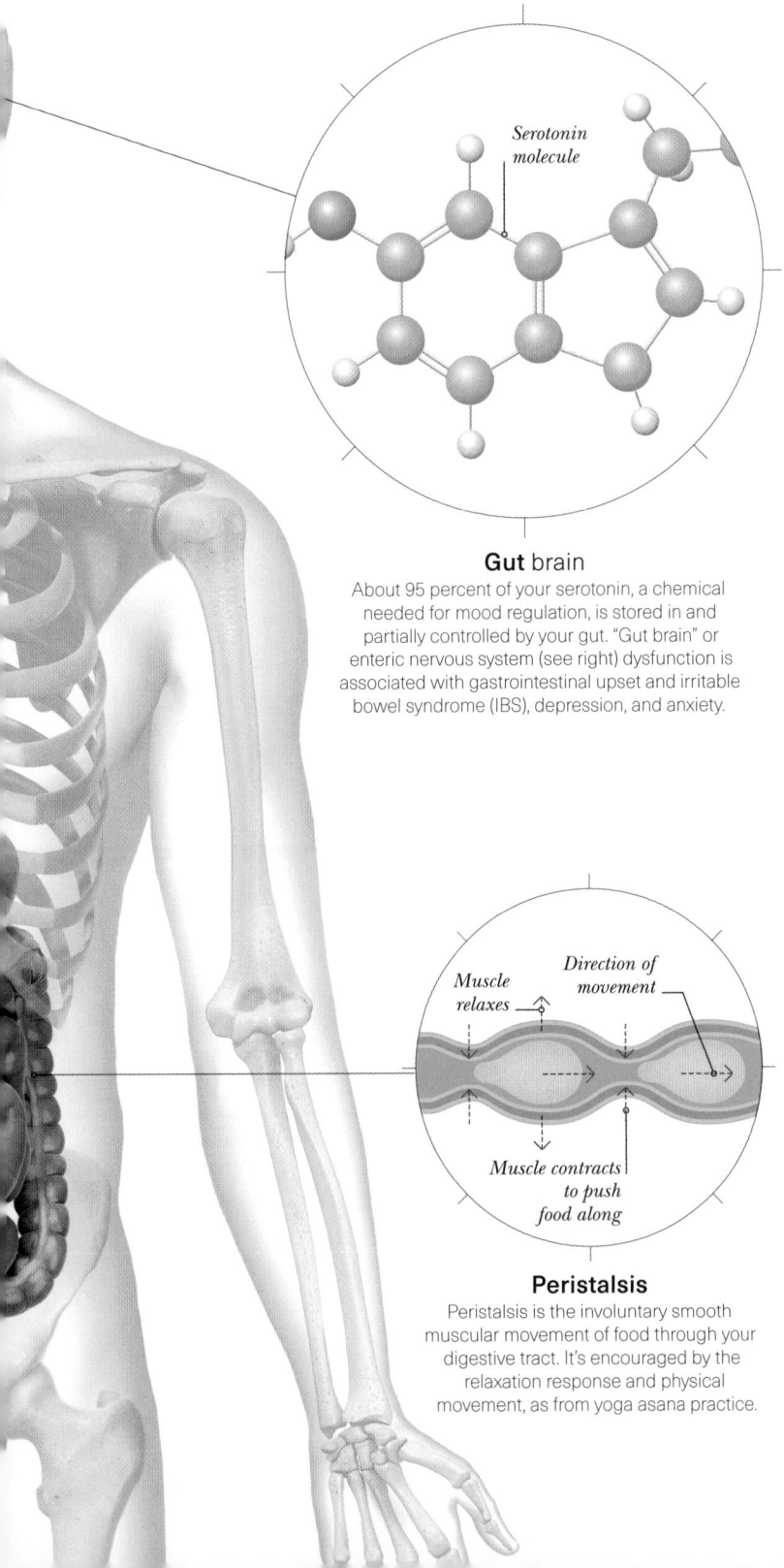

Gut brain

About 95 percent of your serotonin, a chemical needed for mood regulation, is stored in and partially controlled by your gut. "Gut brain" or enteric nervous system (see right) dysfunction is associated with gastrointestinal upset and irritable bowel syndrome (IBS), depression, and anxiety.

Peristalsis

Peristalsis is the involuntary smooth muscular movement of food through your digestive tract. It's encouraged by the relaxation response and physical movement, as from yoga asana practice.

Enteric nervous system (ENS)

Scientists have recently discovered the semi-independent enteric nervous system (ENS). These 100 million neurons may be responsible for you feeling butterflies in your stomach from love or having an intuitive gut feeling. Yoga enhances your mind-body connection, so you can feel what is going on in your gut clearly. This interconnection may explain how yoga can improve both your digestion and mood significantly.

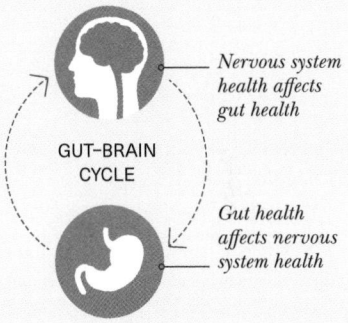

Ahimsa diet

Yogis often make conscious choices about what they put into their body. An ahimsa diet is one of nonharm. For many, this means being a vegetarian to reduce the suffering of other animals. A largely plant-based diet reduces your risk of heart disease, cancer, and related major killers. Scientists project that a mostly vegetarian diet may reduce global mortality by 6–10 percent and cut food-based greenhouse gas emissions by 29–70 percent—a huge impact on the environment. Even small dietary changes like a Meatless Monday can make a big difference.

NONHARM FOOD

URINARY SYSTEM

The urinary system filters out waste and excess fluids to maintain correct blood volume. This, in turn, affects blood pressure, which yoga has also been shown to help regulate.

SYSTEM OVERVIEW

Your kidneys process waste from blood into urine, which is then stored in your bladder. Urine release is voluntary in adults but some people lose this control, leading to urinary incontinence. A recent study showed that yoga classes may help manage urinary incontinence.

Inferior vena cava
Returns blood from kidneys and lower body

Abdominal aorta
Delivers blood to kidneys and lower body

Adrenal gland
Regulates fluid volume

Kidney
Filters blood to make urine

Ureter
Carries urine from kidneys to bladder

Bladder
Stores urine

Prostate gland
Surrounds male urethra

Urethra
Carries urine from bladder out of body

Urethra
Shorter urethra increases chance of infections

FEMALE

MALE

Pelvic floor muscles

Your pelvic floor muscles are vital for bladder control. Common issues such as frequent, urgent, or painful urination, or slight leaking—such as when sneezing or laughing—may be helped by yoga exercises. For example, a gentle version of *mula bandha* (see p.153) and relaxation practices could improve pelvic floor health.

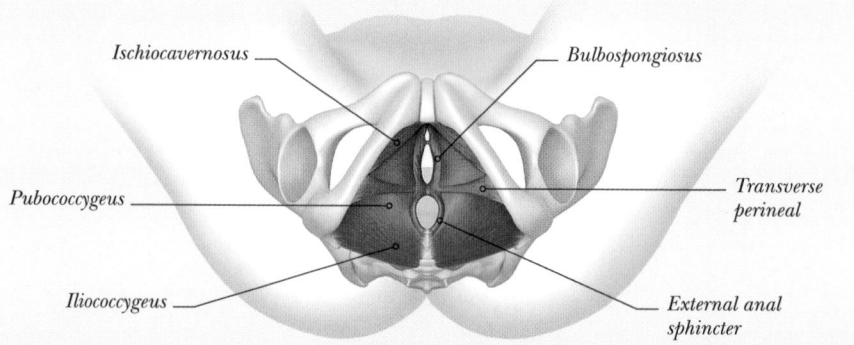

INFERIOR VIEW OF FEMALE

REPRODUCTIVE SYSTEM

The reproductive system functions to help continue our species by sexual reproduction. Yoga may help aspects of reproductive health, including pelvic floor health. This may improve sexual satisfaction, and labor and delivery.

SYSTEM OVERVIEW

Yoga seems to indirectly address aspects of pelvic health, both urinary and reproductive, partly by promoting optimal breathing. It is also feasible that, because yoga helps manage stress, it can improve fertility and conception; although we need more research to confirm this.

Milk ducts
Carry milk from glands to nipple

Nipple
Opening where a baby sucks milk

Fallopian tube
Connects ovary to uterus

Uterus
Where a fertilized egg develops

Ovary
Where eggs are stored and released

Endometrium
Uterus lining that thickens to receive an egg

Cervix
Opening to uterus

Vagina
Muscular tube

Penis
Transfers sperm into the vagina

Testes
Manufacture sperm and testosterone

MALE

FEMALE

Pelvic floor motion

A healthy pelvic floor is able to move through its full range of motion with your breath, following the movement of your diaphragm. Yoga practice may enhance neurological awareness, along with increasing strength, flexibility, and the relaxation of these muscles. This may improve your bladder, bowel, sexual, and reproductive health.

BREATHING
Your pelvic floor muscles descend as you inhale and ascend as you exhale.

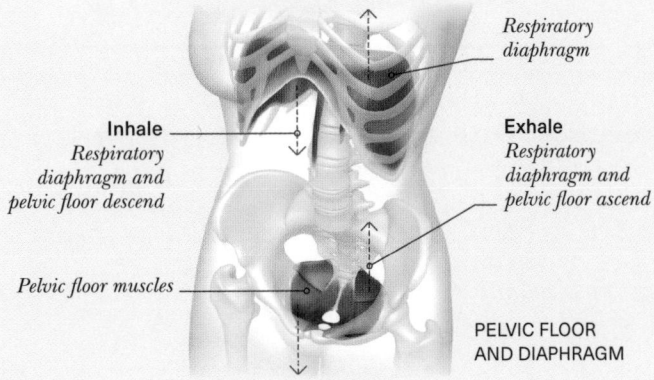

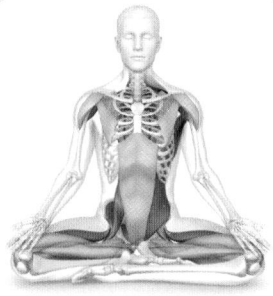

SEATED
Pages 44–83

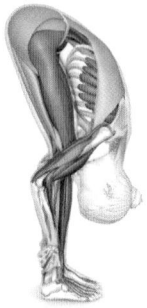

STANDING
Pages 84–121

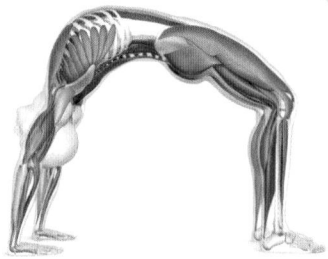

INVERSIONS
Pages 122–143

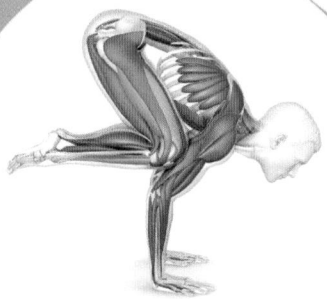

FLOOR
Pages 144–173

THE ASANAS

Allow this section to guide a meditative exploration of your inner world. Visualize, physically touch, and become curious about how your body feels. Studying these 30 asanas can be an engaging way to memorize the muscles and better understand the basics of anatomy, physiology, and kinesiology. I hope these poses, or any variation of them, help you to become more connected to yourself.

SEATED ASANAS

Seated and kneeling poses can be grounding and meditative, often forming the starting and ending points of yoga sessions. The asanas presented here show how the body can benefit physically from yoga in a range of ways. Use variations and modify to find stability and ease in body and mind, and remember: if you can breathe, you can do yoga.

ACCOMPLISHED

Siddhasana

This seated pose is so called because the traditional purpose of all the other poses is to prepare your body physically for this meditative posture. The neutral spine and engaged abdominals should make this pose steady and comfortable; if it isn't, try other options.

VARIATION
The common variation *Sukhasana,* or "easy pose," has the legs crossing at the shins. For many, this may not be so "easy;" find support by sitting on a prop to elevate your hips.

THE BIG PICTURE

Your back muscles and abdominals engage, while stretching muscles on the outside of your hips. You may feel this minimally, but for many people it can be challenging to maintain a neutral spine and pelvis, using muscles in ways your body isn't used to.

KEY
- *Joints*
- *Muscles*
- Engaging
- Engaging while stretching
- Stretching

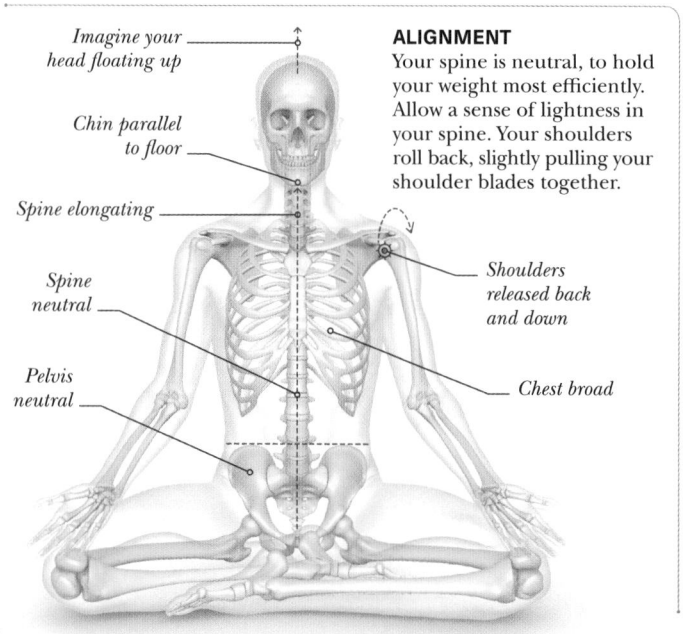

ALIGNMENT
Your spine is neutral, to hold your weight most efficiently. Allow a sense of lightness in your spine. Your shoulders roll back, slightly pulling your shoulder blades together.

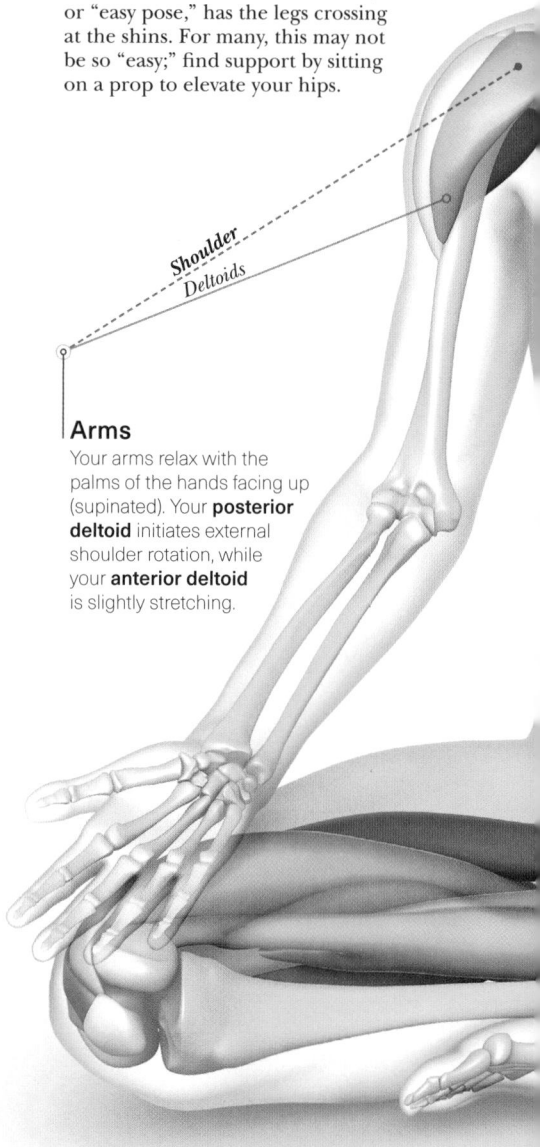

Arms
Your arms relax with the palms of the hands facing up (supinated). Your **posterior deltoid** initiates external shoulder rotation, while your **anterior deltoid** is slightly stretching.

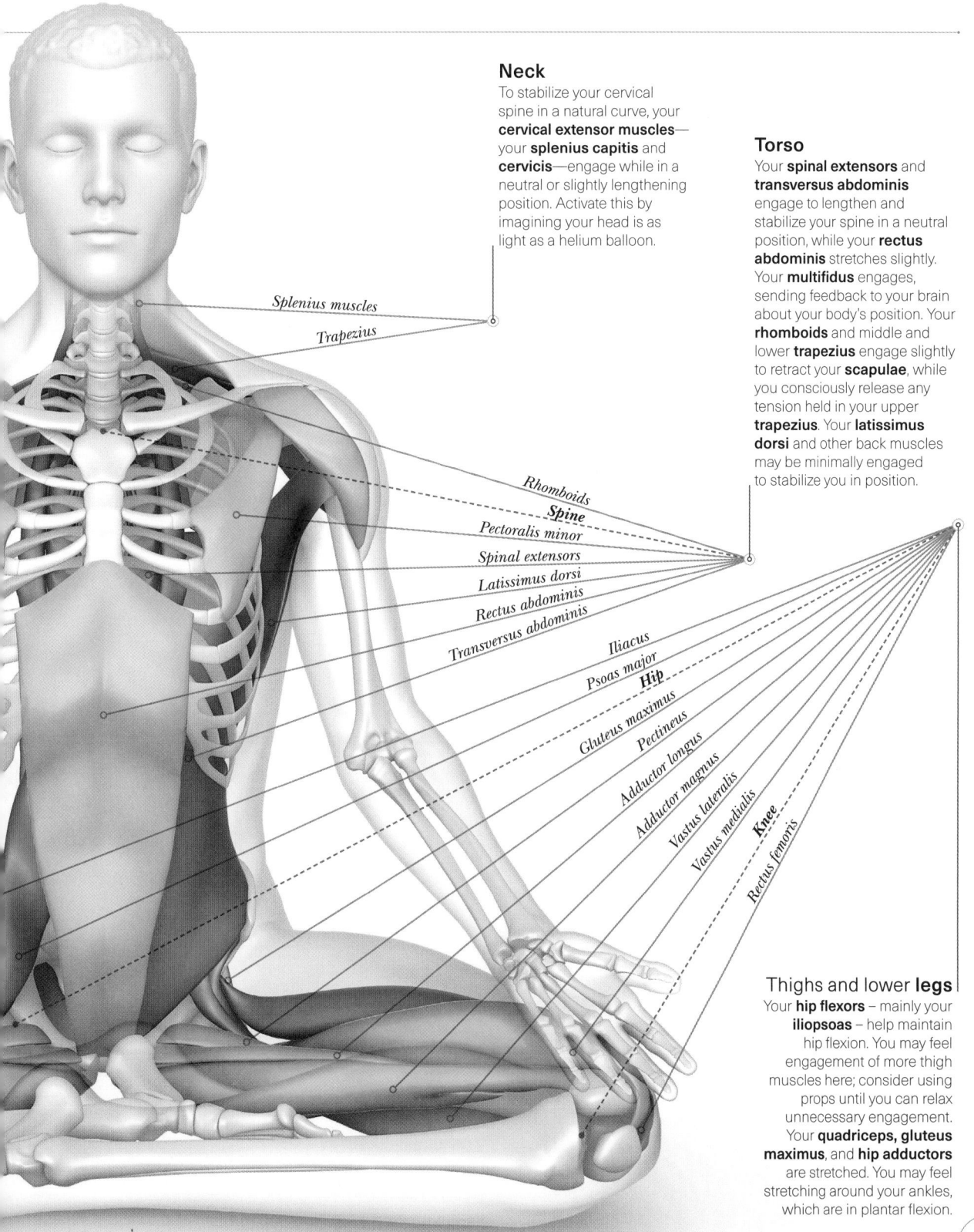

Neck

To stabilize your cervical spine in a natural curve, your **cervical extensor muscles**—your **splenius capitis** and **cervicis**—engage while in a neutral or slightly lengthening position. Activate this by imagining your head is as light as a helium balloon.

Torso

Your **spinal extensors** and **transversus abdominis** engage to lengthen and stabilize your spine in a neutral position, while your **rectus abdominis** stretches slightly. Your **multifidus** engages, sending feedback to your brain about your body's position. Your **rhomboids** and middle and lower **trapezius** engage slightly to retract your **scapulae**, while you consciously release any tension held in your upper **trapezius**. Your **latissimus dorsi** and other back muscles may be minimally engaged to stabilize you in position.

Thighs and lower **legs**

Your **hip flexors** – mainly your **iliopsoas** – help maintain hip flexion. You may feel engagement of more thigh muscles here; consider using props until you can relax unnecessary engagement. Your **quadriceps, gluteus maximus**, and **hip adductors** are stretched. You may feel stretching around your ankles, which are in plantar flexion.

» CLOSER LOOK

In Accomplished pose, your intervertebral disks are stacked on top of each other, creating the natural curves of the neutral spine. As you breathe, your ribcage expands and releases efficiently, which is facilitated by sitting tall with good posture.

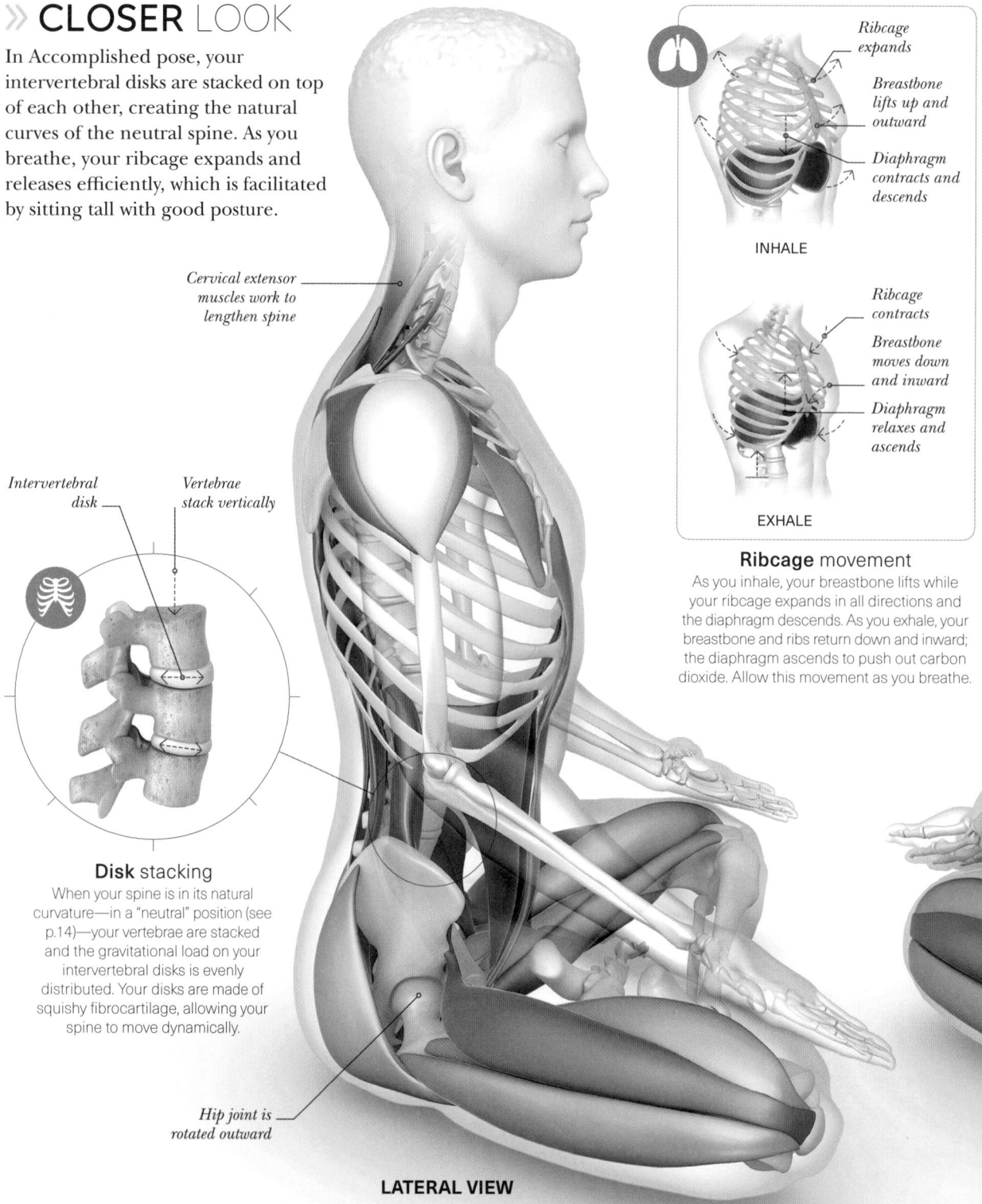

Disk stacking

When your spine is in its natural curvature—in a "neutral" position (see p.14)—your vertebrae are stacked and the gravitational load on your intervertebral disks is evenly distributed. Your disks are made of squishy fibrocartilage, allowing your spine to move dynamically.

Ribcage movement

As you inhale, your breastbone lifts while your ribcage expands in all directions and the diaphragm descends. As you exhale, your breastbone and ribs return down and inward; the diaphragm ascends to push out carbon dioxide. Allow this movement as you breathe.

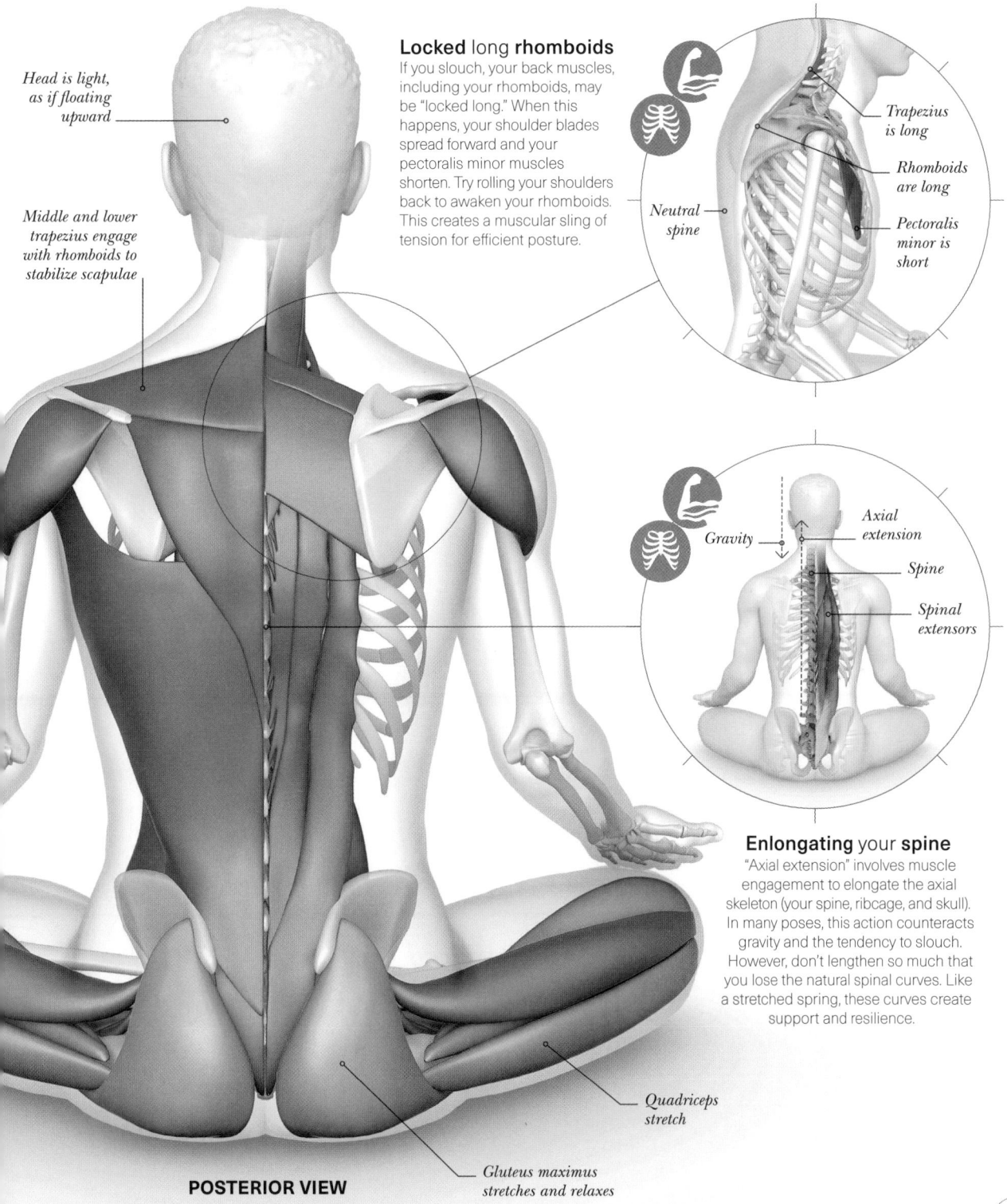

Locked long **rhomboids**

If you slouch, your back muscles, including your rhomboids, may be "locked long." When this happens, your shoulder blades spread forward and your pectoralis minor muscles shorten. Try rolling your shoulders back to awaken your rhomboids. This creates a muscular sling of tension for efficient posture.

Enlongating your **spine**

"Axial extension" involves muscle engagement to elongate the axial skeleton (your spine, ribcage, and skull). In many poses, this action counteracts gravity and the tendency to slouch. However, don't lengthen so much that you lose the natural spinal curves. Like a stretched spring, these curves create support and resilience.

BOUND ANGLE

Baddha Konasana

Bound angle pose is a seated hip opener and groin stretch. It can relieve pelvic cramping, and this version of the pose also improves your ankle flexibility and awareness, which will come in handy in balancing poses.

THE BIG PICTURE

Your inner thighs stretch, particularly around your groin. If you can reach, this is also an opportunity to stretch your ankle muscles by opening your feet like a book revealing its pages.

KEY

- *Joints*
- *Muscles*
- Engaging
- Engaging while stretching
- Stretching

Arms

As you reach toward your feet with flexed elbows, your **brachialis** flexes your elbow with the help of the **biceps brachii** and **brachioradialis**.

Shoulder
Biceps brachii
Brachialis
Brachioradialis
Elbow

Lower **legs**

Your **tibialis anterior** muscles dorsiflex your ankles, and your **extensor digitorum** muscles extend your toes. If you are using your hands to manually invert your feet, your **fibularis muscles** are stretching.

Ankle
Fibularis muscles
Extensor digitorum longus
Tibialis anterior

ALIGNMENT

Your spine is stabilized into neutral and, in this version of the pose, your pelvis is also neutral. Your thighs rest in a rotated outward position.

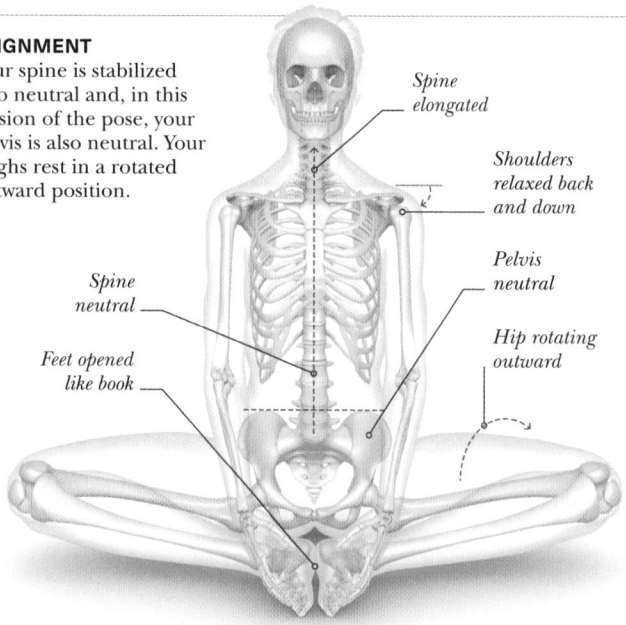

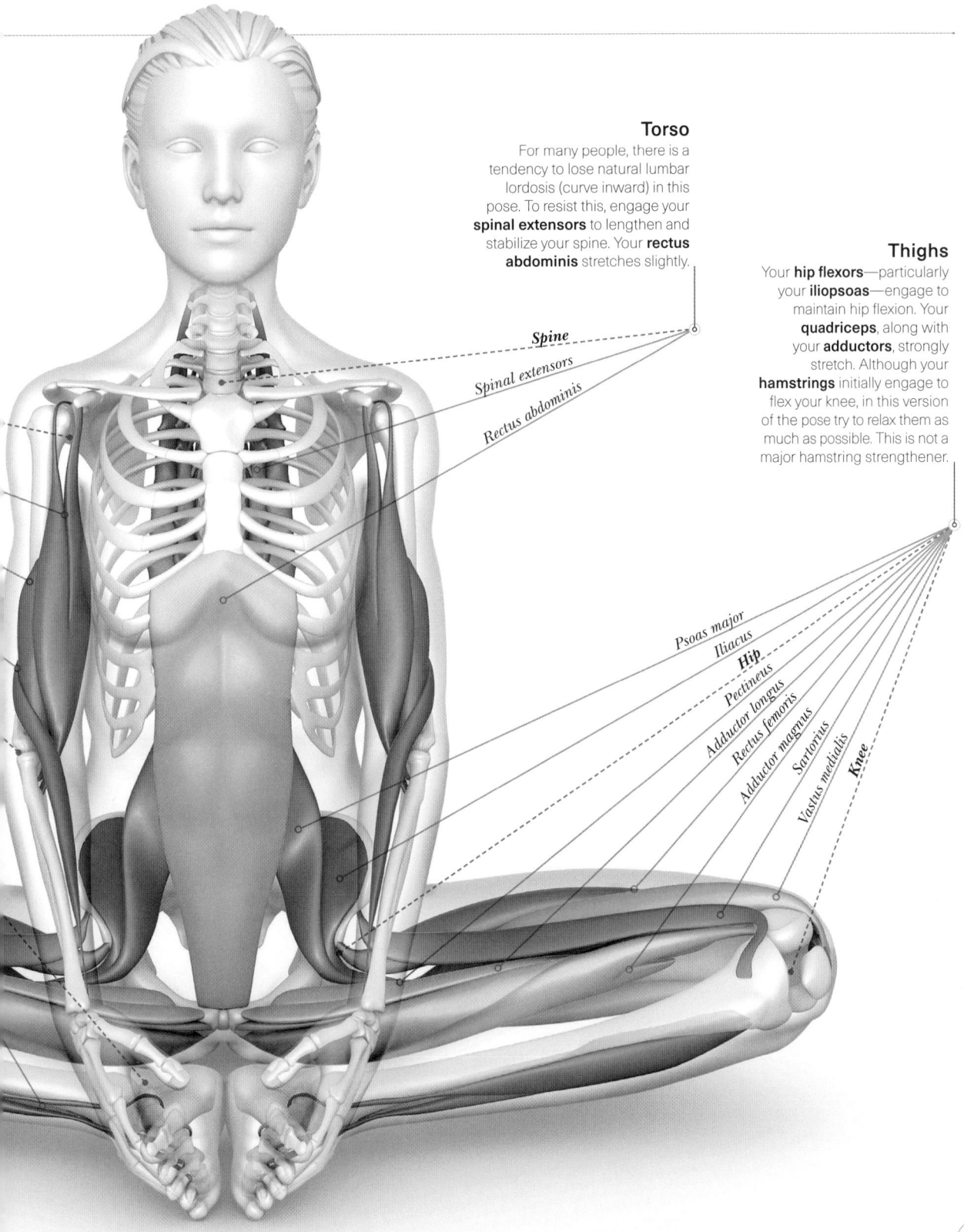

Torso

For many people, there is a tendency to lose natural lumbar lordosis (curve inward) in this pose. To resist this, engage your **spinal extensors** to lengthen and stabilize your spine. Your **rectus abdominis** stretches slightly.

Thighs

Your **hip flexors**—particularly your **iliopsoas**—engage to maintain hip flexion. Your **quadriceps**, along with your **adductors**, strongly stretch. Although your **hamstrings** initially engage to flex your knee, in this version of the pose try to relax them as much as possible. This is not a major hamstring strengthener.

» CLOSER LOOK

Your one-of-a-kind bone shapes and joint structures determine what your bound angle pose looks like. Some people will never be able to bring their knees to the floor and that is okay. Focus on releasing your hips.

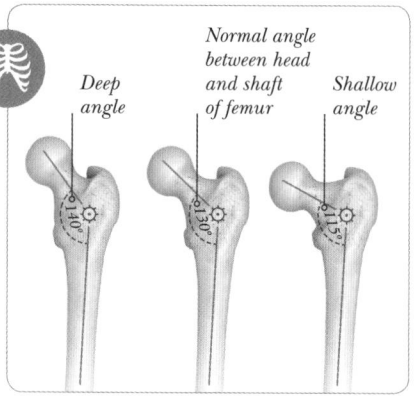

Femur differences

When seated in Bound Angle pose, notice if you feel a "hard" or a "soft" stop. A hard stop is when the bones get in the way, with little stretching sensation felt. A soft stop is when tight muscles limit movement and a stretching is sensation felt. Soft stops can shift from stretching, but you cannot change hard stops. Variance in femur shape and angle can limit certain poses.

Head is stacked over your spine

Multifidus sends sensory information to your brain, helping body awareness

Erector spinae engage to maintain posture

Lower back is curved inward to neutral, not rounding forward

Head of femur rotates outward in your hip socket

Knee is lowered toward floor

POSTERIOR–LATERAL VIEW

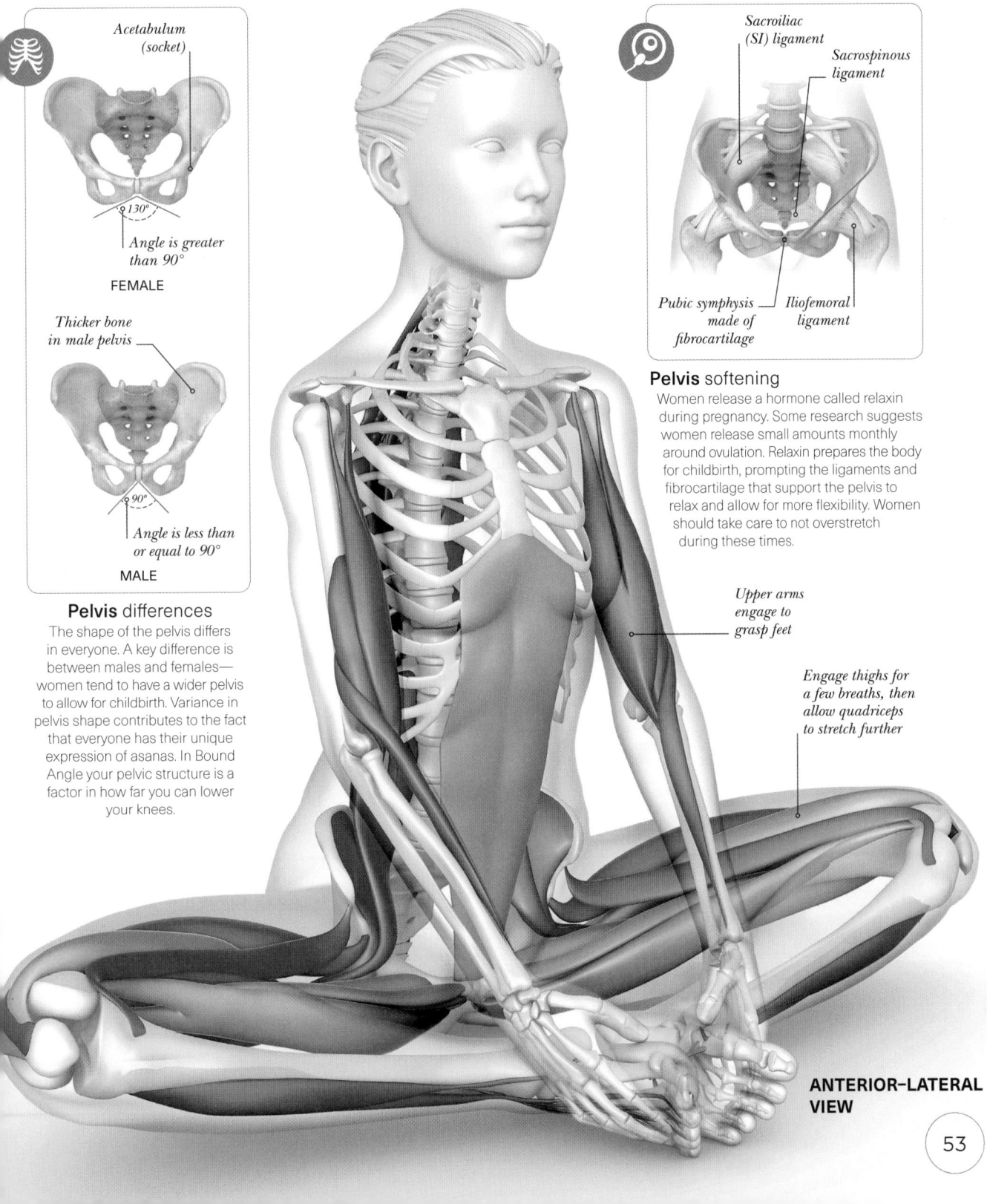

Pelvis differences

The shape of the pelvis differs in everyone. A key difference is between males and females—women tend to have a wider pelvis to allow for childbirth. Variance in pelvis shape contributes to the fact that everyone has their unique expression of asanas. In Bound Angle your pelvic structure is a factor in how far you can lower your knees.

Pelvis softening

Women release a hormone called relaxin during pregnancy. Some research suggests women release small amounts monthly around ovulation. Relaxin prepares the body for childbirth, prompting the ligaments and fibrocartilage that support the pelvis to relax and allow for more flexibility. Women should take care to not overstretch during these times.

ANTERIOR–LATERAL VIEW

CAT

Marjaryasana

This is a gentle kneeling pose that takes the position of a scared cat, warming up joints in your spine, hips, and shoulders. Try exhaling as you move into the pose. This is often done with the next pose, Cow, by flowing from Cat to Cow with the exhale and inhale.

THE BIG PICTURE

Your back muscles stretch while the muscles on the front of your body—including your chest and abdominal muscles—engage. Muscles in your arms work to stabilize you. Your rib cage is compressed, helping to facilitate a deep exhale into the pose.

ALIGNMENT

Your arms and thighs are fixed in place, with your knees directly under your hips and hands under your shoulders (or slightly forward). The rounding of your spine is as even as possible.

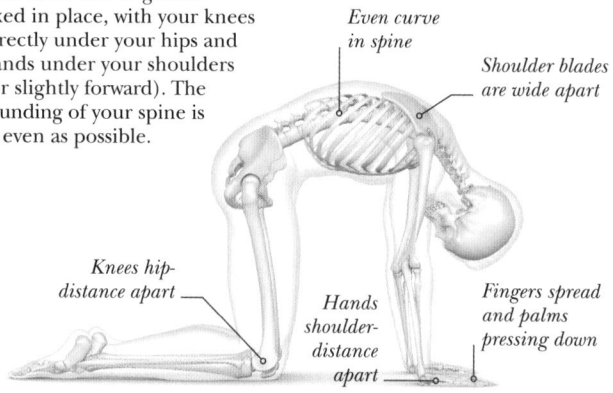

Lower **torso**

Your lumbar spine is in flexion, stretching your **quadratus lumborum**. Your **abdominals** engage to compress your abdomen, squeezing your belly button in toward your spine. Your pelvis is in a posterior pelvic tilt.

Internal obliques
Rectus abdominis
Iliopsoas
Hip

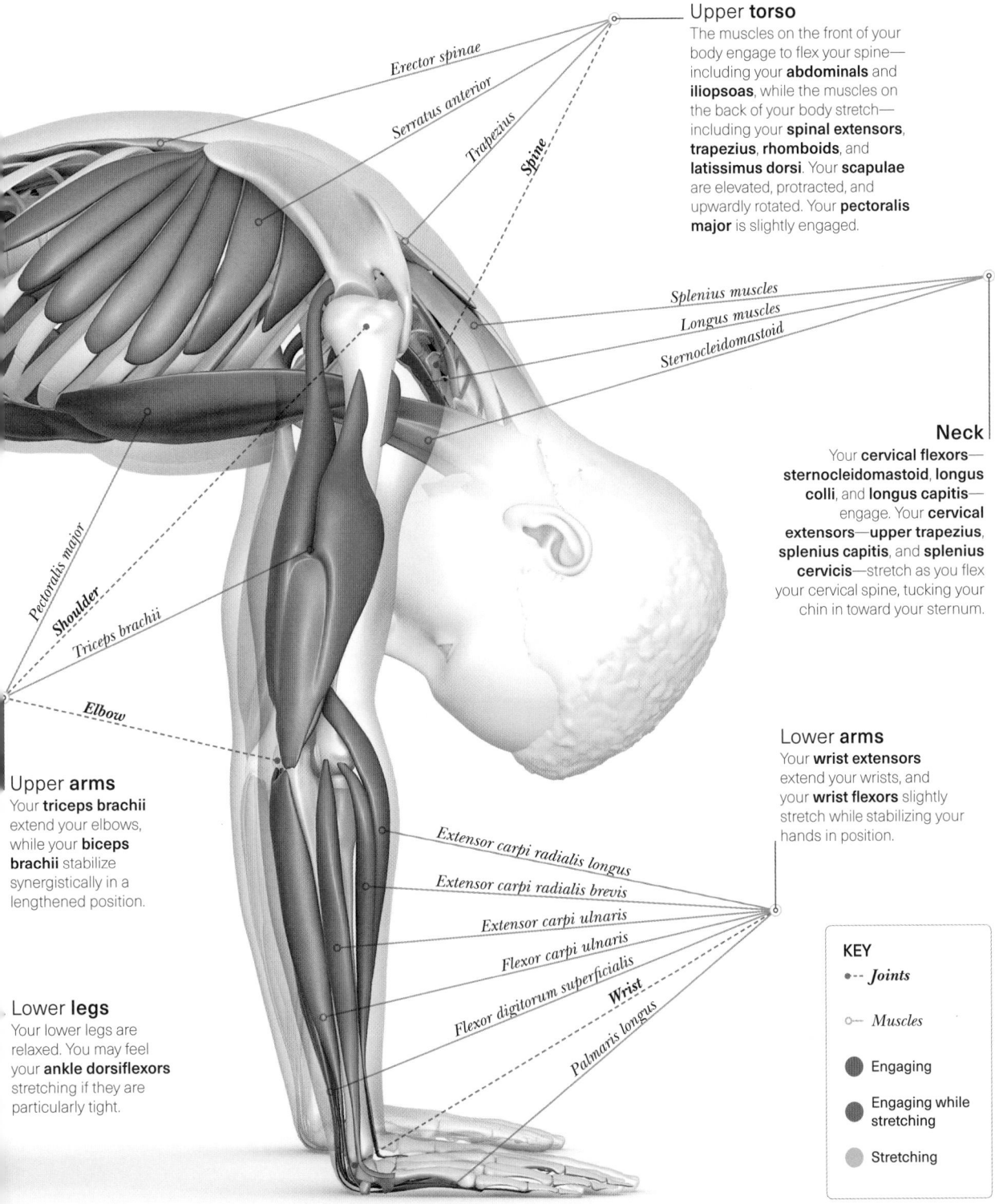

Upper **torso**

The muscles on the front of your body engage to flex your spine—including your **abdominals** and **iliopsoas**, while the muscles on the back of your body stretch—including your **spinal extensors**, **trapezius**, **rhomboids**, and **latissimus dorsi**. Your **scapulae** are elevated, protracted, and upwardly rotated. Your **pectoralis major** is slightly engaged.

Neck

Your **cervical flexors**—**sternocleidomastoid**, **longus colli**, and **longus capitis**—engage. Your **cervical extensors**—**upper trapezius**, **splenius capitis**, and **splenius cervicis**—stretch as you flex your cervical spine, tucking your chin in toward your sternum.

Upper **arms**

Your **triceps brachii** extend your elbows, while your **biceps brachii** stabilize synergistically in a lengthened position.

Lower **arms**

Your **wrist extensors** extend your wrists, and your **wrist flexors** slightly stretch while stabilizing your hands in position.

Lower **legs**

Your lower legs are relaxed. You may feel your **ankle dorsiflexors** stretching if they are particularly tight.

COW

Bitilasana

Mimicking the slightly dipped back of a cow, this gentle kneeling pose incorporates a backbend, and is practiced to warm up the spine, hips, and shoulders. Inhale as you enter the pose; you can also alternate between this and Cat pose, in time with your breath.

THE BIG PICTURE

Your abdominal and chest muscles stretch, while your back muscles—including your spinal extensors—engage. Your rib cage is expanding, making it possible to inhale fully. A subtle, even curve is created by the backbend and raised head.

ALIGNMENT

Your arms and thighs are fixed in place with your knees under your hips and hands under your shoulders (or slightly forward). Your backbend is as even as possible, focusing on lengthening your neck, creating a subtle, even curve.

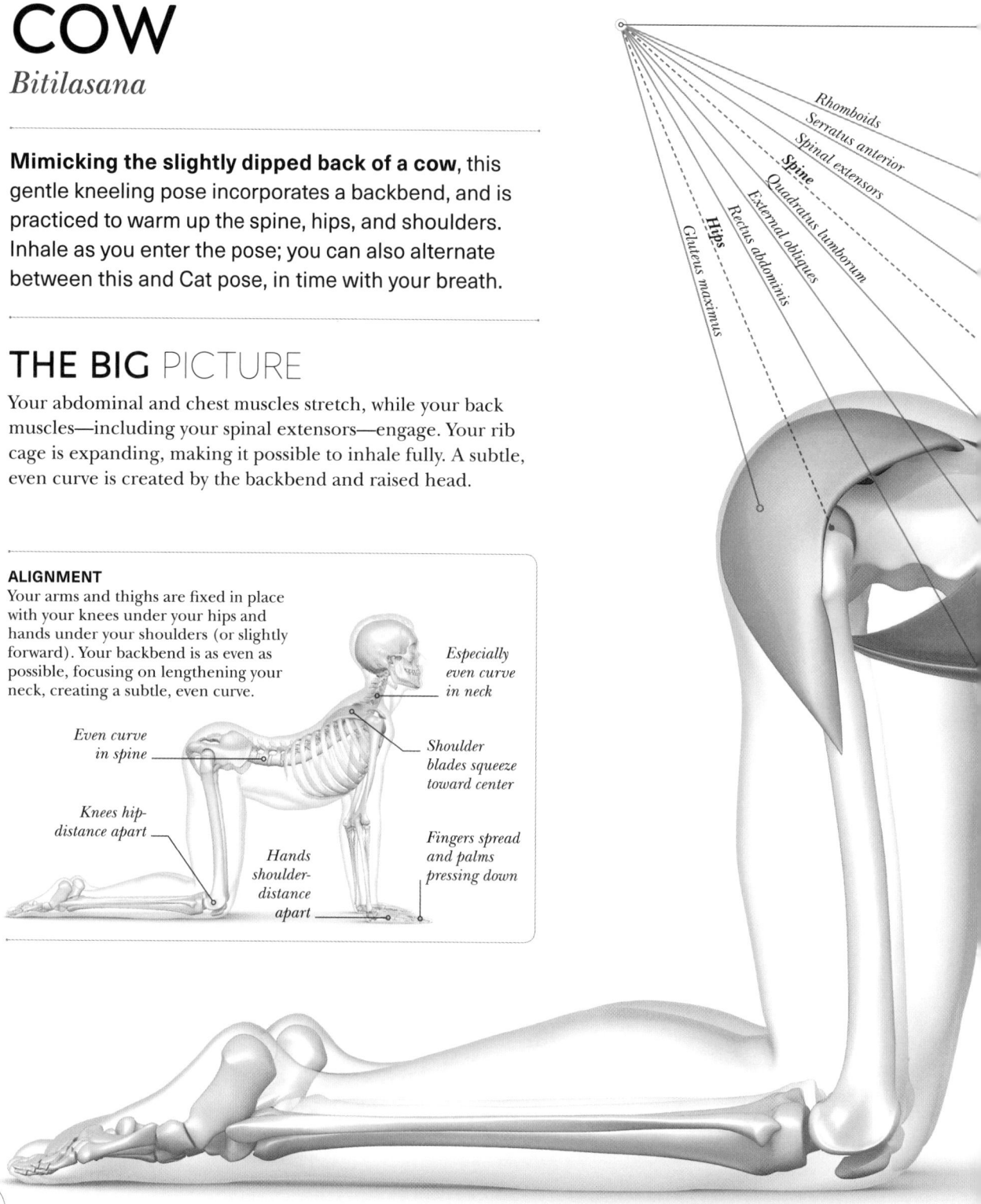

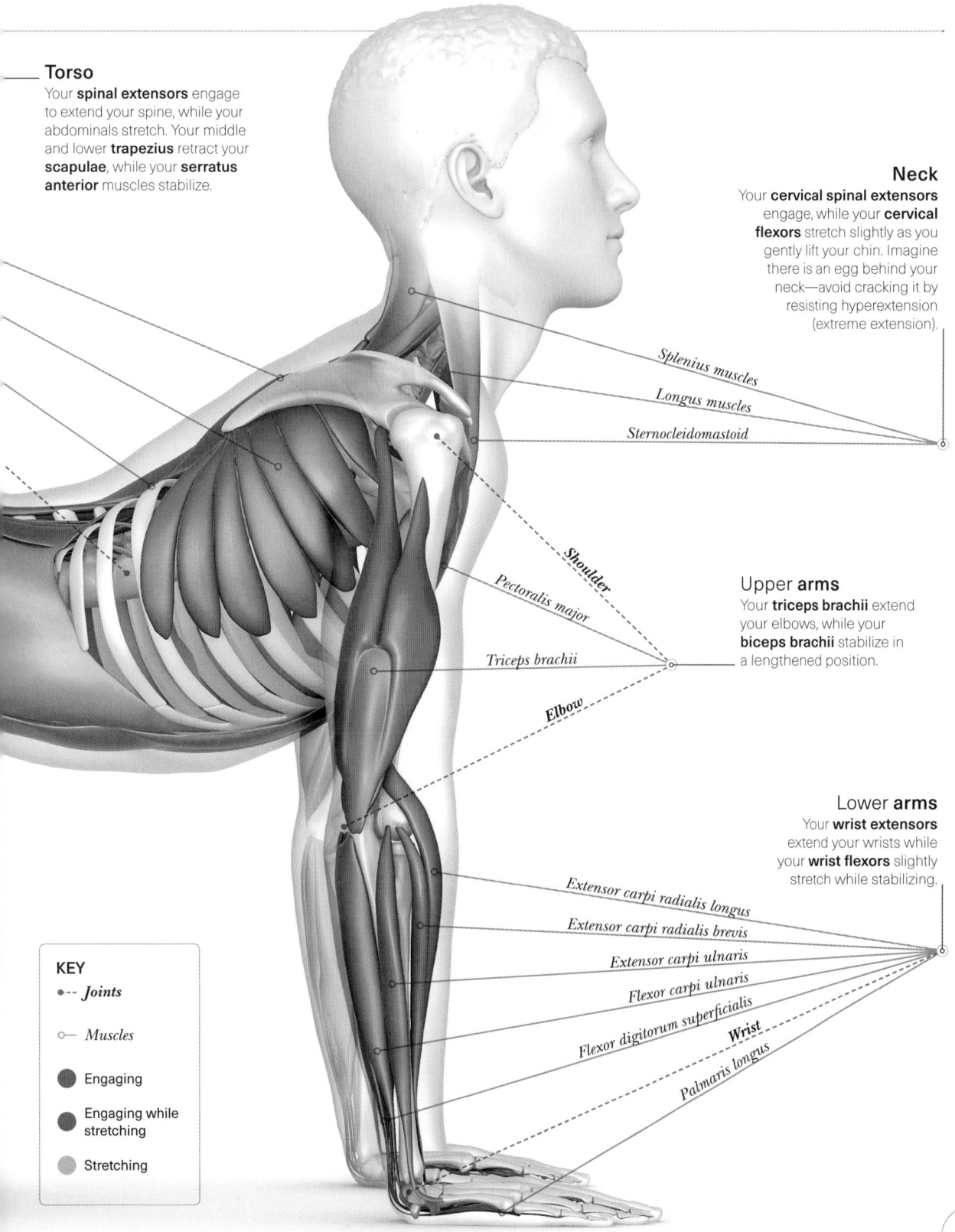
Torso
Your **spinal extensors** engage to extend your spine, while your abdominals stretch. Your middle and lower **trapezius** retract your **scapulae**, while your **serratus anterior** muscles stabilize.
Neck
Your **cervical spinal extensors** engage, while your **cervical flexors** stretch slightly as you gently lift your chin. Imagine there is an egg behind your neck—avoid cracking it by resisting hyperextension (extreme extension).
Splenius muscles
Longus muscles
Sternocleidomastoid
Shoulder
Pectoralis major
Upper arms
Your **triceps brachii** extend your elbows, while your **biceps brachii** stabilize in a lengthened position.
Triceps brachii
Elbow
Lower arms
Your **wrist extensors** extend your wrists while your **wrist flexors** slightly stretch while stabilizing.
Extensor carpi radialis longus
Extensor carpi radialis brevis
Extensor carpi ulnaris
Flexor carpi ulnaris
Flexor digitorum superficialis
Wrist
Palmaris longus
KEY
Joints
Muscles
Engaging
Engaging while stretching
Stretching

» CLOSER LOOK

Flowing from the flexion of Cat to the extension of Cow as you breathe deeply in and out improves your mind–body connection, as well as your sense of body awareness.

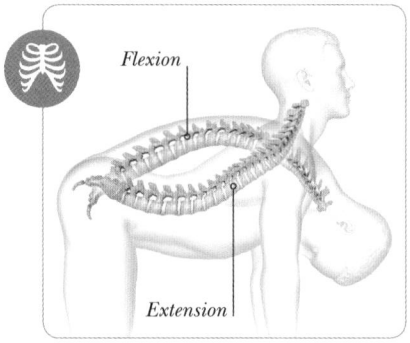

Spine flexion and extension

When your spine flexes, the front of your body engages while the back of your body stretches. When your spine extends, going into a backbend, the back of your body engages while the front of your body stretches. Your spinal extensors are the main players in this extension.

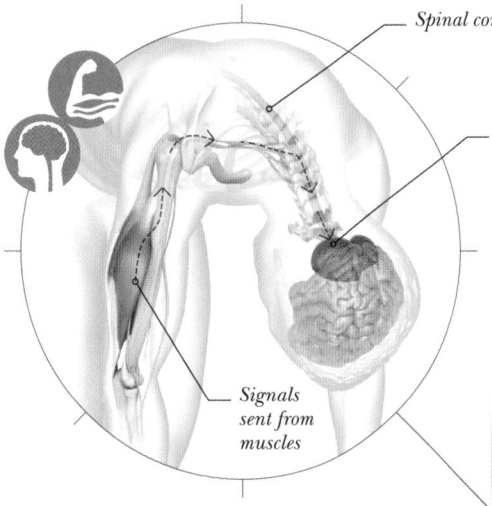

Mind–body connection

We often think of the brain as controlling our muscles. This is true: those motor signals tell your muscles what to do. However, your nervous system is a two-way conversation. Your body sends tons of sensory signals to your brain. Yoga improves the mind–body connection by encouraging you to listen to your body.

Trapezius stretches

Ankles and feet are relaxed (in plantar flexion)

Middle finger is facing forward

ANTERIOR–LATERAL VIEW OF CAT

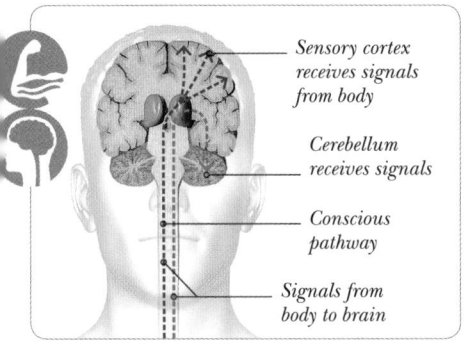

Proprioception

Proprioception is body awareness, particularly while moving through space. Your cerebellum constantly receives unconscious signals from your body about its position, while your cerebral cortex consciously deciphers where you are in space. Mindfully flowing between poses can help develop this awareness and improve balance.

Eyes open or closed

Inhalation and exhalation

Breathe mindfully, coordinating your breath and movement; your nervous system loves this sort of integration. As a general rule, exhale when your ribcage is compressed (as in forward folds or twists), and inhale when your ribcage can expand (as in backbends).

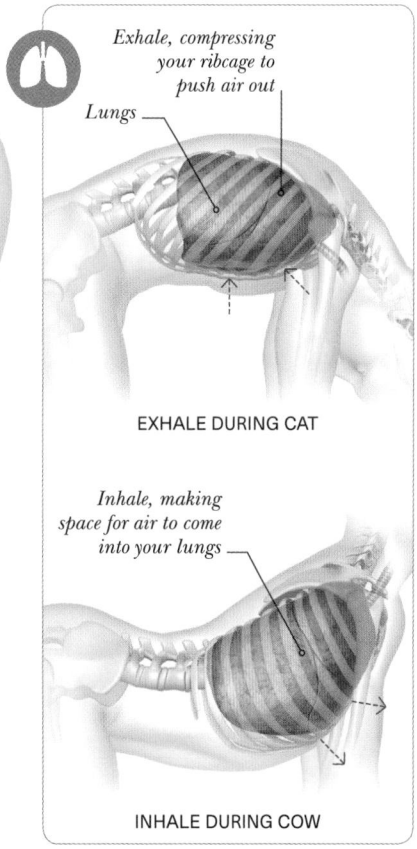

Pectoral muscles stabilize your shoulders

Wrists are extended

Kneecap may need padding, such as a folded blanket

ANTERIOR–LATERAL VIEW OF COW

COW FACE

Gomukhasana

This seated pose involves unique actions of your shoulder joints. This can be helpful in stretching out tight shoulders, particularly if you work at a desk and spend a lot of time typing—but you should avoid this pose if you have a rotator cuff injury. Switch arms and notice if you feel a difference between each side.

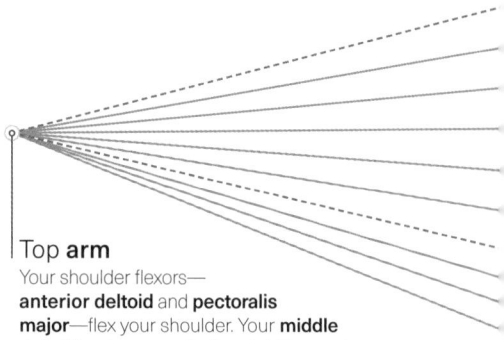

Top arm

Your shoulder flexors—**anterior deltoid** and **pectoralis major**—flex your shoulder. Your **middle deltoid** and **supraspinatus** stabilize and abduct it, and your **infraspinatus**, **teres minor**, and **posterior deltoid** engage to externally rotate. Your elbow flexors engage and **triceps brachii** stretches.

THE BIG PICTURE

In this seated pose, you particularly stretch around your shoulders and the outside of your hips and buttocks. You are also engaging key postural muscles to counteract slouching or rounding forward.

KEY

- *Joints*
- *Muscles*
- Engaging
- Engaging while stretching
- Stretching

Torso

Your **spinal extensors** and **transversus abdominis** engage to slightly extend and stabilize your spine, while your **rectus abdominis** stretches. Your **rhomboids** engage to retract your **scapulae**.

ALIGNMENT

Your knees are stacked in the center while your hands reach toward each other, trying to clasp your fingertips. Your elbows are squeezing in toward the center. Keep your spine neutral or in a slight backbend, trying to twist or lean.

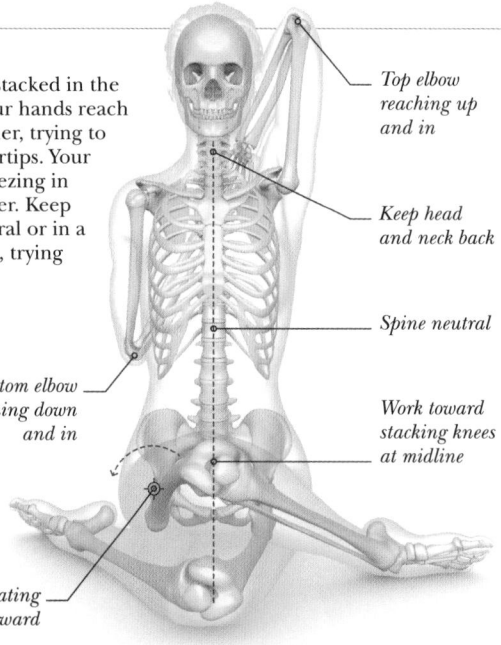

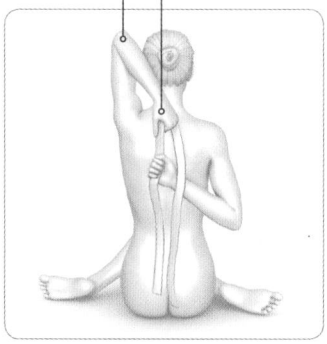

VARIATION

If you cannot reach your hands together, use a strap or towel to extend your reach. If you hold for approximately 10 breaths, you may find you can walk your fingers closer toward each other.

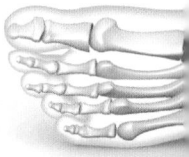

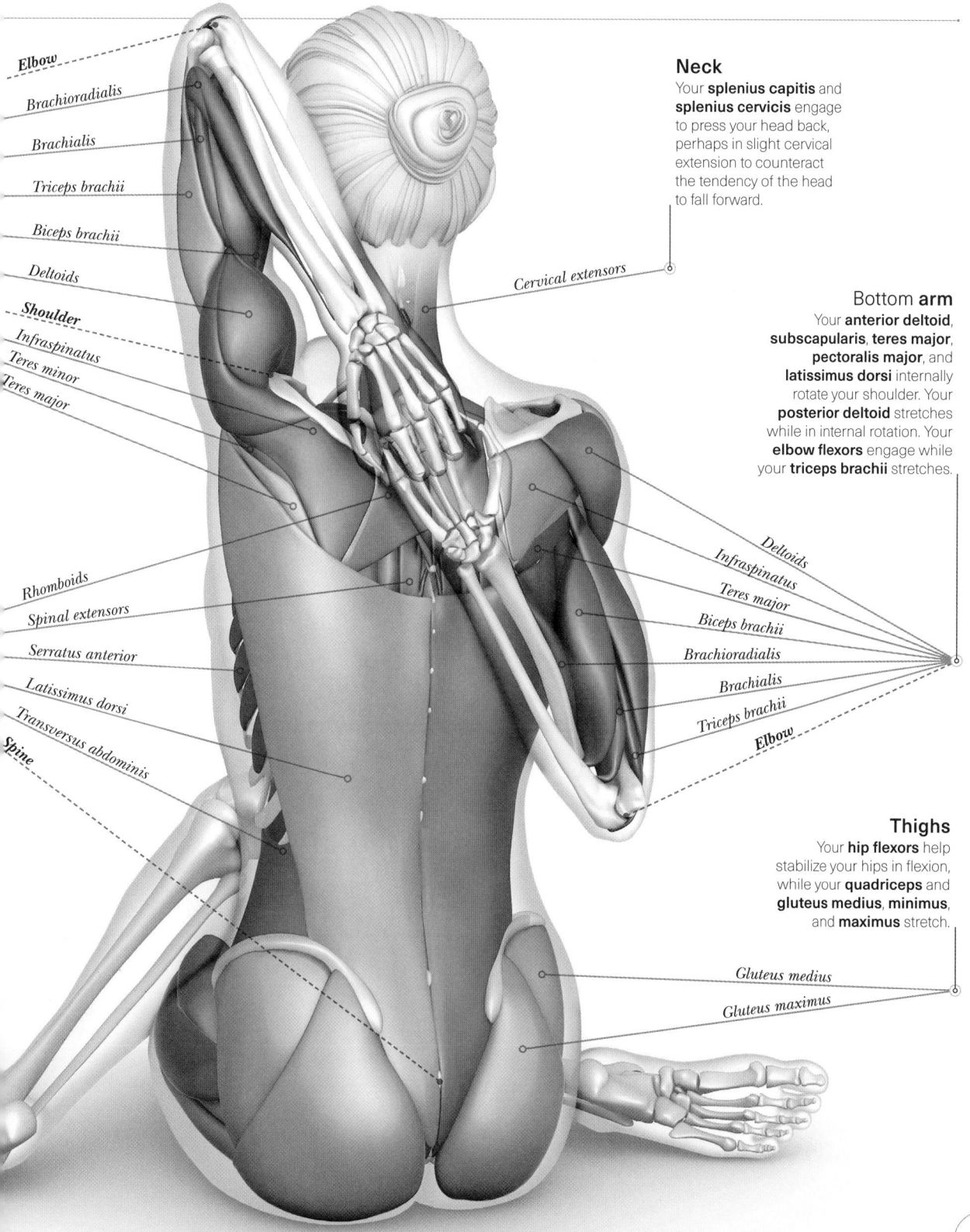

Neck

Your **splenius capitis** and **splenius cervicis** engage to press your head back, perhaps in slight cervical extension to counteract the tendency of the head to fall forward.

Bottom **arm**

Your **anterior deltoid**, **subscapularis**, **teres major**, **pectoralis major**, and **latissimus dorsi** internally rotate your shoulder. Your **posterior deltoid** stretches while in internal rotation. Your **elbow flexors** engage while your **triceps brachii** stretches.

Thighs

Your **hip flexors** help stabilize your hips in flexion, while your **quadriceps** and **gluteus medius**, **minimus**, and **maximus** stretch.

» CLOSER LOOK

Cow Face works your shoulders dynamically—including your deltoid muscles. Compression at your shoulder joint can also lead to cardiovascular shifts both system-wide and in local blood vessels.

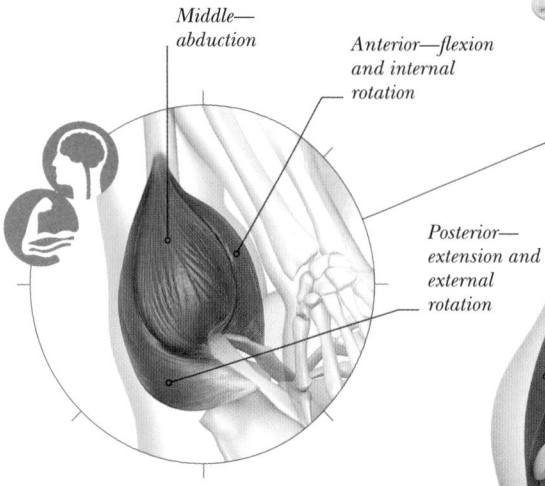

Dynamic **deltoids**

Your deltoids are split into three parts, or heads, which have opposing actions when engaged. Some research suggests that there are 19 parts filled with muscle fibers that can be controlled independently by your nervous system. This pose dynamically engages and stretches each part of your deltoids.

Triceps brachii stretches strongly on your top arm

Fingers gently clasp, if accessible

Pectoralis major stretches while flexing your shoulder

Latissimus dorsi stretches on this side

All of your gluteal muscles stretch, including your gluteus maximus

LATERAL VIEW

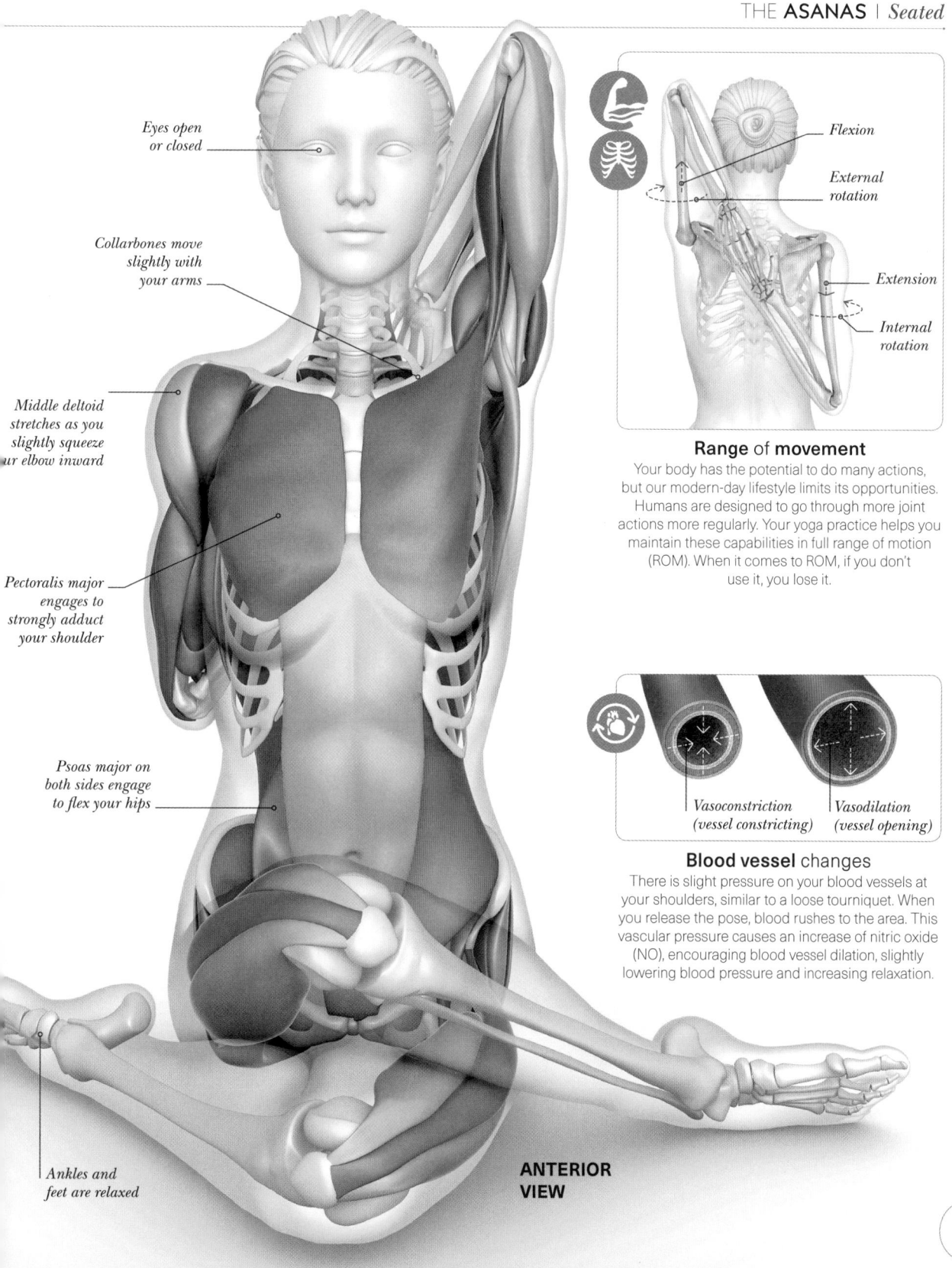

Range of **movement**

Your body has the potential to do many actions, but our modern-day lifestyle limits its opportunities. Humans are designed to go through more joint actions more regularly. Your yoga practice helps you maintain these capabilities in full range of motion (ROM). When it comes to ROM, if you don't use it, you lose it.

Blood vessel changes

There is slight pressure on your blood vessels at your shoulders, similar to a loose tourniquet. When you release the pose, blood rushes to the area. This vascular pressure causes an increase of nitric oxide (NO), encouraging blood vessel dilation, slightly lowering blood pressure and increasing relaxation.

SIDE BEND

Parivrtta Janu Sirsasana

This seated, lateral side stretch allows you to mobilize your spine in a way that you probably don't often move it in everyday life. The novel movement that this pose involves benefits your intervertebral discs, nervous system, and fascia.

THE BIG PICTURE

As you bend deeply to the side, muscles along your spine stretch and strengthen. Your shoulder muscles engage to reach your arms over your head, and your thigh muscles on both sides stretch in different ways.

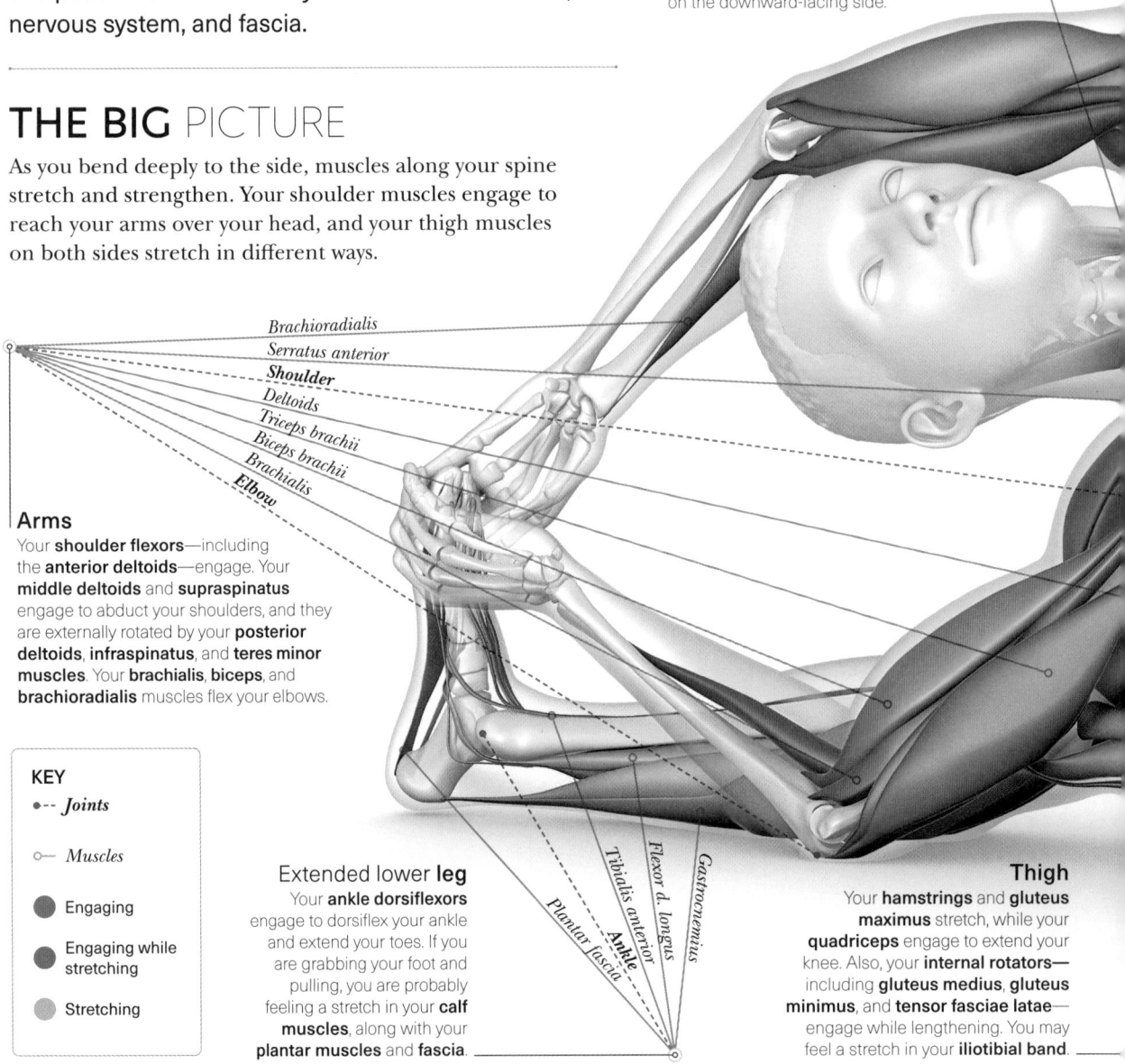

Neck

To rotate your neck, your **rotatores**, **multifidus**, **sternocleidomastoid**, and **semispinalis cervicis** engage on the side toward the ground (model's right in this image), while stretching on the upward-facing side. Your **splenius capitis** and **splenius cervicis** engage on the upward-facing side (model's left in this image), while stretching on the downward-facing side.

Arms

Your **shoulder flexors**—including the **anterior deltoids**—engage. Your **middle deltoids** and **supraspinatus** engage to abduct your shoulders, and they are externally rotated by your **posterior deltoids**, **infraspinatus**, and **teres minor muscles**. Your **brachialis**, **biceps**, and **brachioradialis** muscles flex your elbows.

KEY

- *Joints*
- *Muscles*
- Engaging
- Engaging while stretching
- Stretching

Extended lower **leg**

Your **ankle dorsiflexors** engage to dorsiflex your ankle and extend your toes. If you are grabbing your foot and pulling, you are probably feeling a stretch in your **calf muscles**, along with your **plantar muscles** and **fascia**.

Thigh

Your **hamstrings** and **gluteus maximus** stretch, while your **quadriceps** engage to extend your knee. Also, your **internal rotators**—including **gluteus medius**, **gluteus minimus**, and **tensor fasciae latae**—engage while lengthening. You may feel a stretch in your **iliotibial band**.

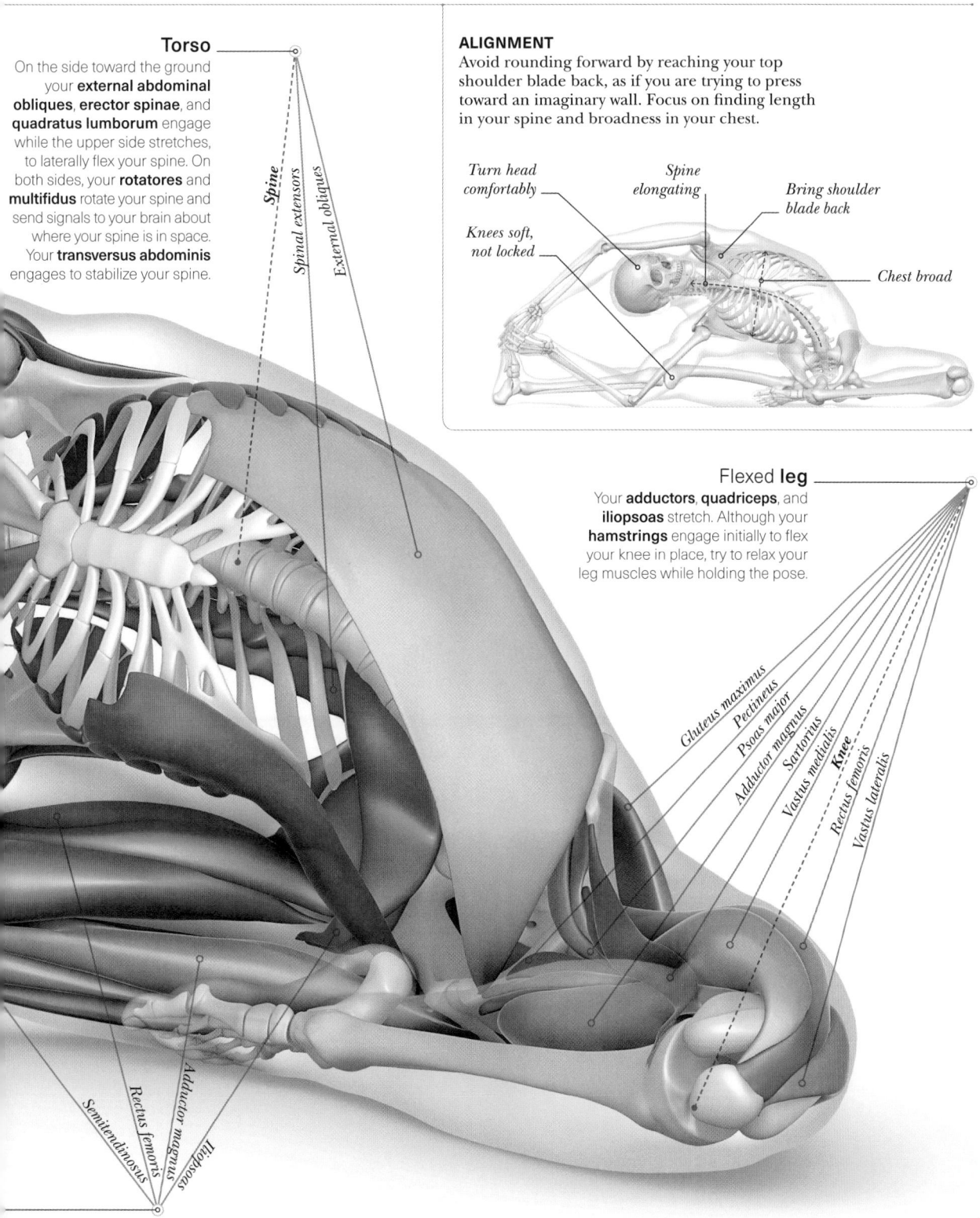
Torso
On the side toward the ground your **external abdominal obliques**, **erector spinae**, and **quadratus lumborum** engage while the upper side stretches, to laterally flex your spine. On both sides, your **rotatores** and **multifidus** rotate your spine and send signals to your brain about where your spine is in space. Your **transversus abdominis** engages to stabilize your spine.
Spine
Spinal extensors
External obliques
ALIGNMENT
Avoid rounding forward by reaching your top shoulder blade back, as if you are trying to press toward an imaginary wall. Focus on finding length in your spine and broadness in your chest.
Turn head comfortably
Spine elongating
Bring shoulder blade back
Knees soft, not locked
Chest broad
Flexed leg
Your **adductors**, **quadriceps**, and **iliopsoas** stretch. Although your **hamstrings** engage initially to flex your knee in place, try to relax your leg muscles while holding the pose.
Gluteus maximus
Pectineus
Psoas major
Adductor magnus
Sartorius
Vastus medialis
Knee
Rectus femoris
Vastus lateralis
Semitendinosus
Rectus femoris
Adductor magnus
Iliopsoas

» CLOSER LOOK

Seated side bend is a one-sided movement that dynamically affects your abdominals, back muscles, and spinal disks. You don't have to be able to reach your foot with either hand to do this pose; your arms can simply reach to the side.

Disk health

When side-bending (lateral flexion of the spine), your intervertebral disks push to the sides. As you bend to the right, your disks shift to the left (and vice versa). The cartilage in your spine allows for this natural process.

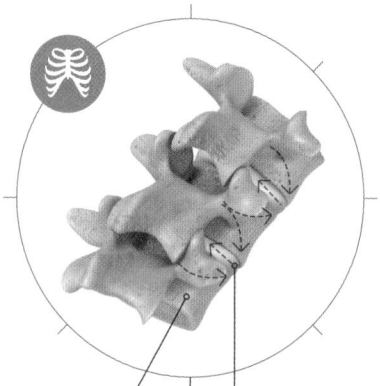

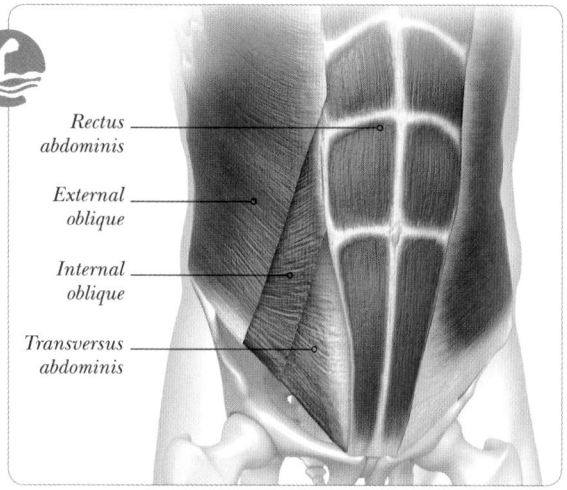

Abdominal structure

Your criss-crossing abdominal muscles provide multilayered support for your internal organs and allow your torso to move. Legend has it that in 1888, Dr. Dunlop, a surgeon, was watching his son bouncing on his tricycle due to the poor design of the wheels, causing a headache. Inspired by the structure of the abdominals, he designed a tire for a smoother ride and fewer flats.

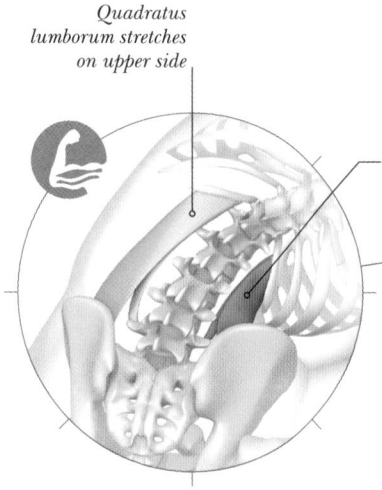

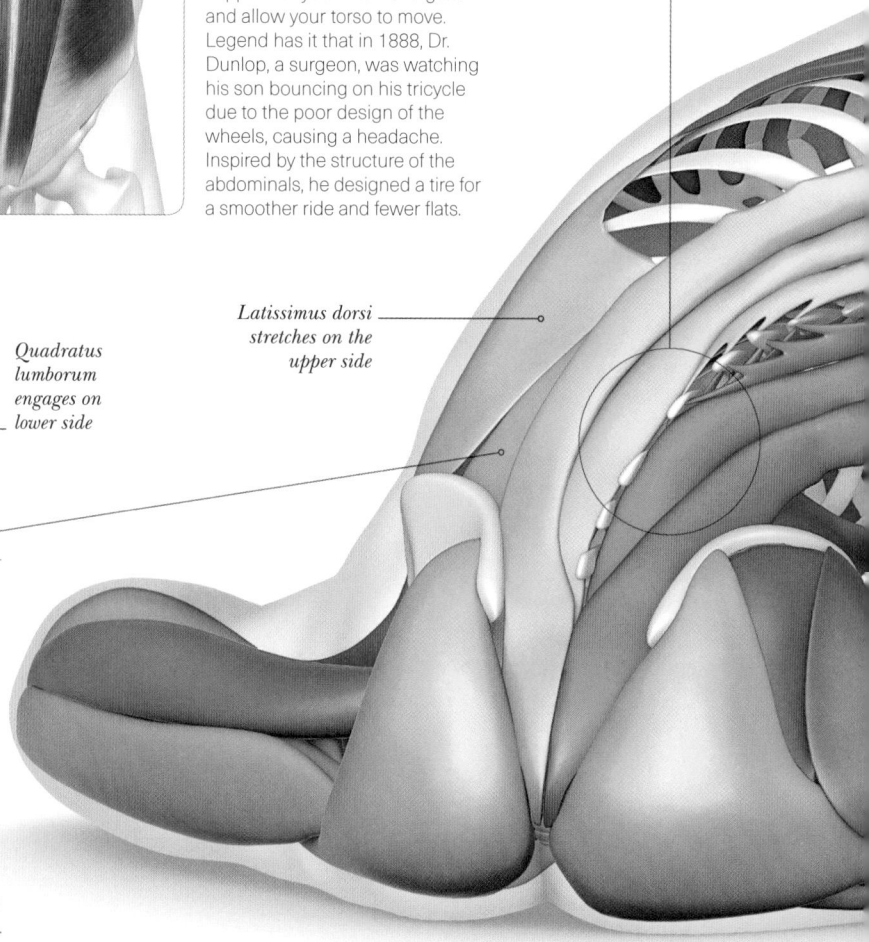

Unilateral movement

Your quadratus lumborum (QL) is important for holding posture. When the erector spinae are weak it picks up the slack. Keeping your spine erect is a big job for this little muscle, leading to muscle fatigue and even pain. This pose helps by stretching and engaging the QL.

Pressure and **balance**

Notice and feel the point of contact of your body on the floor in this pose. It is a little different for everybody. Notice how the pressure points shift as you transition in and out of the pose.

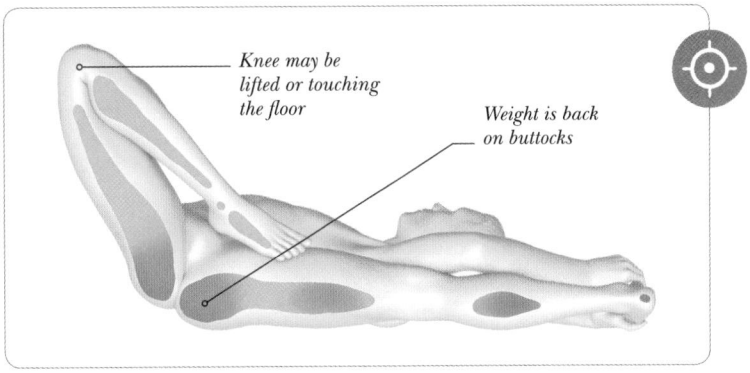

Posterior deltoid stretches and engages to help externally rotate shoulder

Erector spinae stretch on the upper side

Elbow flexors engage

If hands don't reach foot, place one hand on your shin and reach the other out and over toward it

Ankle dorsiflexors engage to flex your ankle

Erector spinae engage on the lower side

POSTERIOR VIEW

SEATED TWIST

Ardha Matsyendrasana

This seated twist will wake up small muscles along your spine and stimulate digestion. Practicing twists mindfully in yoga can help prevent injury from twists you do in everyday life. Take care not to twist too far if you have spinal disk issues or osteoporosis.

THE BIG PICTURE

Your back muscles and abdominals dynamically engage and stretch as you rotate your spine. Your thighs and hips—particularly around your buttocks—are stretching as they rotate outward. Your lowered arm presses down to help you find more length along your spine.

KEY

- *Joints*
- *Muscles*
- Engaging
- Engaging while stretching
- Stretching

Neck

To rotate your neck, on the contralateral side of axial rotation (side you are rotating away from, model's left side on this image), your **rotatores**, **multifidus**, **sternocleidomastoid**, and **semispinalis cervicis** engage while stretching on the ipsilateral side (the side you are rotating toward). Your **splenius capitis** and **splenius cervicis** engage on the ipsilateral side, and stretch on the contralateral.

Sternocleidomastoid

Shoulder
Teres minor
Teres major
Triceps brachii
Biceps brachii
Brachialis
Elbow
Brachioradialis

Arms

On your extended arm, your **teres minor** engages to stabilize and externally rotate your shoulder, while your **teres major** extends your shoulder. Your **elbow flexors** and **triceps** are dynamically engaging to help hold your arm in place, pushing down into the ground to help elongate your spine. On your flexed arm, your **elbow flexors** engage while your **triceps** stretches slightly.

ALIGNMENT

Prioritize elongating your spine over rotating more or leaning. If you do decide to rotate more deeply, try to use your core muscles instead of pulling with the external force of your arms.

Spine elongating

Keep the rotation as even as possible

Pelvis rotates slightly with you

Arm presses down

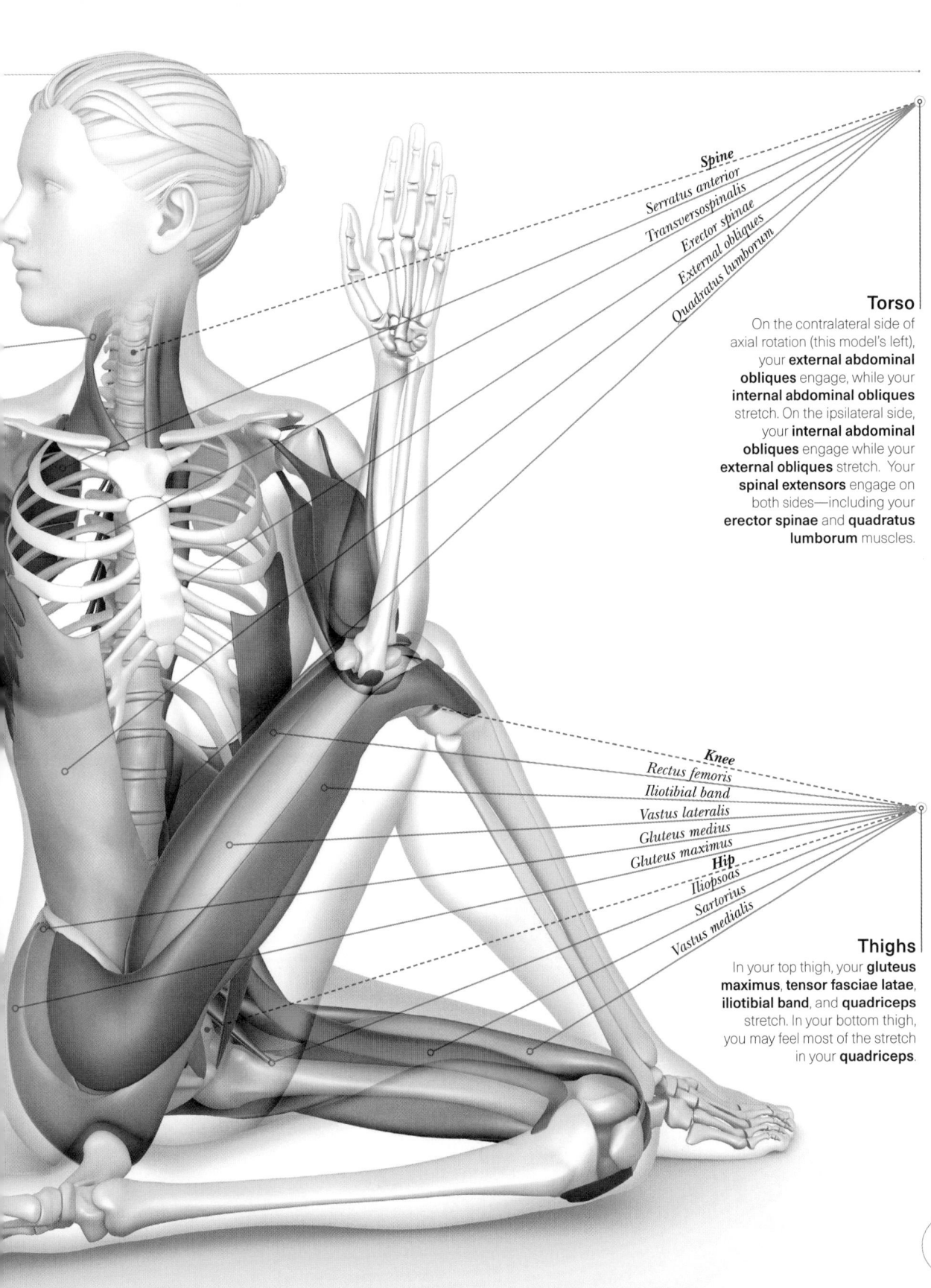

Torso

On the contralateral side of axial rotation (this model's left), your **external abdominal obliques** engage, while your **internal abdominal obliques** stretch. On the ipsilateral side, your **internal abdominal obliques** engage while your **external obliques** stretch. Your **spinal extensors** engage on both sides—including your **erector spinae** and **quadratus lumborum** muscles.

Thighs

In your top thigh, your **gluteus maximus**, **tensor fasciae latae**, **iliotibial band**, and **quadriceps** stretch. In your bottom thigh, you may feel most of the stretch in your **quadriceps**.

» CLOSER LOOK

Spinal twists affect the disks between your vertebrae and your sacroiliac joint. Although this action may not "wring out toxins" as is sometimes claimed, it does encourage healthy digestive movement in your intestines, known as peristalsis.

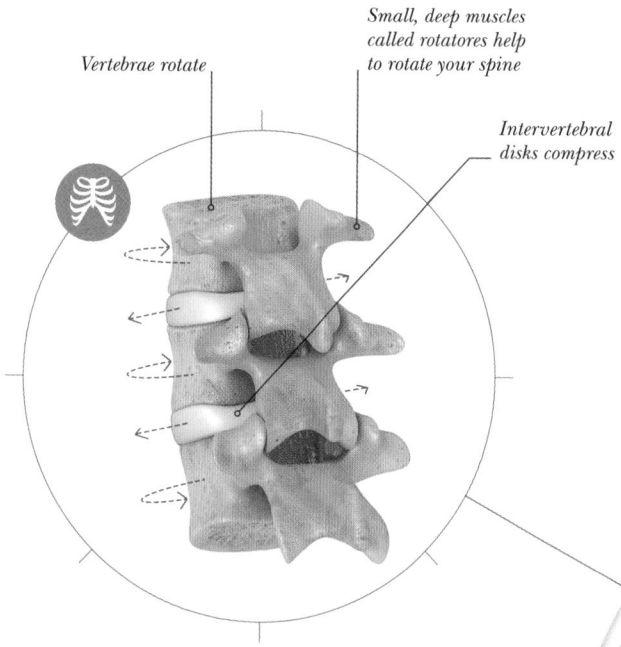

Spinal rotation

With rotation of your spine (spinal twists), your intervertebral disks naturally compress. For optimal safety and benefit, first consciously lengthen your spine into axial extension as much as possible, then twist only as far as your muscles bring you. Use your arms to help you maintain length in your spine. Breathe.

External obliques engage as you twist away

Gluteal muscles stretch

Foot is relaxed

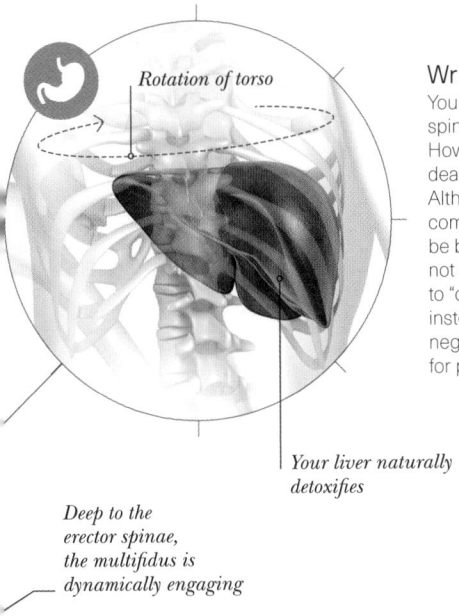

Wringing out **toxins**

You may have heard that spinal twists "wring out toxins." However, your liver efficiently deals with toxins automatically. Although mechanically compressing the organs may be beneficial, evidence does not show that this contributes to "detoxification." You can instead visualize wringing out negative energy as you twist, for psychological benefit.

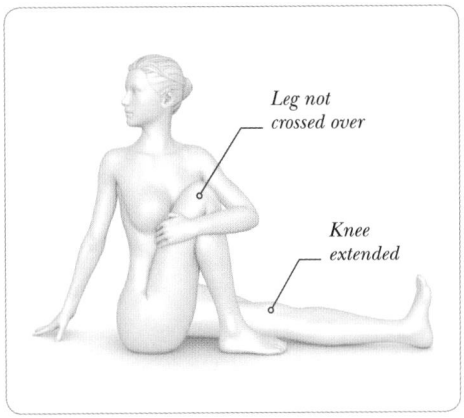

VARIATION

For a gentler twist, keep one leg extended and consider not crossing the lifted leg over the midline. Use your arm wrapped around the leg to help you sit tall as you twist.

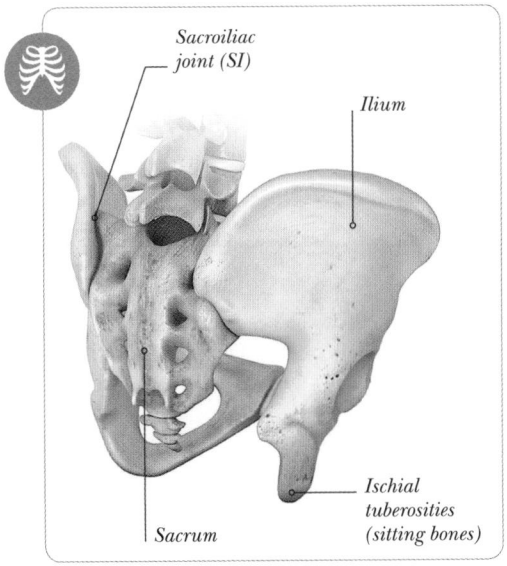

Sacroiliac joint

Allow your sitting bones to move slightly on the floor with the twist. If you anchor them down, the twist puts a lot of pressure on the structure of the SI joint, which can cause aches. Alternatively, allowing too much movement in your SI joint can also lead to achiness. Find the middle way for your body.

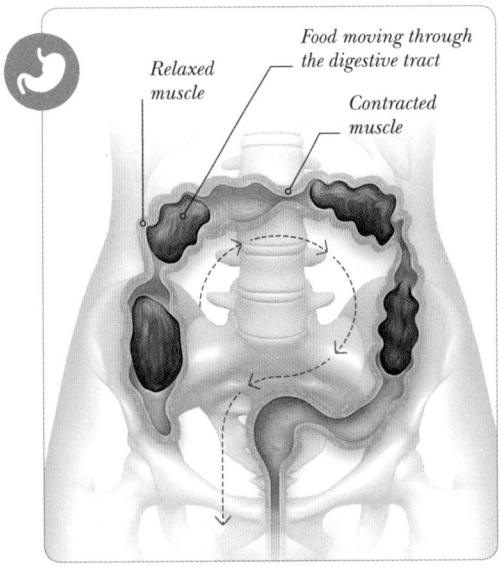

Stimulating **peristalsis**

In your digestive tract, peristalsis is the involuntary engagement of smooth muscles to move digesting food (see p.39). Thankfully, you don't have to consciously tell your stomach to empty into the small intestines. Stress and a sedentary lifestyle can affect peristalsis and lead to digestive issues. Twisting can stimulate healthy peristalsis.

POSTERIOR–LATERAL VIEW

CHILD'S POSE

Balasana

Reminiscent of the fetal position and with your weight supported by the floor, the restorative forward bend of Child's pose can be a deeply relaxing, restful posture for many. It provides a deep but gentle stretch for your back muscles, calming both body and mind.

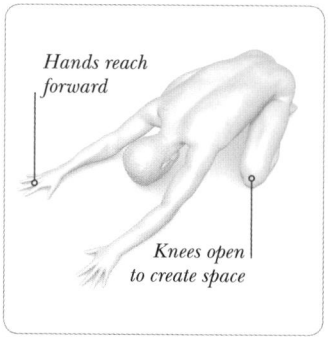

VARIATION
Another option is to separate your knees and bring your hands forward. This allows more space for the torso and is a common resting pose during sequences such as sun salutations.

THE BIG PICTURE

With as little muscle engagement as possible, your body releases down. In particular your back muscles, buttocks, and ankles stretch out. As you breathe deeply, the muscles in and around your ribcage dynamically engage and stretch with each breath.

Neck and upper **arms**
Your neck muscles are passive with your **splenius capitis** and **splenius cervicis** stretching. Your **posterior deltoids** slightly stretch while your shoulders are in internal rotation. Your **arm muscles** are passive with your forearms pronated, allowing the backs of your hands to rest on the floor.

Deltoids
Splenius cervicis
Splenius capitis

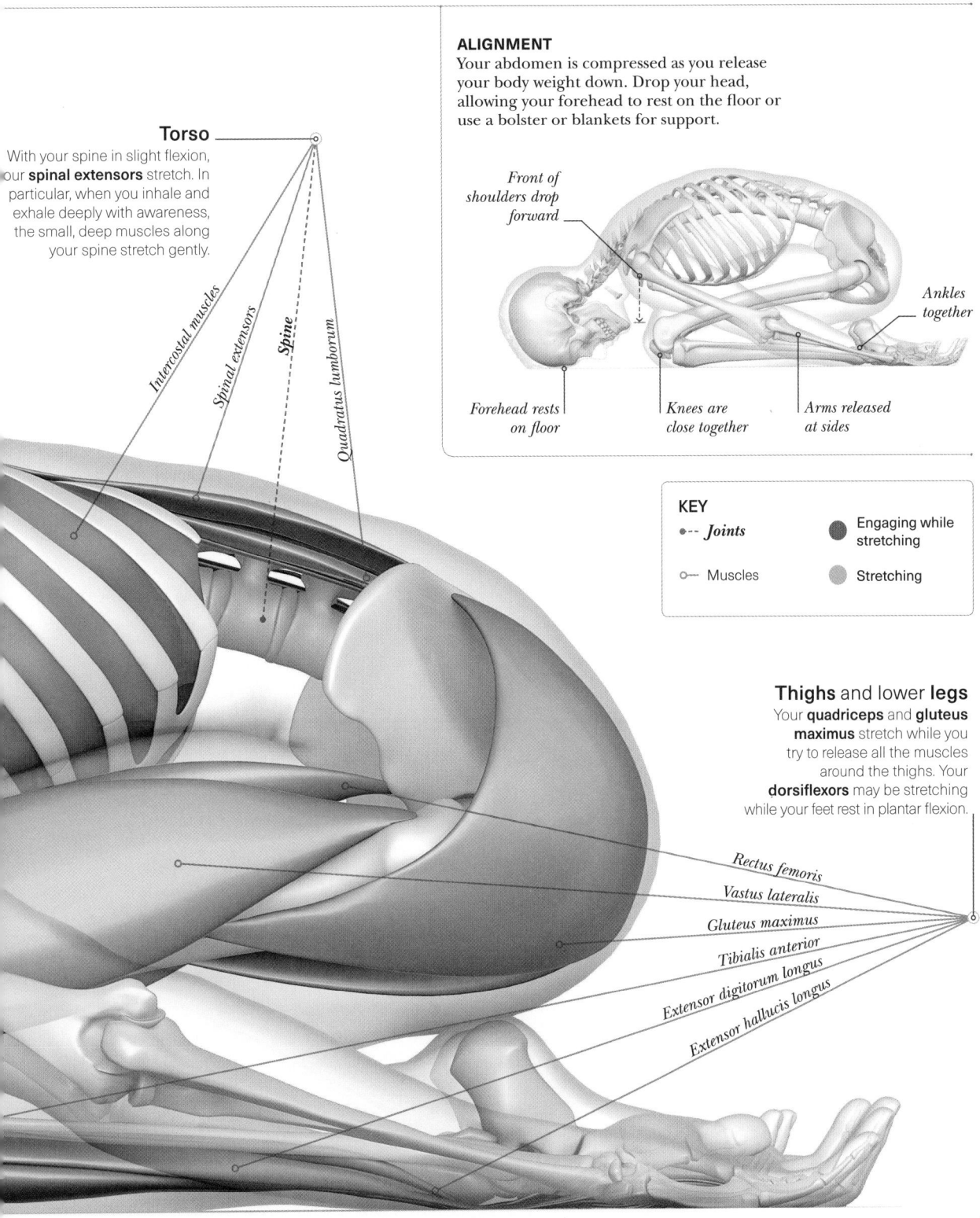
Torso
With your spine in slight flexion, your spinal extensors stretch. In particular, when you inhale and exhale deeply with awareness, the small, deep muscles along your spine stretch gently.
Intercostal muscles
Spinal extensors
Spine
Quadratus lumborum
ALIGNMENT
Your abdomen is compressed as you release your body weight down. Drop your head, allowing your forehead to rest on the floor or use a bolster or blankets for support.
Front of shoulders drop forward
Ankles together
Forehead rests on floor
Knees are close together
Arms released at sides
KEY
Joints
Muscles
Engaging while stretching
Stretching
Thighs and lower legs
Your quadriceps and gluteus maximus stretch while you try to release all the muscles around the thighs. Your dorsiflexors may be stretching while your feet rest in plantar flexion.
Rectus femoris
Vastus lateralis
Gluteus maximus
Tibialis anterior
Extensor digitorum longus
Extensor hallucis longus

» CLOSER LOOK

Child's pose can be an opportunity to rest, take deep breaths, relax tired muscles, and access a primal sense of safety. If comfortable, you can use this pose as a place of rest and rejuvenation between challenging poses.

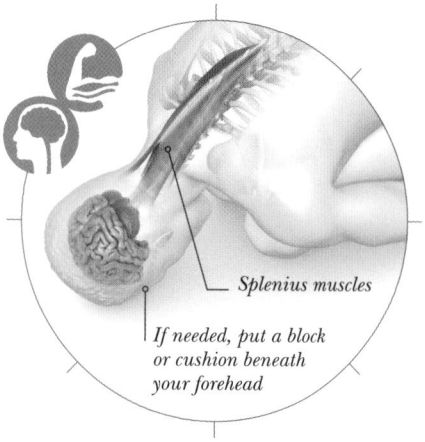

Head rest

Throughout the day, your neck muscles have the job of holding up the 11lb (5kg) bowling ball that is your head. This muscular activity keeps your nervous system on slight alert. Letting the muscles of your neck and head completely relax lets your nervous system know it is safe to rest.

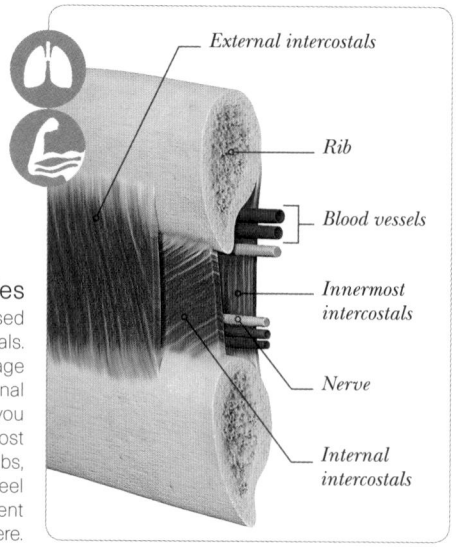

Intercostal muscles

Your intercostals are crisscrossed and layered, like your abdominals. Your external intercostals engage to help you inhale. Your internal intercostals engage to help you forcefully exhale. Your innermost intercostals stabilize your ribs, stretching when you inhale. Feel how dynamic your rib movement is while you take deep breaths here.

Neck muscles are completely relaxed

Quadriceps stretch

SUPERIOR–LATERAL VIEW

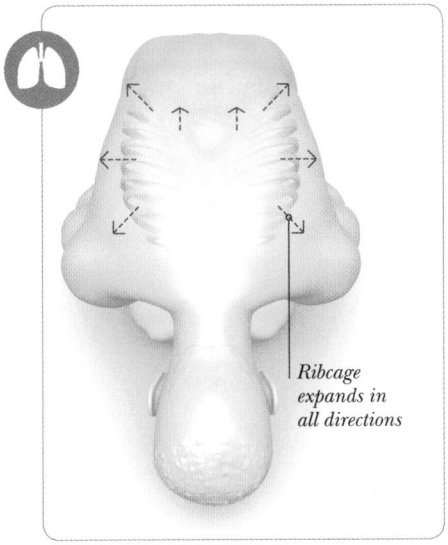

Fetal position

This pose may evoke comfort as it is reminiscent of being in your mother's womb. In the fetal position, most of your joints are in flexion, protecting your abdominal organs from harm. Notice how your body moves with each breath: your torso rising and broadening with each inhale and releasing back as you exhale.

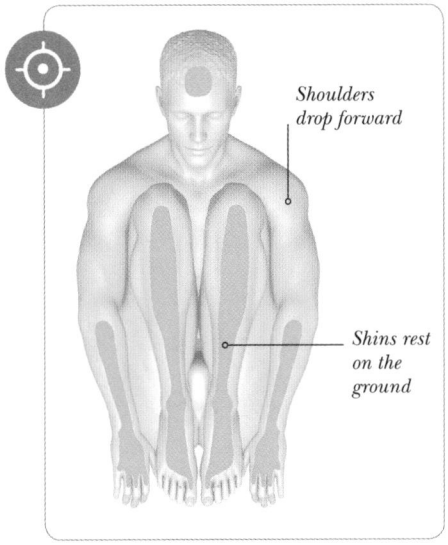

Pressure points

Allow your body to release down completely, with your shins, feet, forearms, hands, and forehead all resting on the ground. If your body doesn't make this shape, use blankets and props to find support.

Spinal extensors stretch while in relaxed flexion

Intercostal muscles dynamically stretch and engage with deep breaths

Gluteal muscles stretch

Arms are relaxed and rest down at your sides

A blanket under the ankles can relieve pressure for some

POSTERIOR VIEW

CAMEL

Ustrasana

Camel is an energetic backbend that can leave you feeling confident and ready to take on the day. This pose counteracts our flexion-driven modern lifestyles by broadening the chest. It is challenging but can be adapted for anyone who struggles to reach their feet.

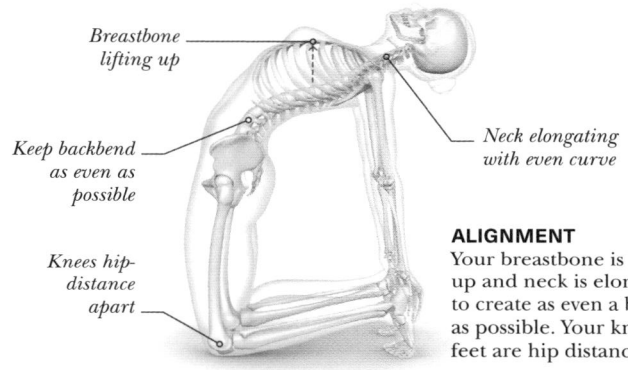

ALIGNMENT
Your breastbone is lifting up and neck is elongating to create as even a backbend as possible. Your knees and feet are hip distance apart.

THE BIG PICTURE

The front of your body—including your abdominals and thighs—stretches, while the back of your body—including your back muscles, buttocks, and thighs—engages. You may also feel stretching on the soles of your feet as you tuck your toes under.

Neck
Your **cervical extensors** engage to extend your neck, while your **cervical flexors** stabilize, preventing your head from being thrown back, and creating an even, controlled curve.

Torso
Your **spinal extensors** engage to extend your spine while your **abdominals** stretch. Your **pectorals** stretch as you broaden your chest. Your **middle** and **lower trapezius** work with your **rhomboids** to retract and stabilize your scapulae, while your **serratus anterior** stretches.

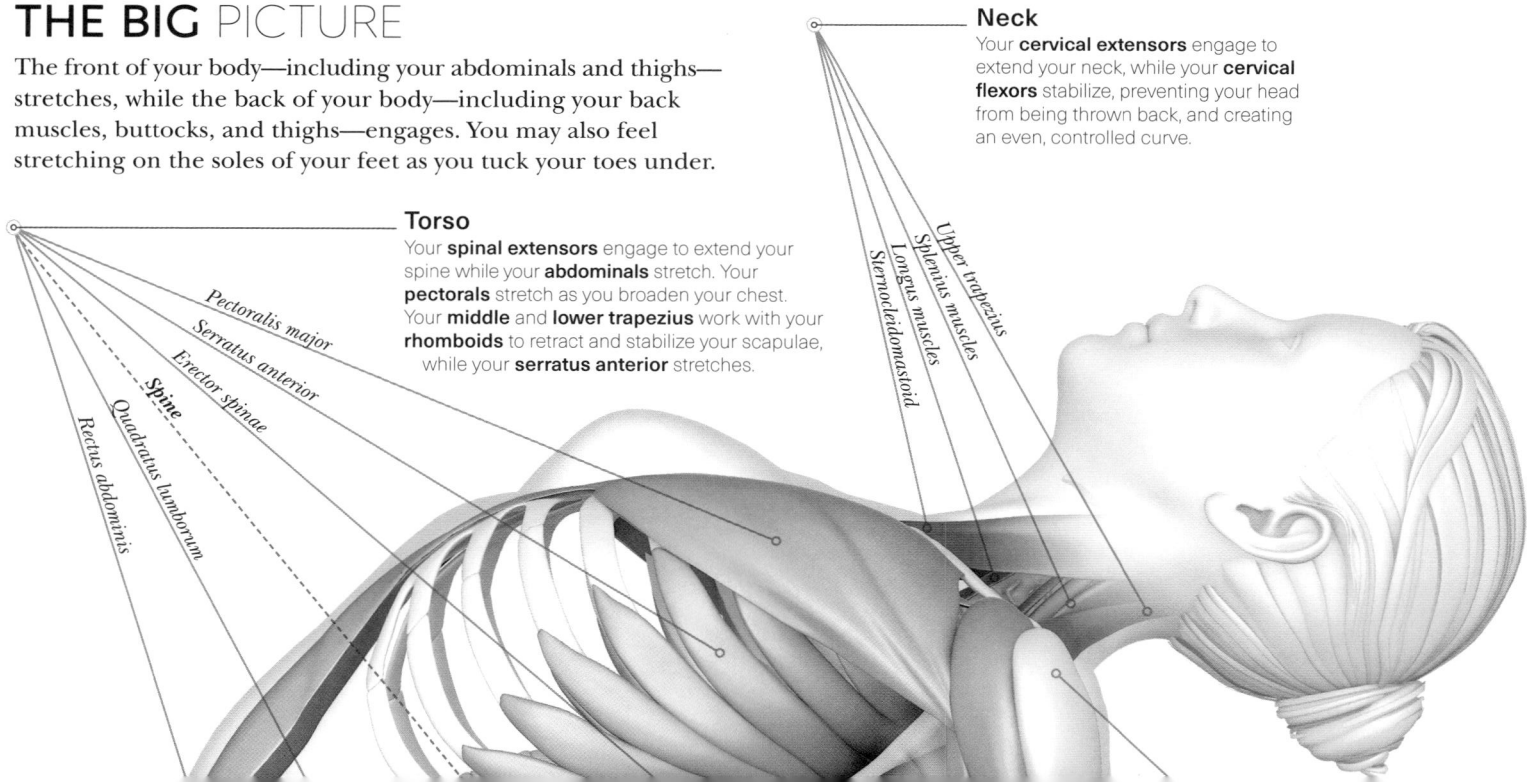

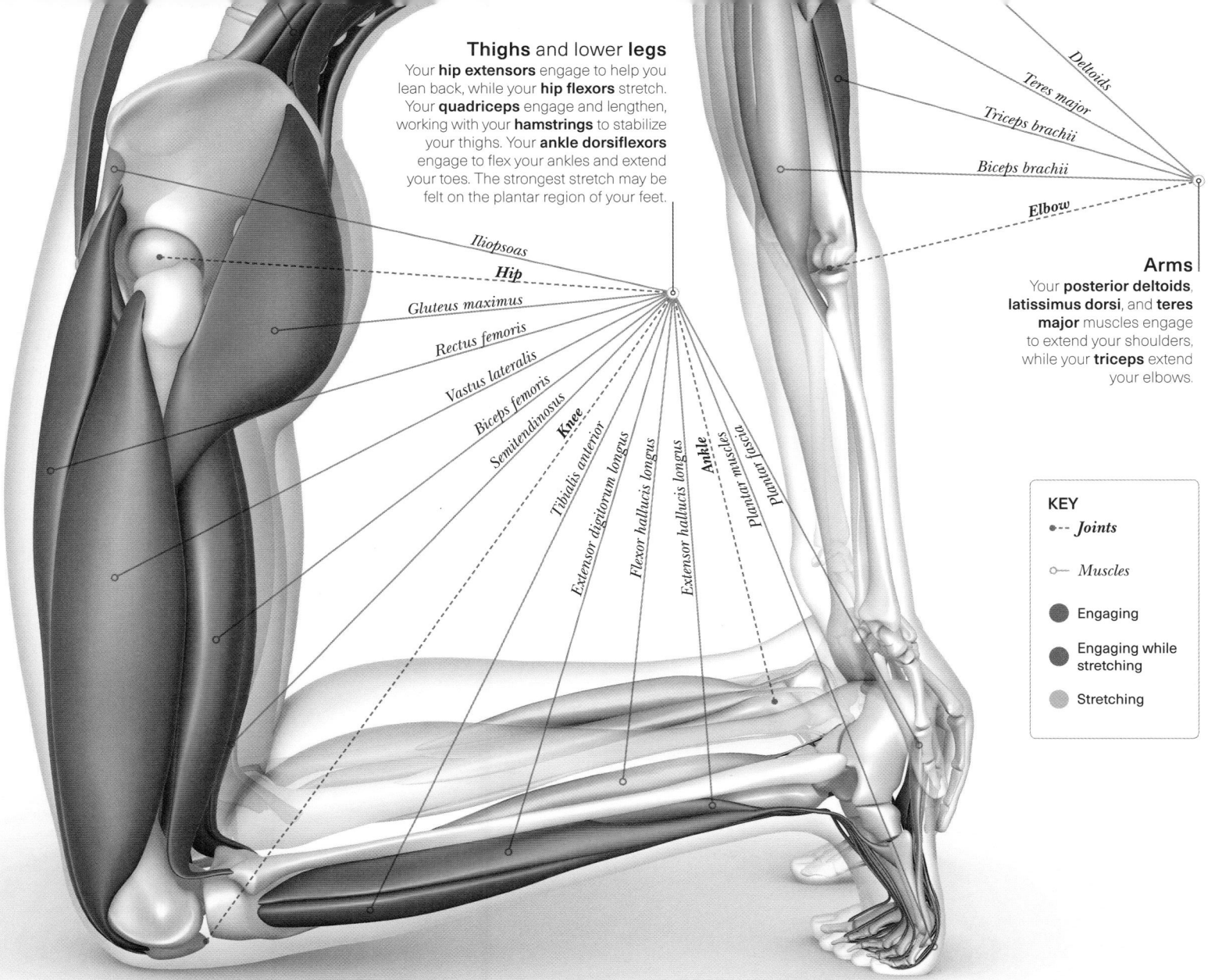
Thighs and lower legs
Your hip extensors engage to help you lean back, while your hip flexors stretch. Your quadriceps engage and lengthen, working with your hamstrings to stabilize your thighs. Your ankle dorsiflexors engage to flex your ankles and extend your toes. The strongest stretch may be felt on the plantar region of your feet.
Iliopsoas
Hip
Gluteus maximus
Rectus femoris
Vastus lateralis
Biceps femoris
Semitendinosus
Knee
Tibialis anterior
Extensor digitorum longus
Flexor hallucis longus
Extensor hallucis longus
Ankle
Plantar muscles
Plantar fascia
Deltoids
Teres major
Triceps brachii
Biceps brachii
Elbow
Arms
Your posterior deltoids, latissimus dorsi, and teres major muscles engage to extend your shoulders, while your triceps extend your elbows.
KEY
Joints
Muscles
Engaging
Engaging while stretching
Stretching

CLOSER LOOK

Camel can be great for your posture and spinal disk health. However, make sure you warm up first, and take care with the position of your neck.

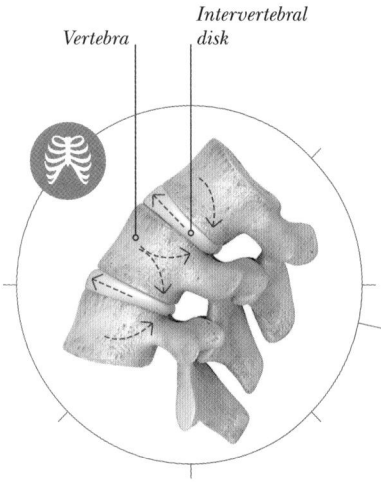

Spinal extension

Backbends (spinal extension) push your intervertebral disks slightly forward, while strengthening your back muscles. This is great for the health of your disks, and can be applied therapeutically for disk issues. Consult your healthcare team and a qualified yoga professional about your unique conditions first.

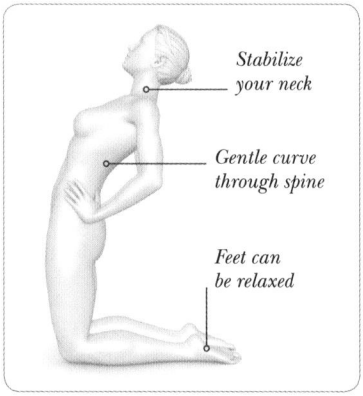

VARIATION

For a gentler backbend, press your hands into your hips as you lean back slightly into the pose. You could also place your hands on blocks set alongside your shins.

This subtle, controlled neck extension protects the small, complex joint structures

Abdominals mainly stretch, perhaps with some engagement

Toes can be relaxed or curled under

ANTERIOR–LATERAL VIEW

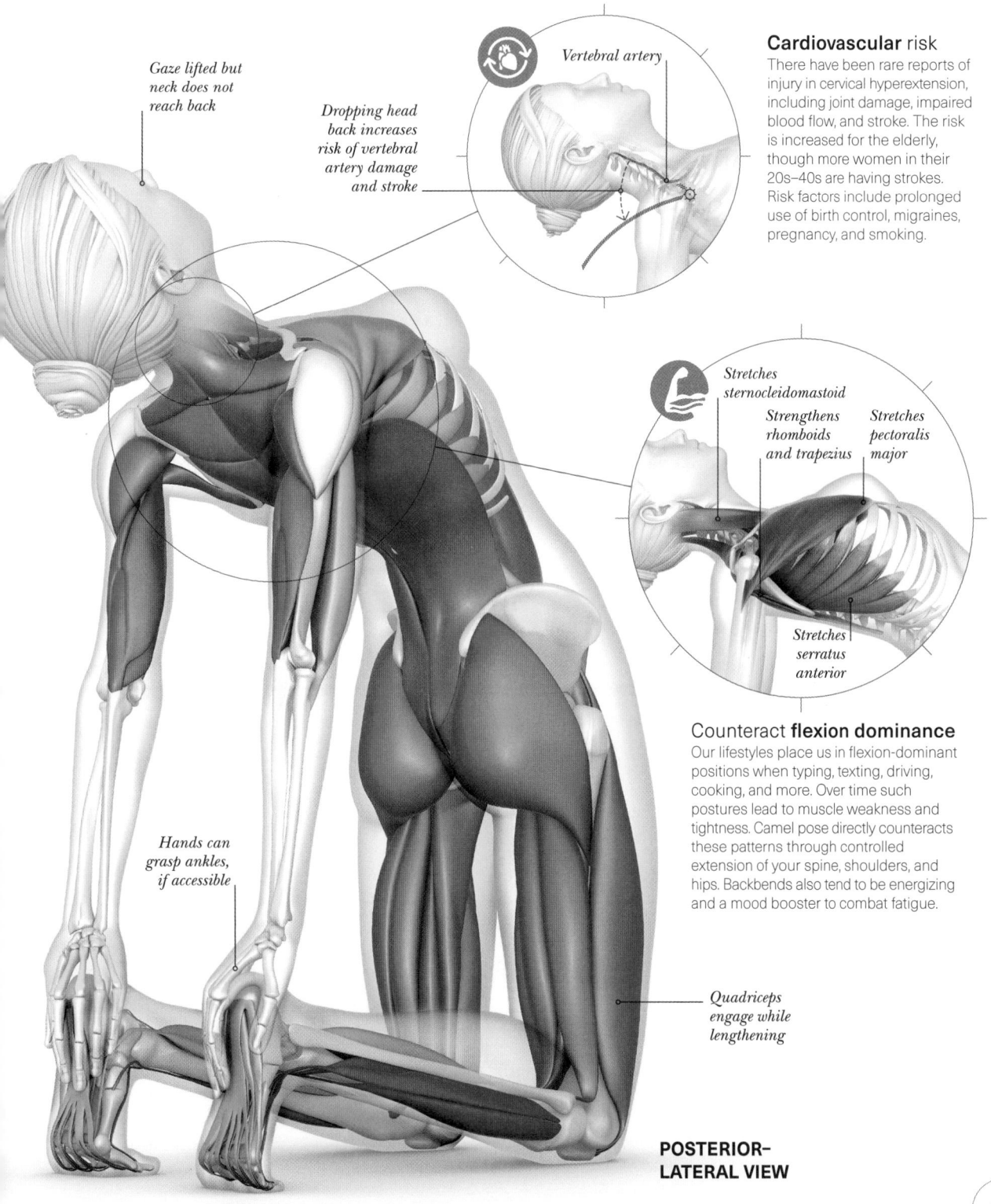

Cardiovascular risk

There have been rare reports of injury in cervical hyperextension, including joint damage, impaired blood flow, and stroke. The risk is increased for the elderly, though more women in their 20s–40s are having strokes. Risk factors include prolonged use of birth control, migraines, pregnancy, and smoking.

Counteract **flexion dominance**

Our lifestyles place us in flexion-dominant positions when typing, texting, driving, cooking, and more. Over time such postures lead to muscle weakness and tightness. Camel pose directly counteracts these patterns through controlled extension of your spine, shoulders, and hips. Backbends also tend to be energizing and a mood booster to combat fatigue.

POSTERIOR-LATERAL VIEW

KING PIGEON

Eka Pada Rajakapotasana

Pigeon pose, as practiced today, is not a traditional yoga asana. This modern kneeling backbend can be modified to offer therapeutic benefits for sciatica and back pain, with suitable options for everyone. Make sure you are warmed up and move slowly into this pose.

THE BIG PICTURE

This version of the pose deeply stretches your hips, buttocks, thighs, abdomen, chest, and shoulders. Muscles in your arms, back, and hips engage to hold you in the pose, preventing you from toppling over.

KEY

- *Joints*
- *Muscles*
- Engaging
- Engaging while stretching
- Stretching

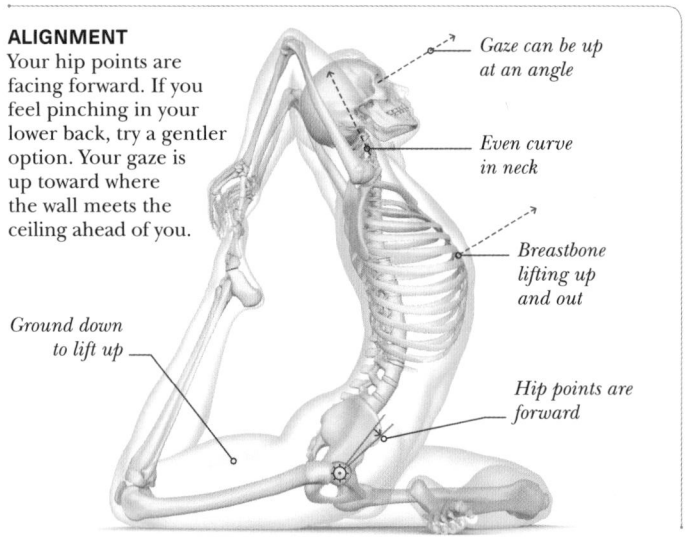

ALIGNMENT
Your hip points are facing forward. If you feel pinching in your lower back, try a gentler option. Your gaze is up toward where the wall meets the ceiling ahead of you.

Arms

Your **shoulder flexors** engage. Your **deltoids** dynamically engage to bring you into the position and then help pull your leg in. Your **brachialis**, **biceps**, and **brachioradialis** engage to flex your elbow, while your **triceps** stretch.

Brachioradialis
Brachialis
Triceps brachii
Biceps brachii
Deltoids
Shoulder

Back thigh

Your **hip extensors** are working to extend your hip, while your **quadriceps** maintain knee extension. Your **hip flexors** stretch strongly.

Gluteus maximus
Tensor fasciae latae
Biceps femoris
Semitendinosus
Rectus femoris
Knee

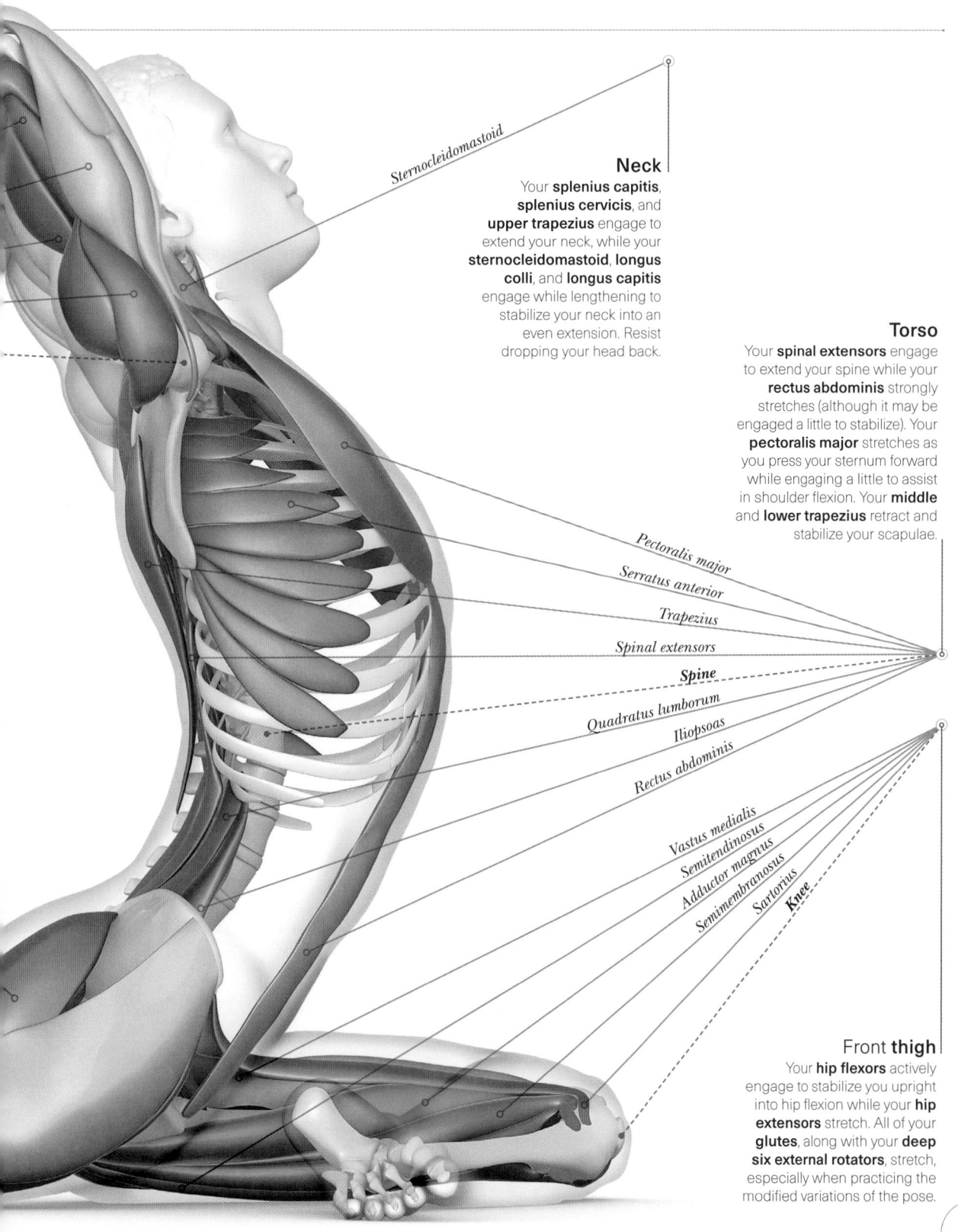

Neck

Your **splenius capitis**, **splenius cervicis**, and **upper trapezius** engage to extend your neck, while your **sternocleidomastoid**, **longus colli**, and **longus capitis** engage while lengthening to stabilize your neck into an even extension. Resist dropping your head back.

Torso

Your **spinal extensors** engage to extend your spine while your **rectus abdominis** strongly stretches (although it may be engaged a little to stabilize). Your **pectoralis major** stretches as you press your sternum forward while engaging a little to assist in shoulder flexion. Your **middle** and **lower trapezius** retract and stabilize your scapulae.

Front **thigh**

Your **hip flexors** actively engage to stabilize you upright into hip flexion while your **hip extensors** stretch. All of your **glutes**, along with your **deep six external rotators**, stretch, especially when practicing the modified variations of the pose.

» CLOSER LOOK

King Pigeon is challenging for some, but you can find a relaxed variation by lying down or using props. These options may relieve pressure on your joints.

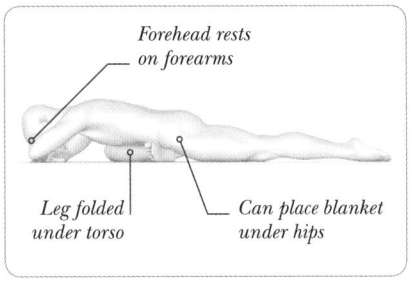

VARIATION

For a more passive version, release forward. You may feel enough of a stretch on your hands or forearms. Consider using blankets or a bolster under your hips. You can get similar benefits by lying on your back and placing your legs in a figure 4 position.

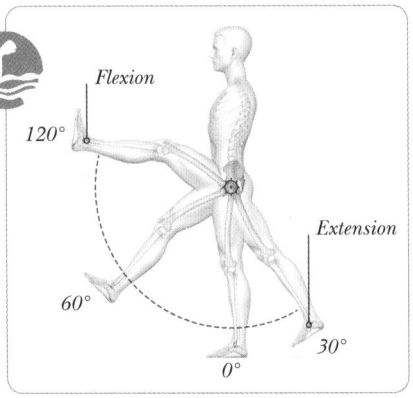

Piriformis

Your piriformis normally externally rotates your hip. However, when your hip is flexed past 60°, your piriformis transforms action to an internal rotator. This means it stretches deeply when in external rotation and flexion, as in the front hip of many versions of Pigeon.

Breastbone reaches forward and up

Rectus abdominis stretches

Hip flexors on extended leg side stretch

Place leg at angle that is comfortable for your knee

ANTERIOR VIEW

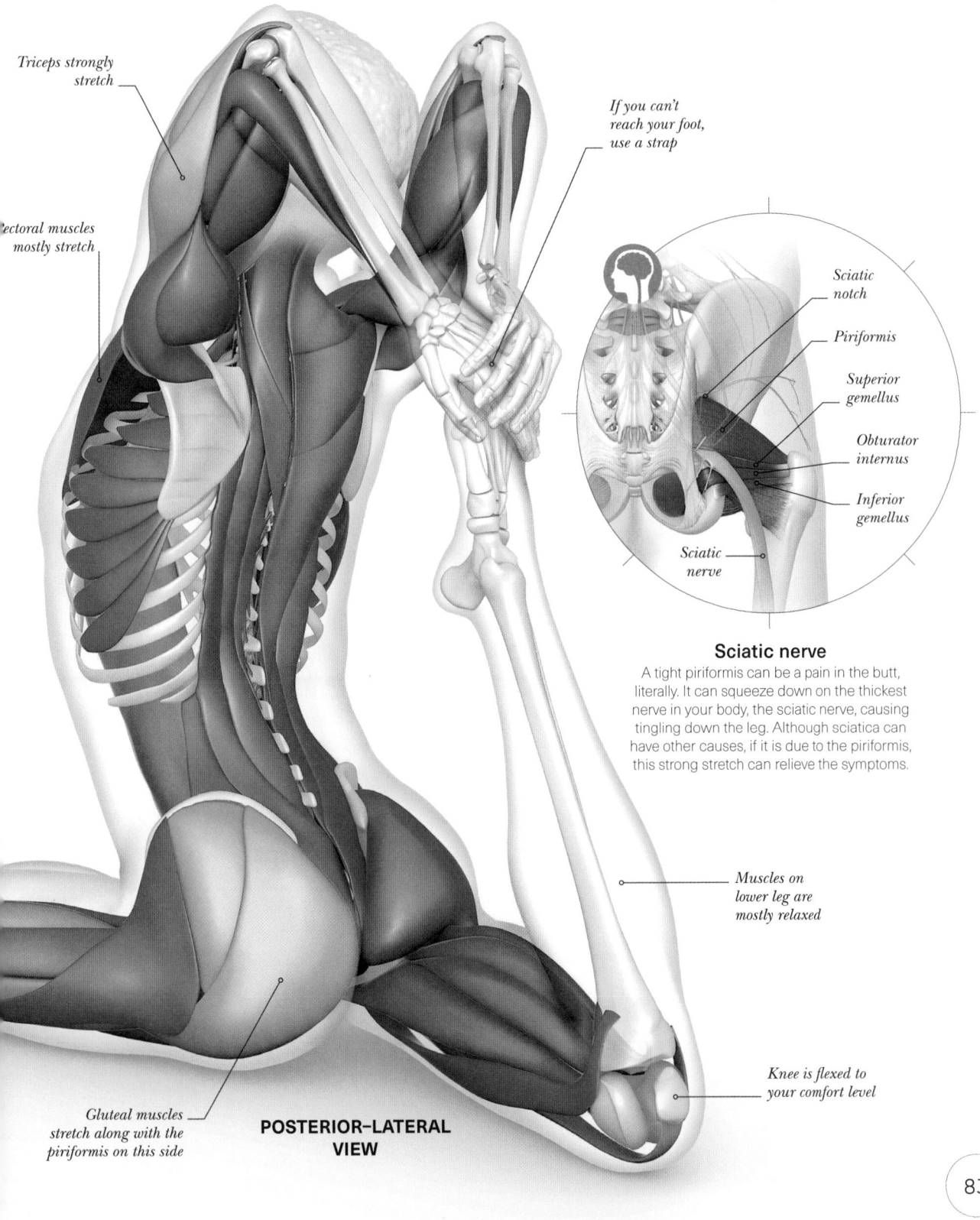

Sciatic nerve

A tight piriformis can be a pain in the butt, literally. It can squeeze down on the thickest nerve in your body, the sciatic nerve, causing tingling down the leg. Although sciatica can have other causes, if it is due to the piriformis, this strong stretch can relieve the symptoms.

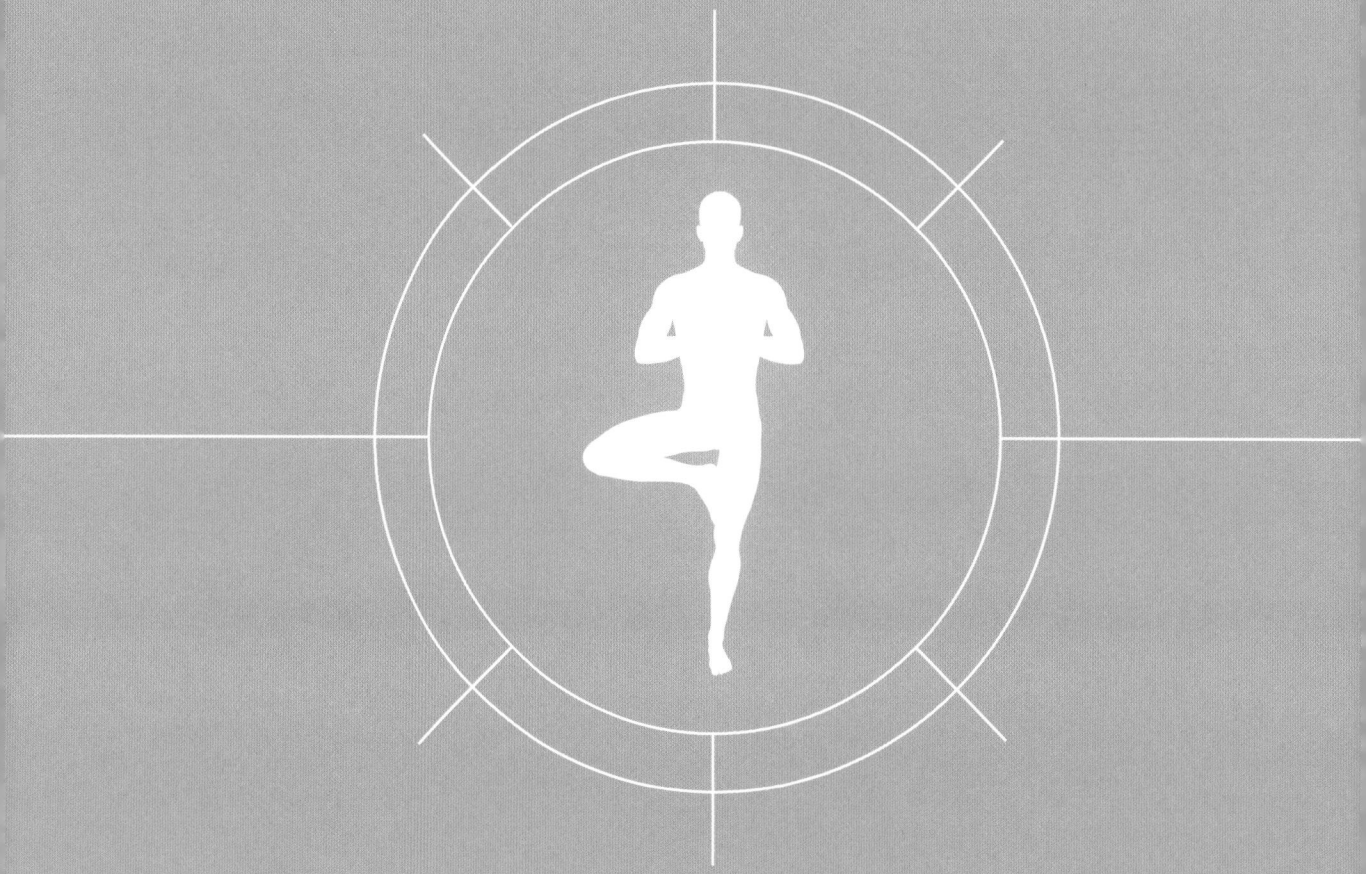

STANDING ASANAS

These standing asanas were specifically chosen to help improve posture and balance. How you hold your body affects all the systems of your anatomy, as well as your energy levels, your cognition, and your confidence. The intention behind these poses is less pain, fewer injuries, improved posture, and optimal movement in everything you do.

MOUNTAIN

Tadasana

This standing pose is essentially the anatomical position. It represents how you hold yourself in the world—your postural alignment. The pose creates a stable connection to the earth. Many muscles are slightly engaged to support you upright, resisting gravity.

THE BIG PICTURE

Although the aim is to activate as few muscles as little as possible, a lot of muscles in your body engage subtly in a neutral or lenthening position to prevent you from leaning or falling in any direction. Your lower legs, thighs, hips, back muscles, and abdominals may all be felt buzzing with this slight engagement.

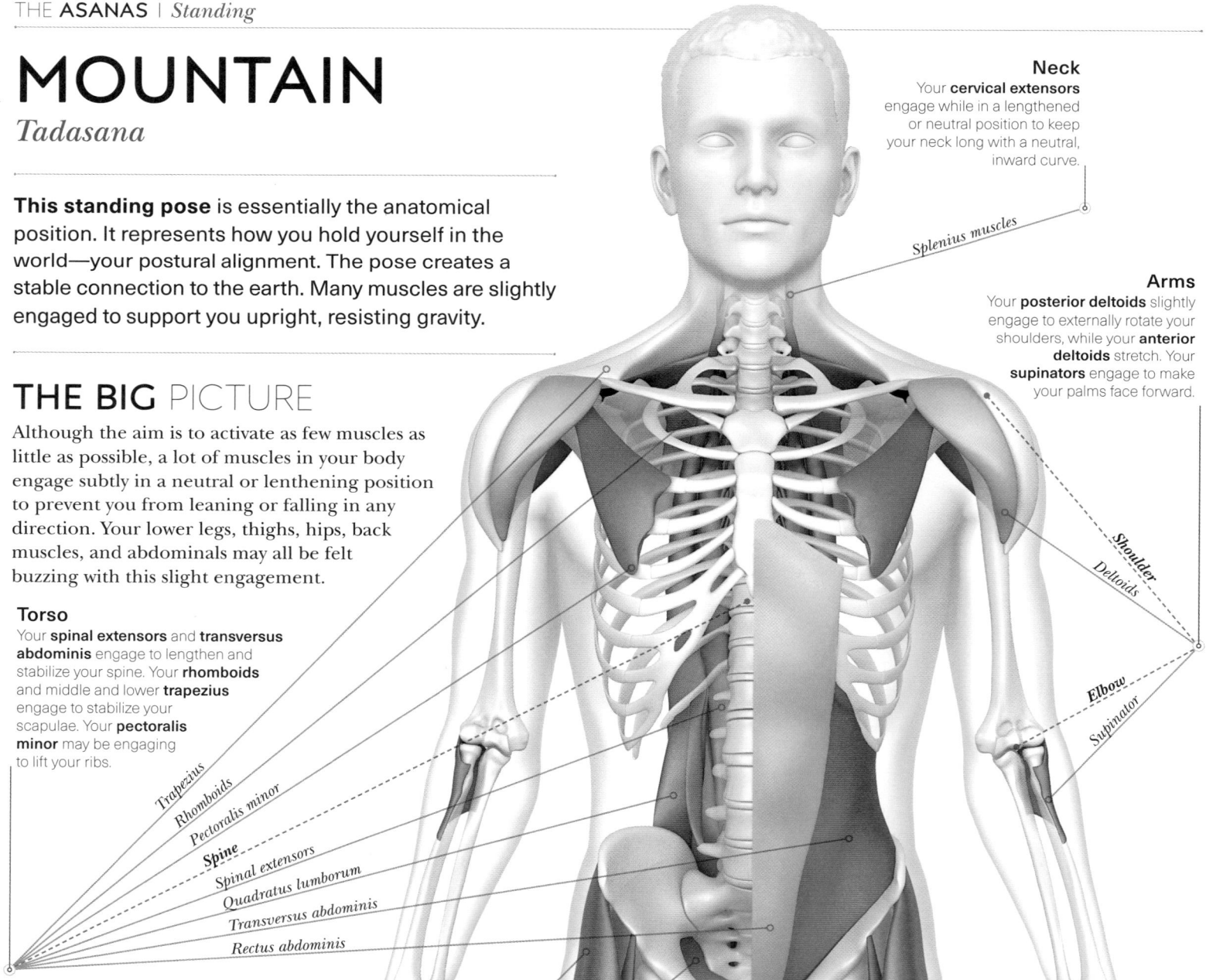

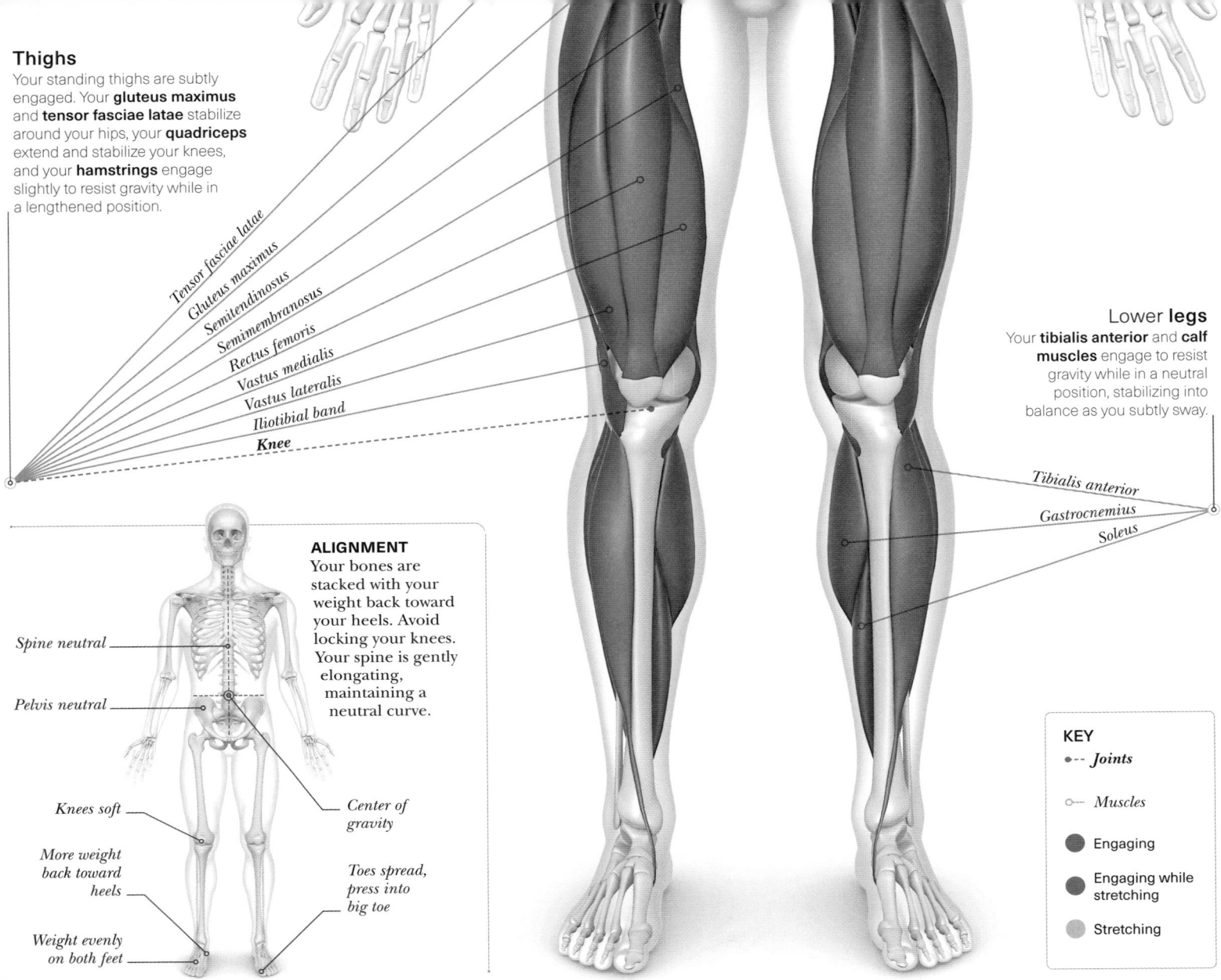
Thighs
Your standing thighs are subtly engaged. Your **gluteus maximus** and **tensor fasciae latae** stabilize around your hips, your **quadriceps** extend and stabilize your knees, and your **hamstrings** engage slightly to resist gravity while in a lengthened position.
Tensor fasciae latae
Gluteus maximus
Semitendinosus
Semimembranosus
Rectus femoris
Vastus medialis
Vastus lateralis
Iliotibial band
Knee
Lower legs
Your **tibialis anterior** and **calf muscles** engage to resist gravity while in a neutral position, stabilizing into balance as you subtly sway.
Tibialis anterior
Gastrocnemius
Soleus
ALIGNMENT
Your bones are stacked with your weight back toward your heels. Avoid locking your knees. Your spine is gently elongating, maintaining a neutral curve.
Spine neutral
Pelvis neutral
Knees soft
More weight back toward heels
Weight evenly on both feet
Center of gravity
Toes spread, press into big toe
KEY
Joints
Muscles
Engaging
Engaging while stretching
Stretching

» CLOSER LOOK

Mountain pose is an opportunity to find a stable, structurally sound base. The structure and placement of your feet can facilitate the foundation of that base.

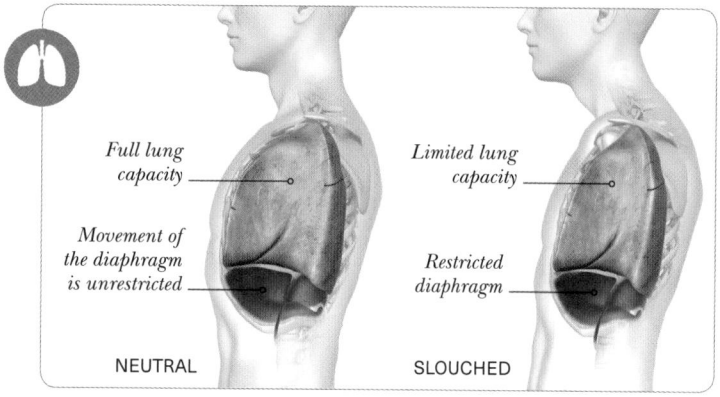

Breathing and posture

When you slouch, you have limited lung capacity, as well as restricted movement of your diaphragm. From a yoga perspective, when you aren't breathing well, your prana, or vital energy, is not flowing properly. From a physiological perspective, when your respiratory system is not efficient, neither are your cardiovascular, digestive, endocrine, or nervous systems. So, stand up tall and let your body function optimally.

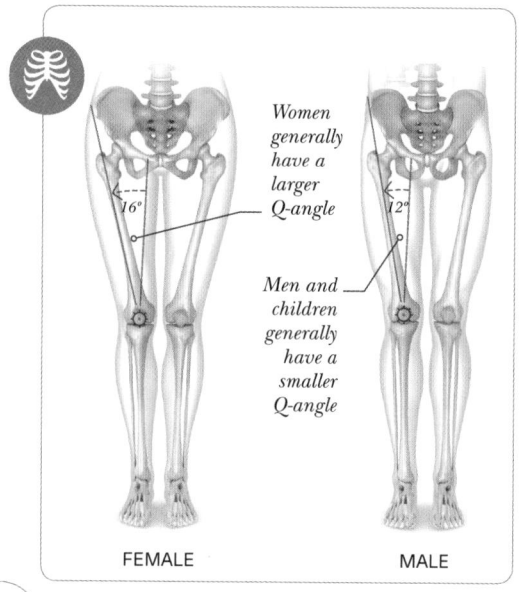

Feet at hip distance

Some styles of yoga bring the feet together in Tadasana. However, while many modern asanas were developed for preadolescent boys in India, who have fairly narrow hips, yoga is now predominantly practiced by adult women, whose hips are wider. For many people, standing with the feet at hip distance is more stable, decreasing the Q-angle (shown left) and reducing stress on the knees.

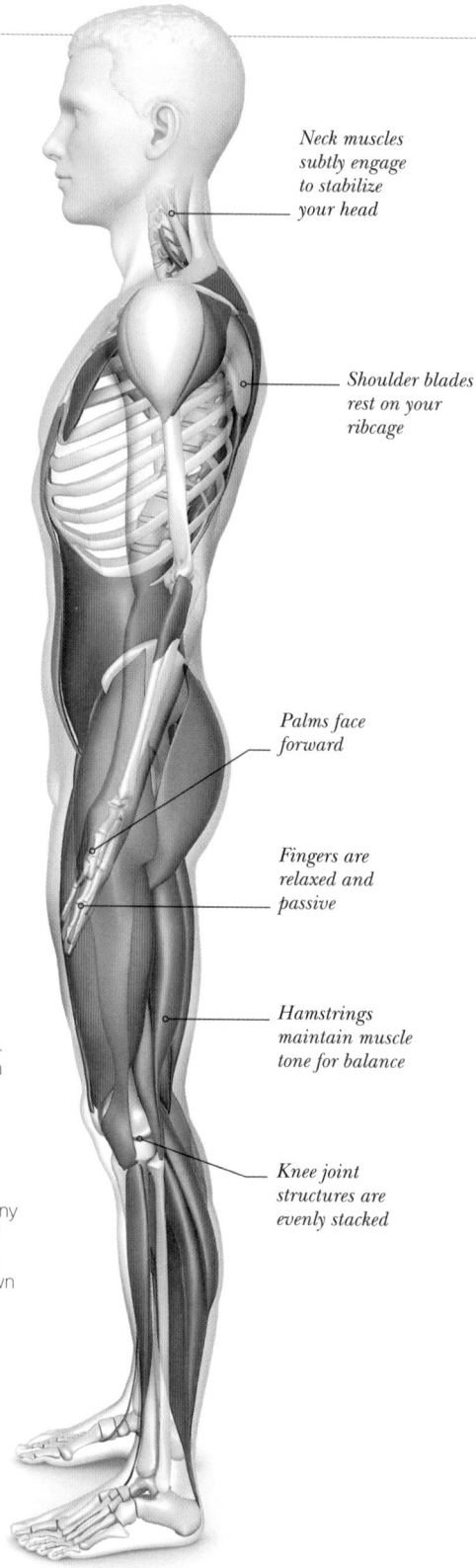

LATERAL VIEW

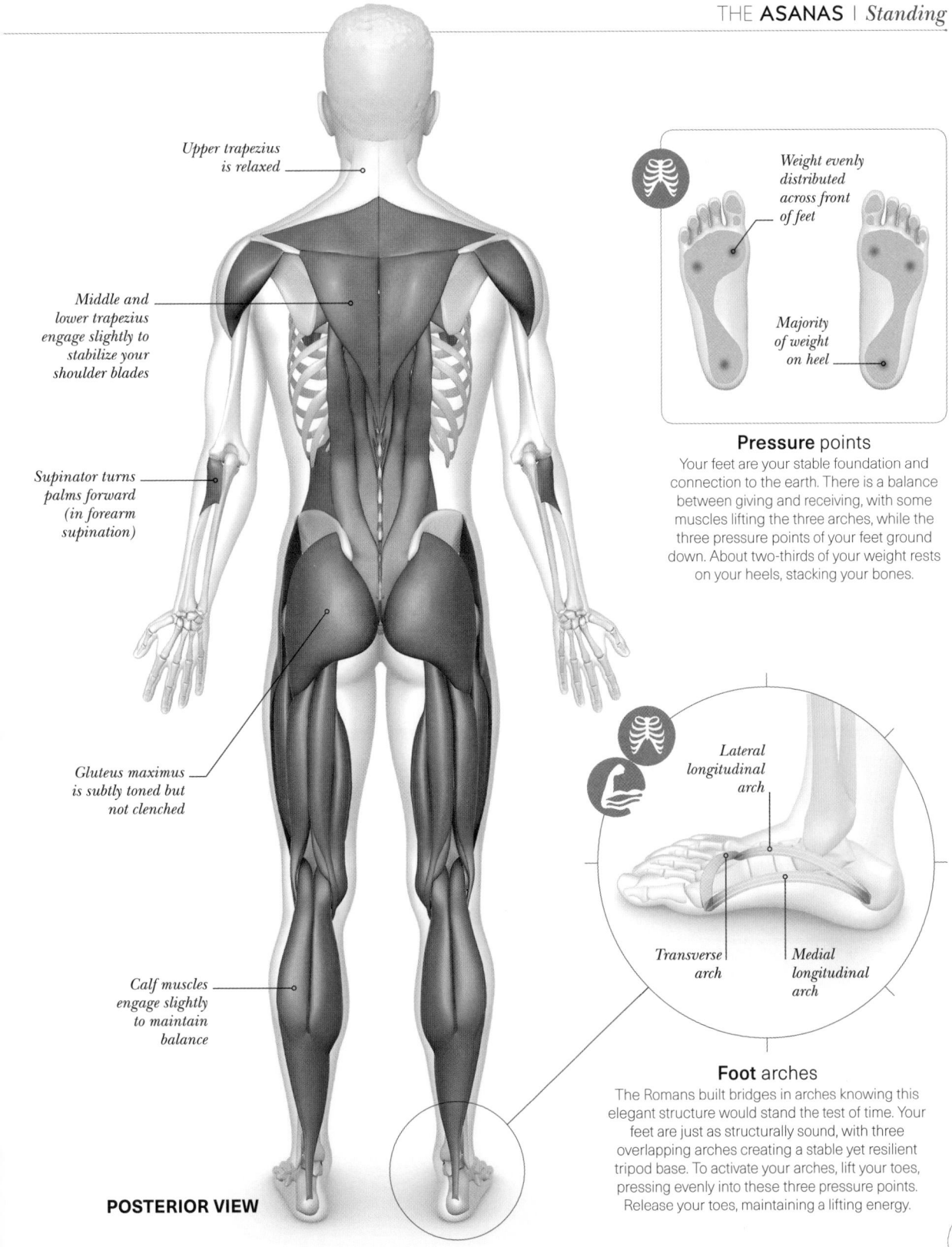

Pressure points

Your feet are your stable foundation and connection to the earth. There is a balance between giving and receiving, with some muscles lifting the three arches, while the three pressure points of your feet ground down. About two-thirds of your weight rests on your heels, stacking your bones.

Foot arches

The Romans built bridges in arches knowing this elegant structure would stand the test of time. Your feet are just as structurally sound, with three overlapping arches creating a stable yet resilient tripod base. To activate your arches, lift your toes, pressing evenly into these three pressure points. Release your toes, maintaining a lifting energy.

FORWARD FOLD

Uttanasana

Forward Fold offers an opportunity to improve flexibility. Transitioning in and out of the pose, such as in Sun salutations, will help prepare you for common functional movements you do throughout the day. This pose can be adapted for all abilities by going into the fold less deeply.

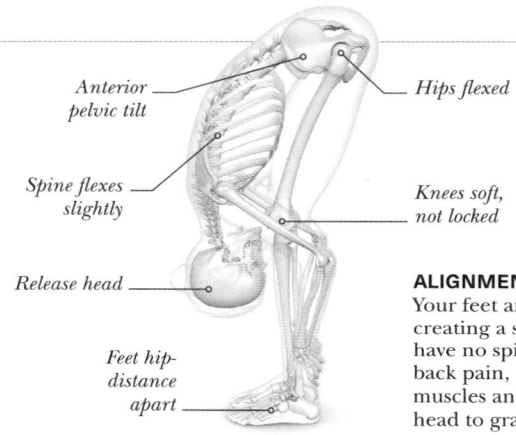

ALIGNMENT
Your feet are at hip distance, creating a stable base. If you have no spinal problems or back pain, relax your neck muscles and release your head to gravity.

THE BIG PICTURE

The back of your whole body is stretching—including your calf muscles, thighs, buttocks, and back muscles. At the front of your body—especially in your legs—your muscles are working to stabilize you in the deep bend.

Thighs
Your **gluteus maximus**, **medius**, and **minimus**, **hamstrings**, and **adductor magnus** stretch strongly in this pose while your **hip flexors** engage. Your **quadriceps** extend the knees and stabilize your base of support.

Neck and **torso**
All of your **spinal extensors** and your **latissimus dorsi** stretch when you release your upper body to gravity.

Spine
Latissimus dorsi
Spinal extensors
Splenius muscle
Gluteus maximus
Gluteus medius
Psoas major
Iliacus
Tensor fasciae latae
Rectus femoris
Vastus lateralis
Biceps femoris
Semitendinosus

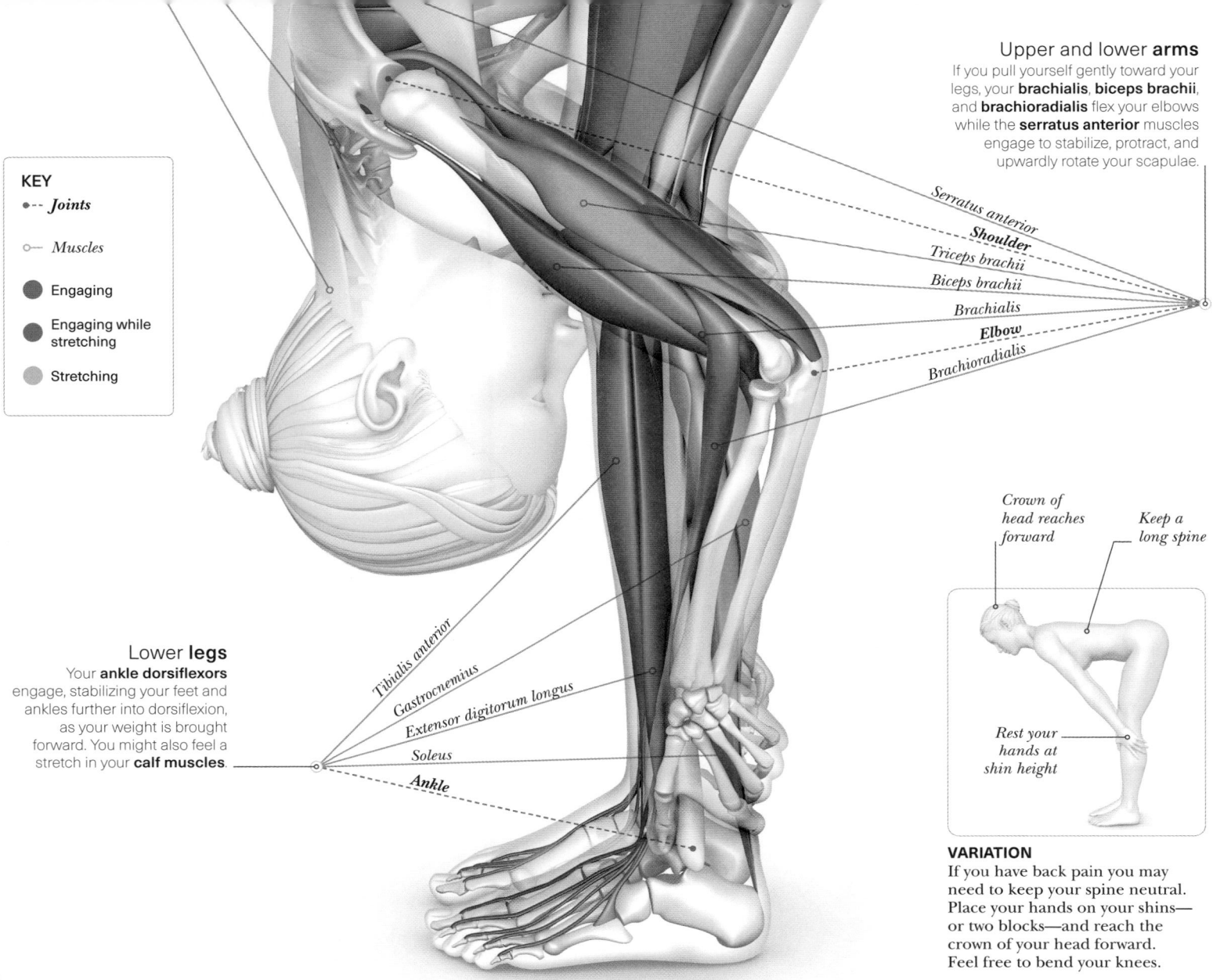

VARIATION
If you have back pain you may need to keep your spine neutral. Place your hands on your shins—or two blocks—and reach the crown of your head forward. Feel free to bend your knees.

» CLOSER LOOK

Forward Fold delivers a deep spinal stretch, which can help to improve back health and reduce back pain. However, care should be taken to reduce the lumbar load for those with intervertebral disk issues.

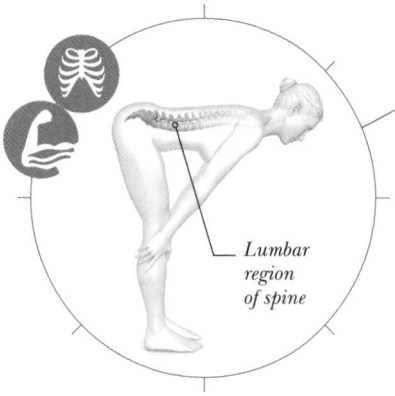

Lumbar load

The load on the lumbar spine in a standing Forward Fold is significant. The lower back is particularly vulnerable during the transition in and out of the pose. If you have any back pain, arthritis, disk issues, osteopenia, or osteoporosis, try keeping your spine neutral and transition in and out of the pose with bent knees and an engaged core.

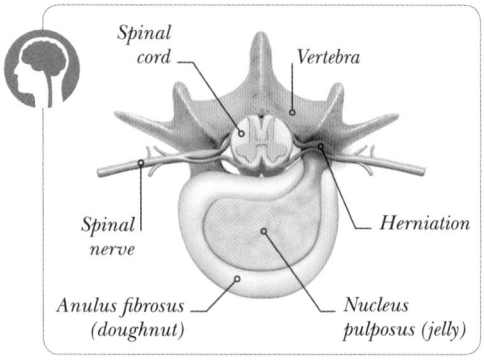

Herniated disk

Intervertebral disks are like jelly doughnuts. In a "slipped" or herniated disk, the "jam" partially leaks out of the tougher fibrocartilage "dough." Since most herniations occur posterior-laterally due to spinal flexion, if you currently have a disk issue, move slowly or avoid flexion by not going into the pose as deeply.

Pelvis tilts anteriorly and hips flex

Latissimus dorsi stretches

Shoulder blades protract and upwardly rotate

Spinal extensors stretch, allowing upper body to release fully

Coming up too quickly can cause dizziness as your blood redistributes

Hands can grasp ankles if comfortable

ANTERIOR VIEW

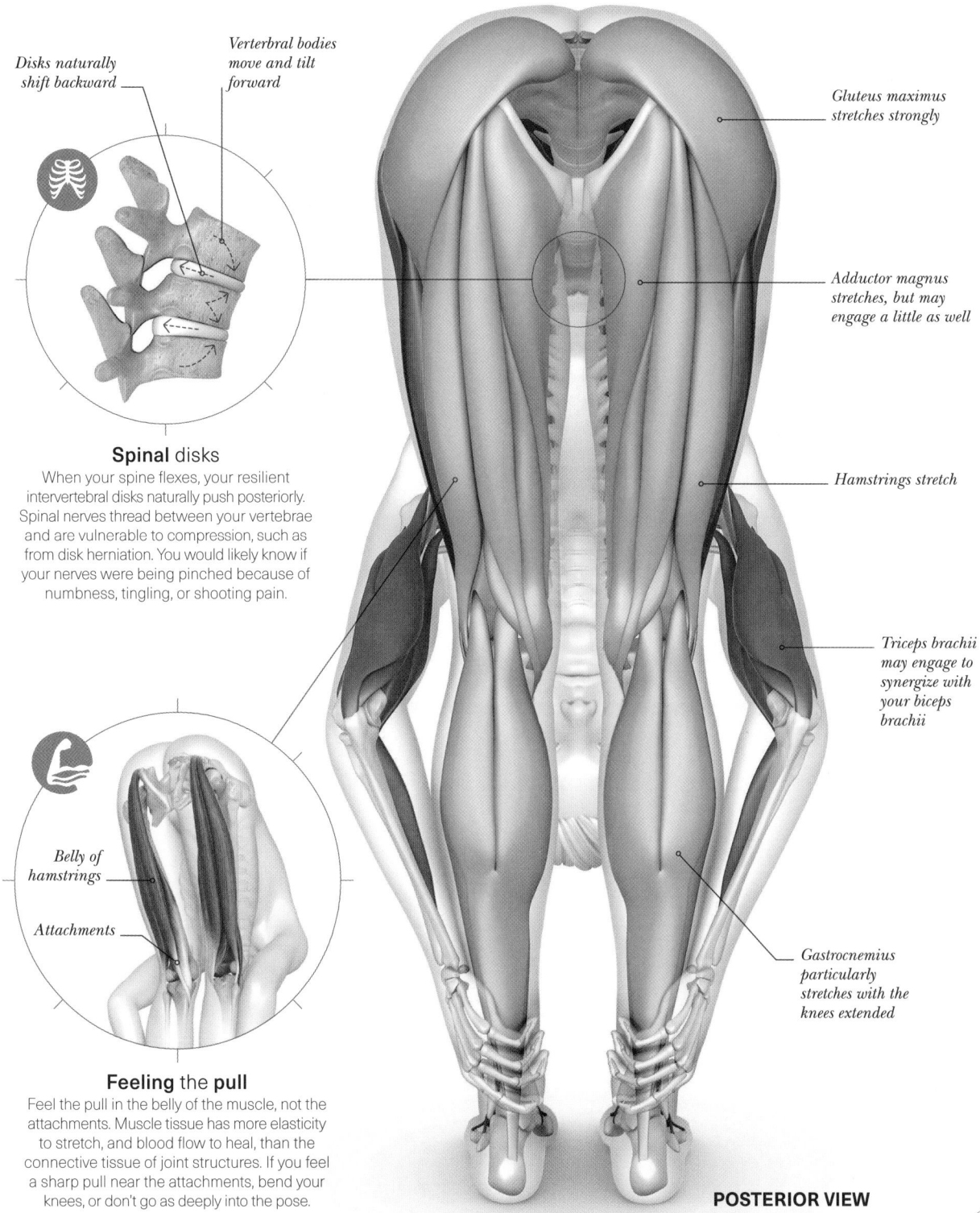

Spinal disks

When your spine flexes, your resilient intervertebral disks naturally push posteriorly. Spinal nerves thread between your vertebrae and are vulnerable to compression, such as from disk herniation. You would likely know if your nerves were being pinched because of numbness, tingling, or shooting pain.

Feeling the **pull**

Feel the pull in the belly of the muscle, not the attachments. Muscle tissue has more elasticity to stretch, and blood flow to heal, than the connective tissue of joint structures. If you feel a sharp pull near the attachments, bend your knees, or don't go as deeply into the pose.

CHAIR

Utkatasana

Chair pose activates the largest muscles in your body, gets your heart pumping, and engages your core strongly. This energizing standing pose improves your thigh strength, which some studies suggest is a key factor in prolonging your life.

THE BIG PICTURE

Muscles around your thighs, hips, and core are engaging strongly to hold you in this squatting position. Lifting your arms overhead further challenges your core strength and engages your shoulder muscles. Alternatively, you can put your hands on your hips to lighten the load.

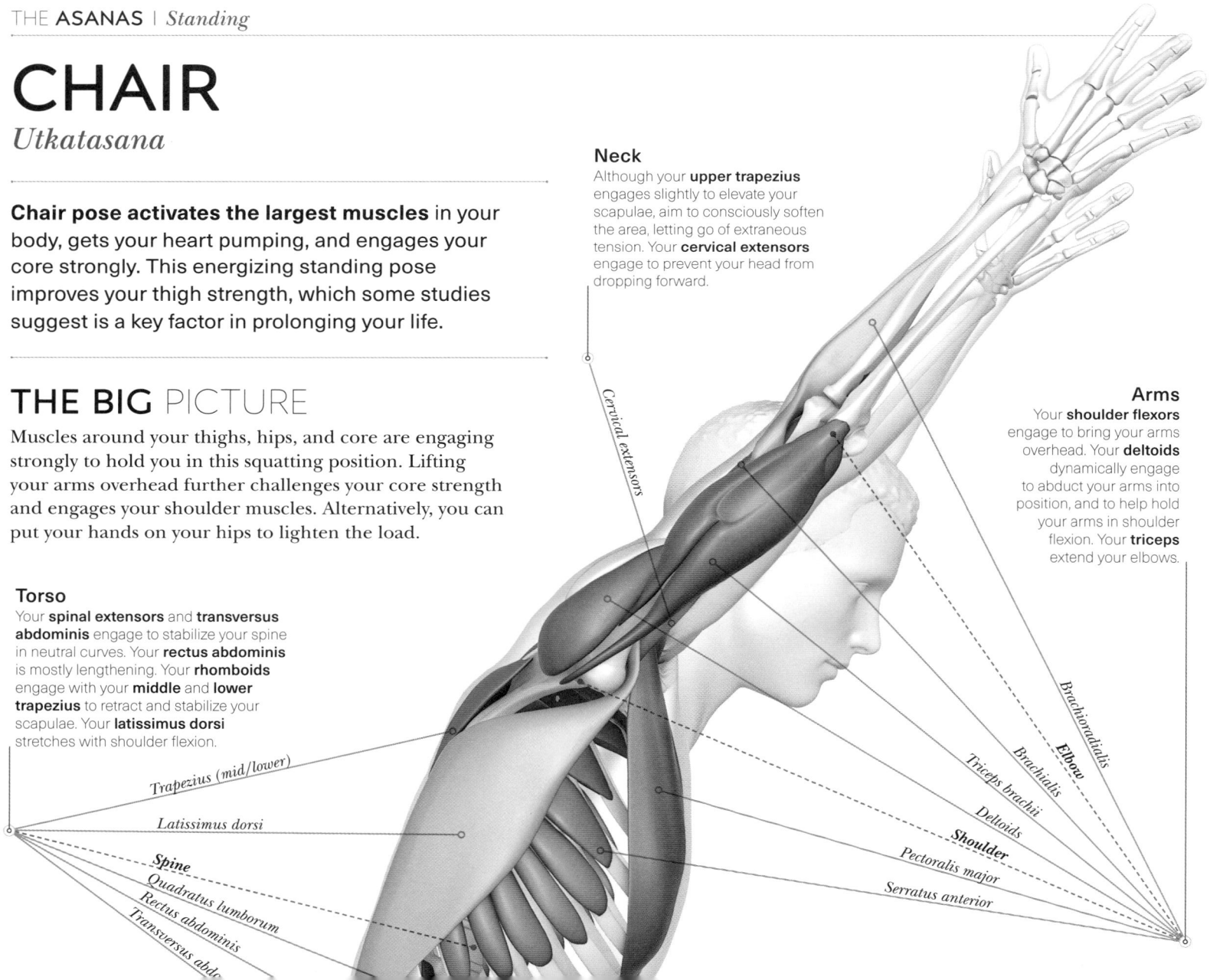

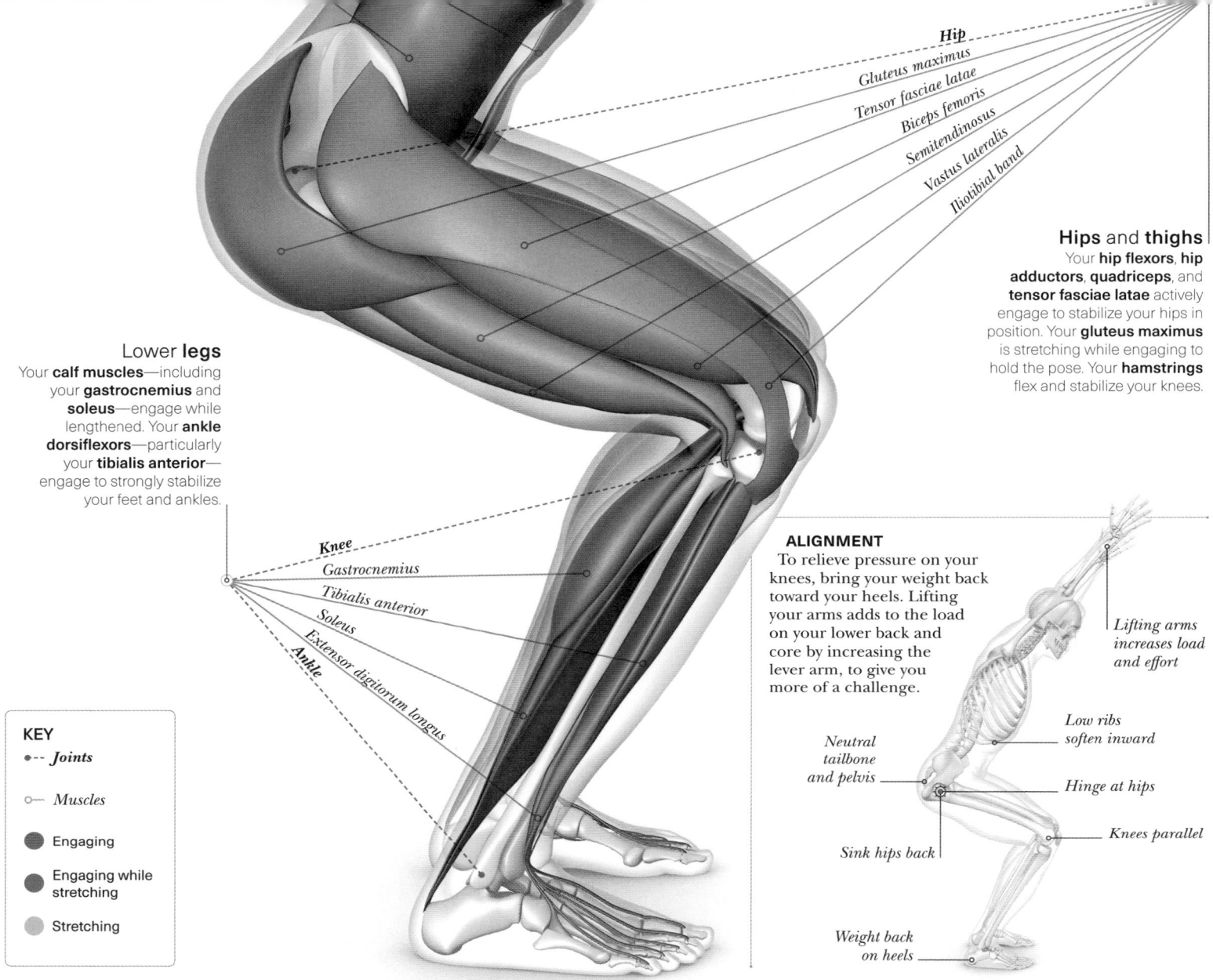
Hip
Gluteus maximus
Tensor fasciae latae
Biceps femoris
Semitendinosus
Vastus lateralis
Iliotibial band
Knee
Gastrocnemius
Tibialis anterior
Soleus
Extensor digitorum longus
Ankle
Lower legs
Your calf muscles—including your gastrocnemius and soleus—engage while lengthened. Your ankle dorsiflexors—particularly your tibialis anterior—engage to strongly stabilize your feet and ankles.
Hips and thighs
Your hip flexors, hip adductors, quadriceps, and tensor fasciae latae actively engage to stabilize your hips in position. Your gluteus maximus is stretching while engaging to hold the pose. Your hamstrings flex and stabilize your knees.
ALIGNMENT
To relieve pressure on your knees, bring your weight back toward your heels. Lifting your arms adds to the load on your lower back and core by increasing the lever arm, to give you more of a challenge.
Lifting arms increases load and effort
Low ribs soften inward
Neutral tailbone and pelvis
Hinge at hips
Knees parallel
Sink hips back
Weight back on heels
KEY
Joints
Muscles
Engaging
Engaging while stretching
Stretching

» CLOSER LOOK

Chair pose leads to body-wide effects. Lifting your arms, for example, raises blood pressure as well as increases the lumbar load, which tests your cardiovascular system and core muscles.

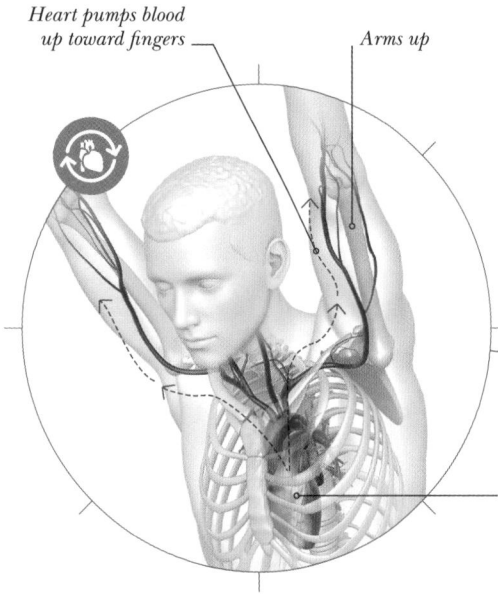

Blood pressure

Raising your arms overhead in any pose quickly increases your heart rate because blood pressure has to increase to pump all the way to your now-higher fingertips. Be conscious of these changes in your body. If you have high blood pressure, consider keeping your arms down with your hands on your hips.

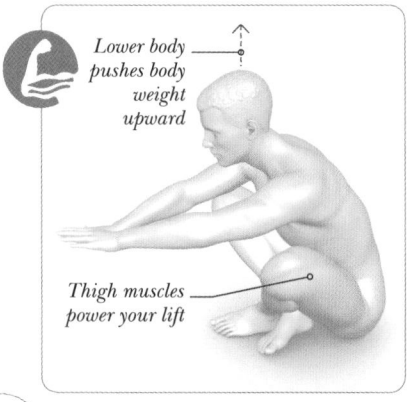

Quadriceps strength

Quadriceps strength is an indicator of longevity. Strengthening your quads in a balanced way can help relieve knee and hip pain (particularly from arthritis) and improve your balance. You can think of your quadriceps as your "independence muscles," as they are vital for getting up from a chair or the floor—researchers often use the sitting–rising test (see left) to determine function and longevity.

Head and neck align with neutral spine

Fingers reach up softly

Shoulder blades are only slightly elevated

Serratus anterior engaging

Abdominals lengthen as you reach up

Hip flexors work to hold hips in pose

Hips are flexed

Ease pressure on knees by keeping weight back

Feet are flexed (in dorsiflexion)

ANTERIOR–LATERAL VIEW

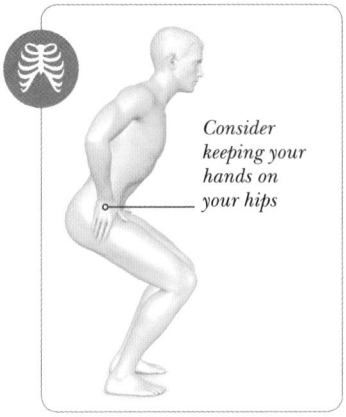

Increased **lumbar load**

Lifting your arms increases the load on your lumbar spine. This can be great to effectively strengthen the core muscles. However, for some people the force is so strong that they lose the integrity of the core and spine, causing lower back strain. If this happens, you can rest your hands on your hips.

Biceps brachii lengthen to reach out arms and engage to stabilize

Brachioradialis is in a stretched position

Spinal extensors stabilize spine in neutral position

Hips are released back and down toward floor

Erector spinae

Diaphragm

Rectus abdominis

Quadratus lumborum

Transversus abdominis

Iliopsoas

Pelvic floor

Core strength

There are many definitions of what makes up the "core," but it generally includes all of the abdominals, the pelvic floor, back muscles like the erector spinae, the respiratory diaphragm, and sometimes the iliopsoas. Asanas like Chair pose help develop core function and awareness, which can improve posture, balance, and functional movements throughout your day.

Try shifting weight back to protect your knees

Pressure and **balance**

Bringing your weight back toward your heels can take pressure off your knees. Try spreading and lifting your toes, feeling your weight shift back. Then, place your toes down gently while keeping your weight back.

Toes are spread and relaxed, without curling or clenching

POSTERIOR–LATERAL VIEW

CRESCENT LUNGE

Anjaneyasana

This lunge is a good antidote to sitting down too much. It is also beneficial for runners or anyone who participates in sports that involve running, because it strengthens the muscles that power your stride and stretches your hip flexors.

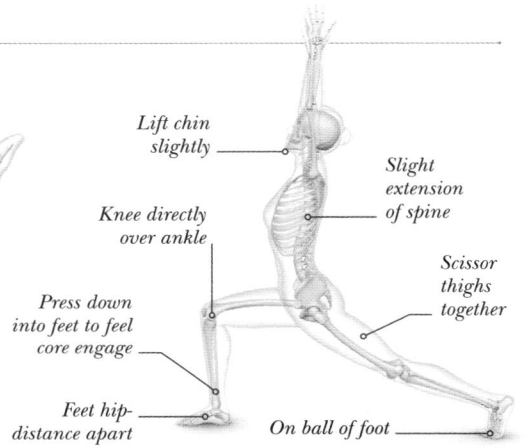

ALIGNMENT
Your feet are hip-width apart to maintain balance. Your front knee is directly over your ankle or behind it for safety.

THE BIG PICTURE

In this pose, the muscles of your hips and your gluteus muscles stretch and activate dynamically to keep you balanced. Your thigh muscles engage strongly to stabilize your hips and knees, while your core muscles stabilize your spine in a slight backbend.

Neck
Your **cervical extensors** engage to extend your cervical spine while your **cervical flexors** engage and lengthen to stabilize your neck, preventing your head from dropping back.

Longus muscles
Sternocleidomastoid
Splenius muscles

Elbow
Brachioradialis
Brachialis
Biceps brachii
Coracobrachialis
Deltoids
Shoulder
Serratus anterior

Arms
Your **shoulder flexors** engage. Your **anterior deltoids** aid in shoulder flexion while your **posterior deltoids** lengthen, yet some fibers engage to stabilize and externally rotate your shoulders. Your **triceps** extend the elbows. Feel space, not stiffness, in your joints as you reach through your fingers.

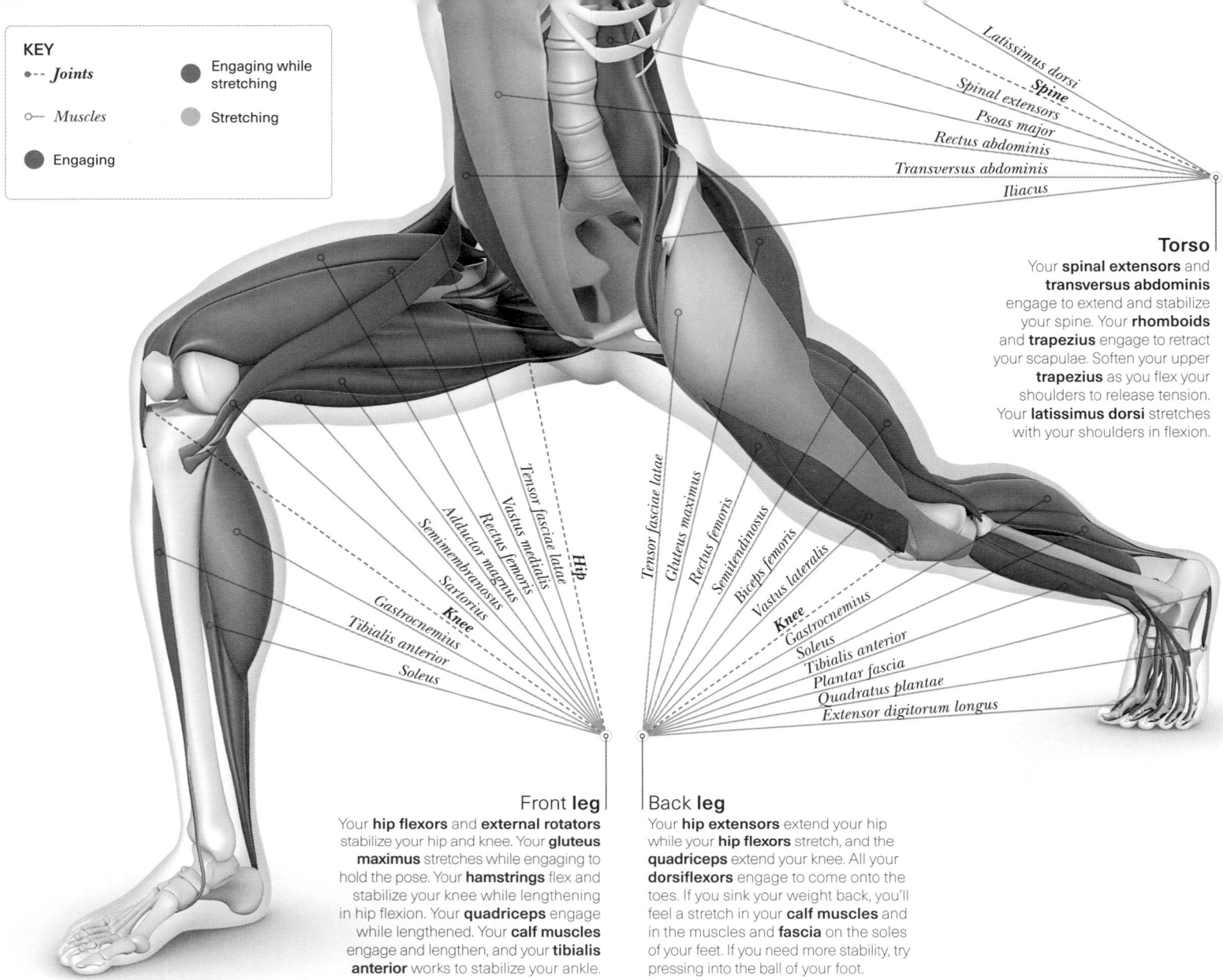

Torso

Your **spinal extensors** and **transversus abdominis** engage to extend and stabilize your spine. Your **rhomboids** and **trapezius** engage to retract your scapulae. Soften your upper **trapezius** as you flex your shoulders to release tension. Your **latissimus dorsi** stretches with your shoulders in flexion.

Front leg

Your **hip flexors** and **external rotators** stabilize your hip and knee. Your **gluteus maximus** stretches while engaging to hold the pose. Your **hamstrings** flex and stabilize your knee while lengthening in hip flexion. Your **quadriceps** engage while lengthened. Your **calf muscles** engage and lengthen, and your **tibialis anterior** works to stabilize your ankle.

Back leg

Your **hip extensors** extend your hip while your **hip flexors** stretch, and the **quadriceps** extend your knee. All your **dorsiflexors** engage to come onto the toes. If you sink your weight back, you'll feel a stretch in your **calf muscles** and in the muscles and **fascia** on the soles of your feet. If you need more stability, try pressing into the ball of your foot.

» CLOSER LOOK

You may try modifications to find comfort and efficient alignment. This pose presents an opportunity to consciously release common "stress" and "fear" muscles like the upper trapezius and psoas major.

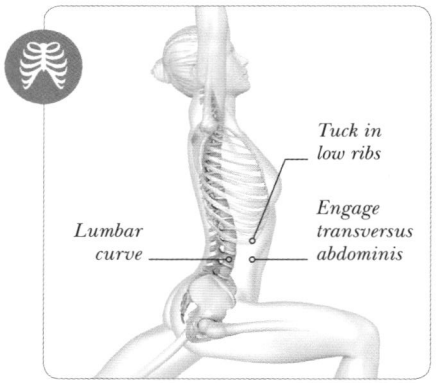

Spinal flexibility

Highly flexible people often allow their pelvic bowl to tilt forward, creating an extreme arch in the spine (see pp.14–15). If this is you, bring in your low ribs and engage your abdominals, particularly your transversus abdominis. However, don't overcompensate by tucking the tailbone and losing the lumbar curve.

Fingers reach up softly

Head is gently tilted up

Deep to the trapezius, your rhomboids stabilize your shoulder blades

Trapezius muscles engage

Spinal extensors engage on both sides

Latissimus dorsi stretches

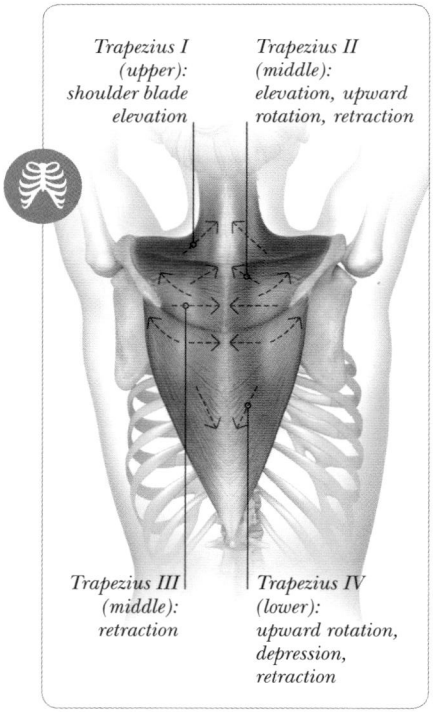

Trapezius recruitment

Your trapezius has three parts, and four fiber directions. When you flex your shoulders, all of the fibers engage to varying degrees. Your upper traps engage slightly to elevate the scapulae, though many overengage these muscles, causing tension. Your middle and lower traps should be the main players.

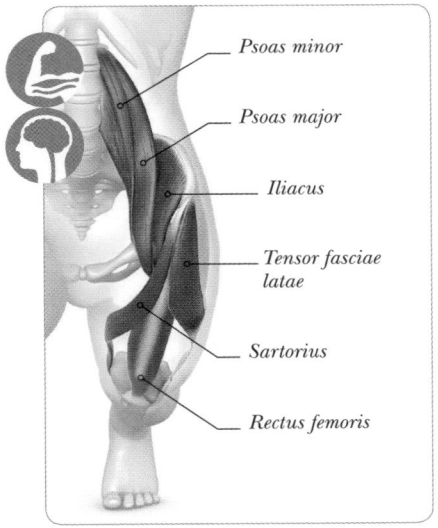

Hip flexors

If you sit a lot, your hip flexors may be tight. You may feel a profound sense of release following this asana, because your psoas is considered a fear reflex muscle. Ancient humans would engage it to run from predators; modern humans unconsciously engage it while stressed, seated at a computer.

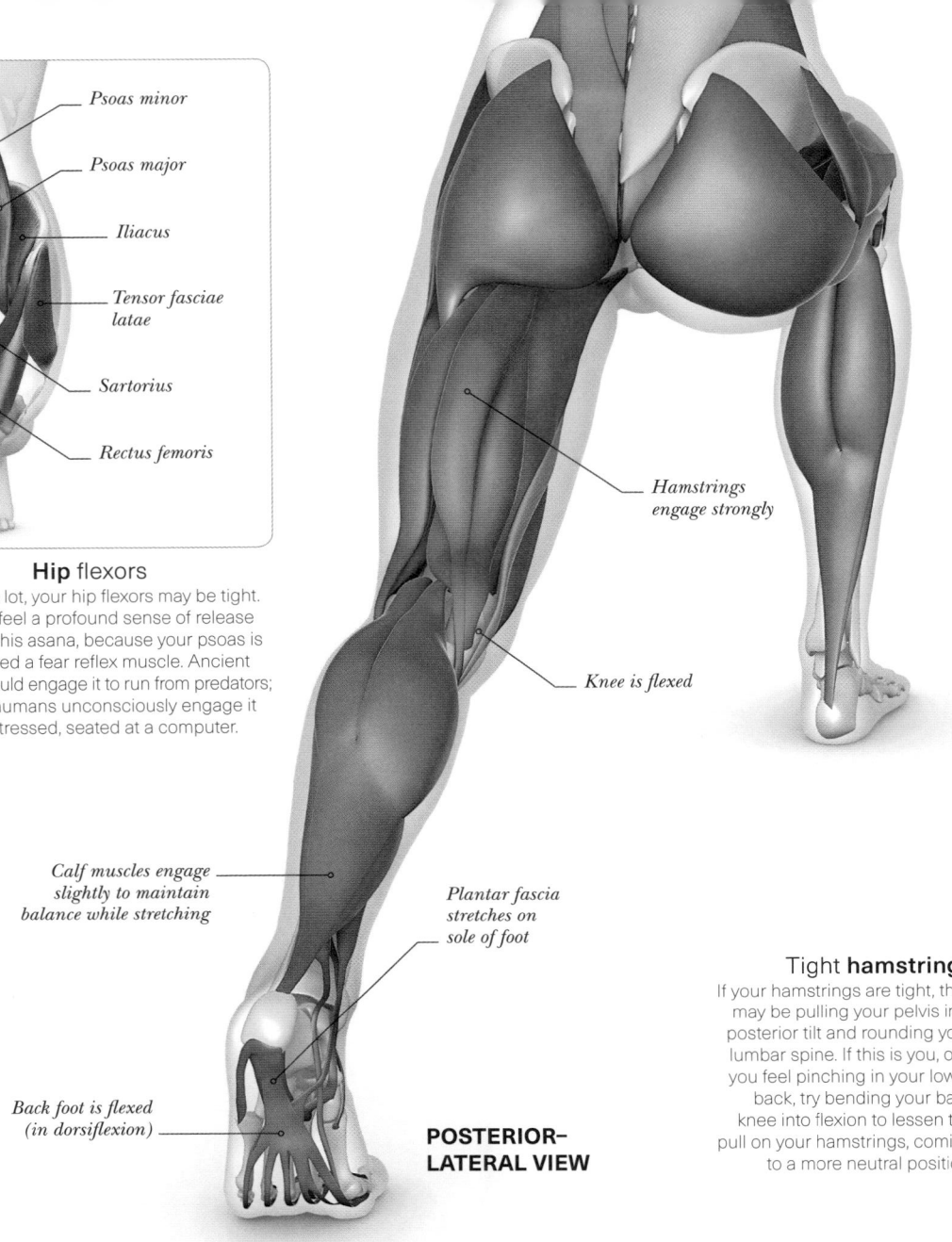

POSTERIOR–LATERAL VIEW

VARIATION

Place your back knee on the floor or on a folded blanket to reduce intensity. You could also place your hands on the floor to take balance out of the equation, allowing you to focus solely on stretching.

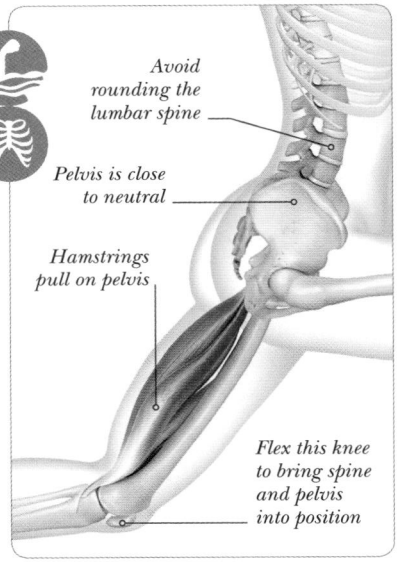

Tight hamstrings

If your hamstrings are tight, they may be pulling your pelvis into posterior tilt and rounding your lumbar spine. If this is you, or if you feel pinching in your lower back, try bending your back knee into flexion to lessen the pull on your hamstrings, coming to a more neutral position.

WARRIOR II

Virabhadrasana II

This strong standing pose is grounding, energizing, and stabilizing. Holding Warrior II for a period of time works on your balance and muscular strength, and provides a great opportunity to observe how your mind reacts during a heated challenge.

THE BIG PICTURE

This pose engages large muscles around your thighs and core. Your arms are reaching in both directions, creating space in the joints, without stiffening or locking your elbows or fingers.

KEY

- *Joints*
- *Muscles*
- Engaging
- Engaging while stretching
- Stretching

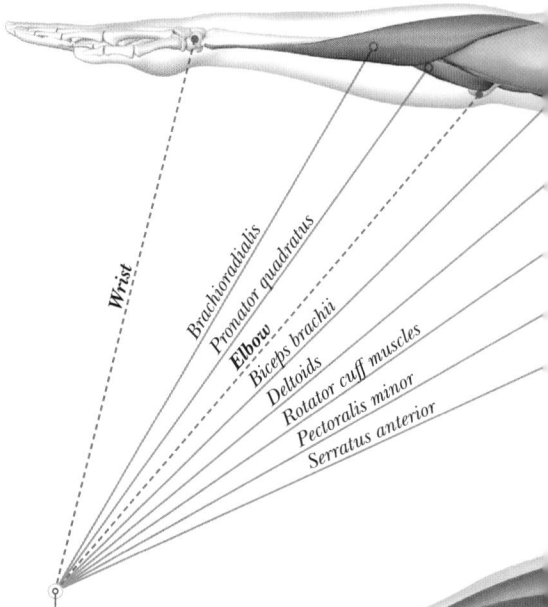

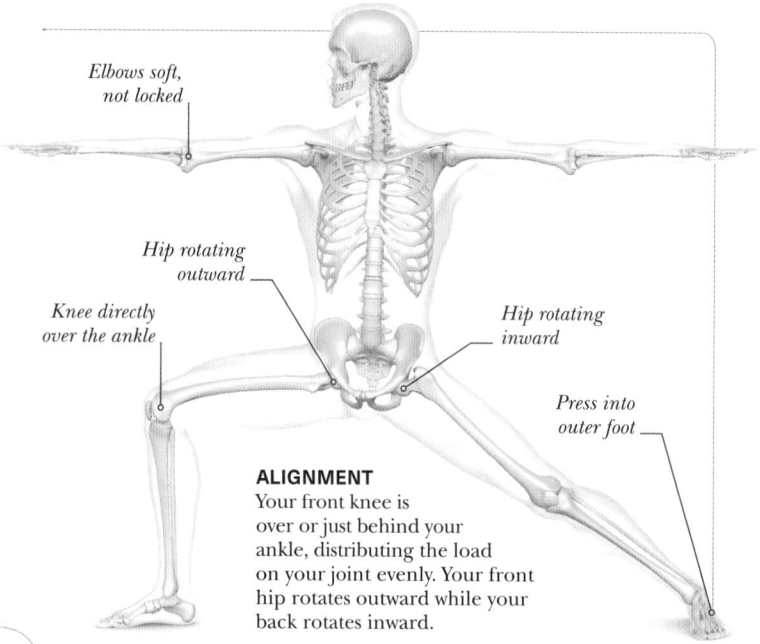

ALIGNMENT
Your front knee is over or just behind your ankle, distributing the load on your joint evenly. Your front hip rotates outward while your back rotates inward.

Arms

Your shoulders are abducted by your **middle deltoid** and **supraspinatus**. While all the **deltoid heads** engage to stabilize your shoulders in place, your **anterior deltoid** helps your **latissimus dorsi** to internally rotate the joints. Your elbows are extended by your **triceps** and your forearms are turned palm face down by your **pronators**. Your **pectorals** stabilize while in a lengthened position on both sides.

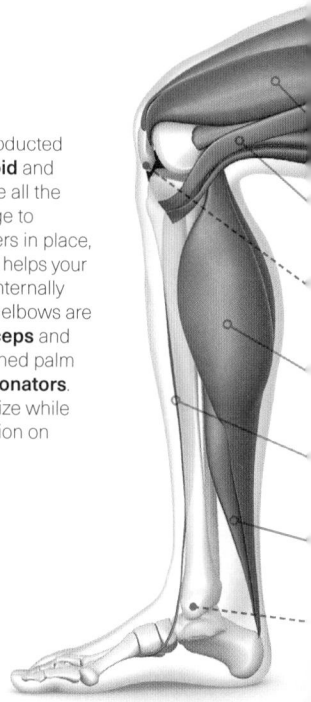

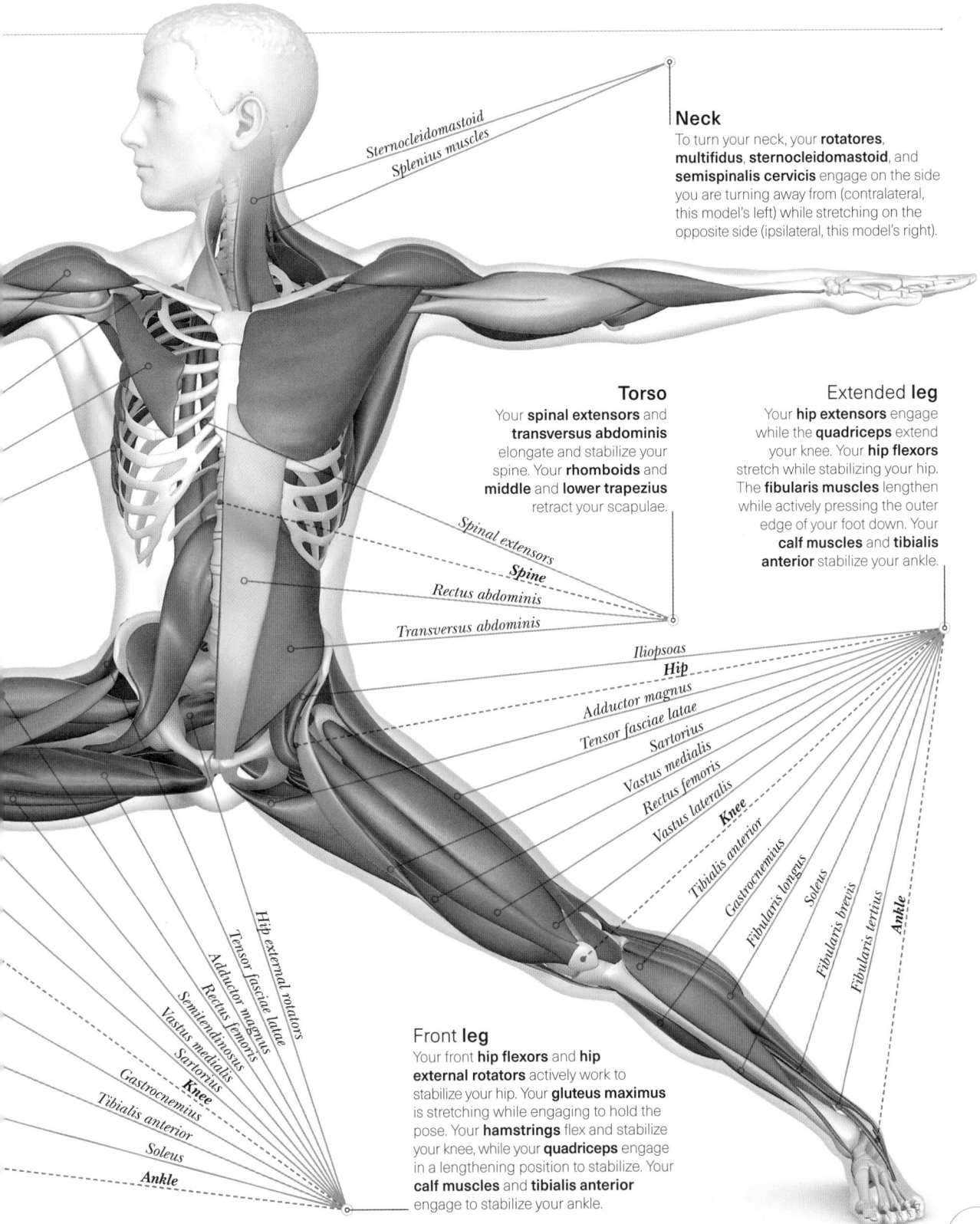
Sternocleidomastoid
Splenius muscles
Neck
To turn your neck, your **rotatores**, **multifidus**, **sternocleidomastoid**, and **semispinalis cervicis** engage on the side you are turning away from (contralateral, this model's left) while stretching on the opposite side (ipsilateral, this model's right).
Torso
Your **spinal extensors** and **transversus abdominis** elongate and stabilize your spine. Your **rhomboids** and **middle** and **lower trapezius** retract your scapulae.
Extended leg
Your **hip extensors** engage while the **quadriceps** extend your knee. Your **hip flexors** stretch while stabilizing your hip. The **fibularis muscles** lengthen while actively pressing the outer edge of your foot down. Your **calf muscles** and **tibialis anterior** stabilize your ankle.
Spinal extensors
Spine
Rectus abdominis
Transversus abdominis
Iliopsoas
Hip
Adductor magnus
Tensor fasciae latae
Sartorius
Vastus medialis
Rectus femoris
Vastus lateralis
Knee
Tibialis anterior
Gastrocnemius
Fibularis longus
Soleus
Fibularis brevis
Fibularis tertius
Ankle
Hip external rotators
Tensor fasciae latae
Adductor magnus
Rectus femoris
Semitendinosus
Vastus medialis
Sartorius
Knee
Gastrocnemius
Tibialis anterior
Soleus
Ankle
Front leg
Your front **hip flexors** and **hip external rotators** actively work to stabilize your hip. Your **gluteus maximus** is stretching while engaging to hold the pose. Your **hamstrings** flex and stabilize your knee, while your **quadriceps** engage in a lengthening position to stabilize. Your **calf muscles** and **tibialis anterior** engage to stabilize your ankle.

» CLOSER LOOK

Proper alignment in Warrior II can prevent damage to joint structures, especially in your knees. This is vital because the knee is one of the most mechanically complex joints in the body.

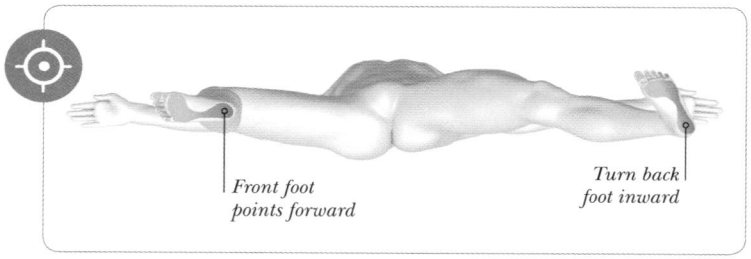

Pressure and **balance**

Try bringing your front heel in line with the center of the arch of your back foot. Distribute weight evenly between feet.

Soft gaze over your middle finger

Neck muscles dynamically stretch and engage

Hands reach in both directions

Biceps stretch

Deep to the pectoralis major, the pectoralis minor also engages

Knee over **ankle**

This pose is traditionally done with the knee stacked over the ankle. Allowing your knee to move forward past your ankle increases the load on joint structures such as your ACL. If you have any increased knee pain, an ACL injury, or knee arthritis, avoid moving your knee past your ankle.

Flexing knee past 90 degrees can put strain on the ACL

Bent knee aligned with ankle

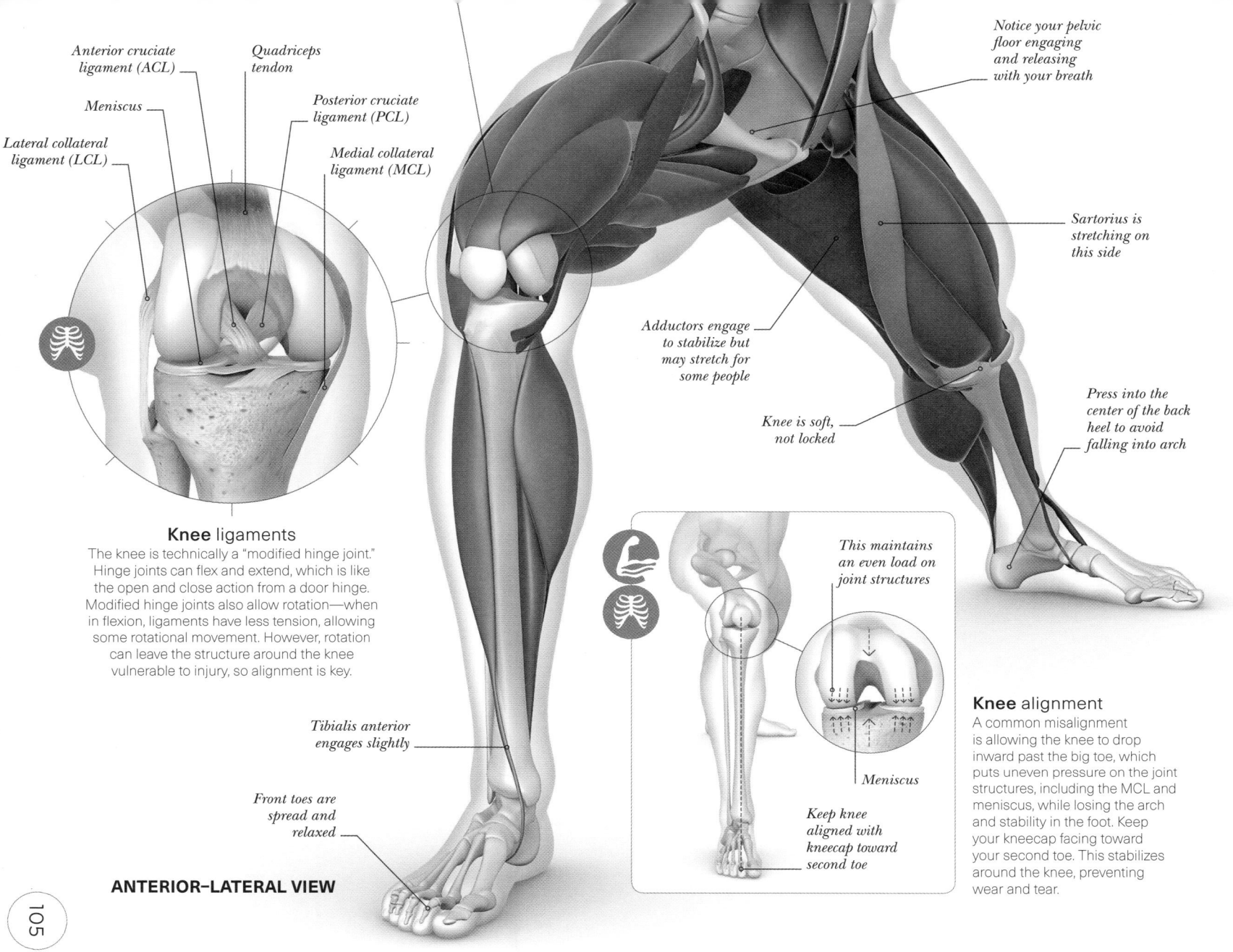

Knee ligaments

The knee is technically a "modified hinge joint." Hinge joints can flex and extend, which is like the open and close action from a door hinge. Modified hinge joints also allow rotation—when in flexion, ligaments have less tension, allowing some rotational movement. However, rotation can leave the structure around the knee vulnerable to injury, so alignment is key.

Knee alignment

A common misalignment is allowing the knee to drop inward past the big toe, which puts uneven pressure on the joint structures, including the MCL and meniscus, while losing the arch and stability in the foot. Keep your kneecap facing toward your second toe. This stabilizes around the knee, preventing wear and tear.

ANTERIOR–LATERAL VIEW

WARRIOR III

Virabhadrasana III

Warrior III is a strong, standing balancing pose that increases your focus and coordination. Your balance is particularly challenged as you bring your head parallel to the ground, affecting structures inside your inner ear that monitor your position and help to keep you upright.

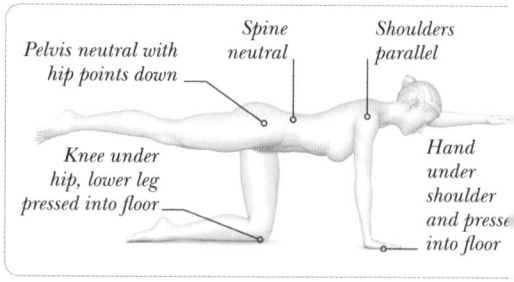

VARIATION
Sunbird challenges your balance but from a more stable base. Start on all fours, then lift an arm at shoulder height and the opposite leg at hip height.

THE BIG PICTURE

The muscles of your thighs, lower legs, and ankles strengthen as you try to maintain your balance on one leg. Muscles around your hips, core, and shoulders work hard to hold the rest of your body horizontal.

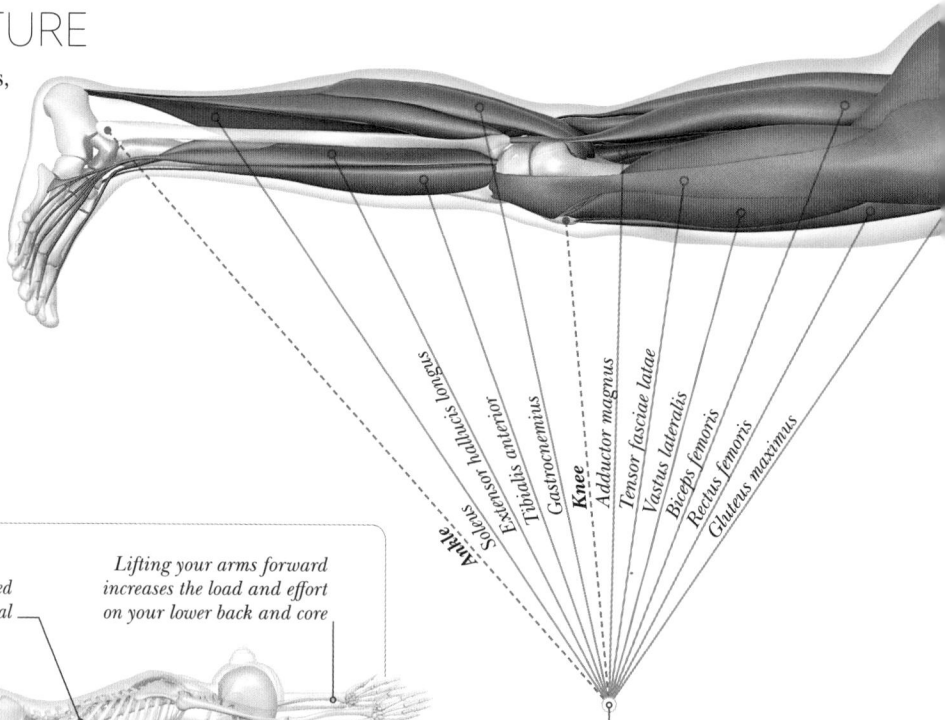

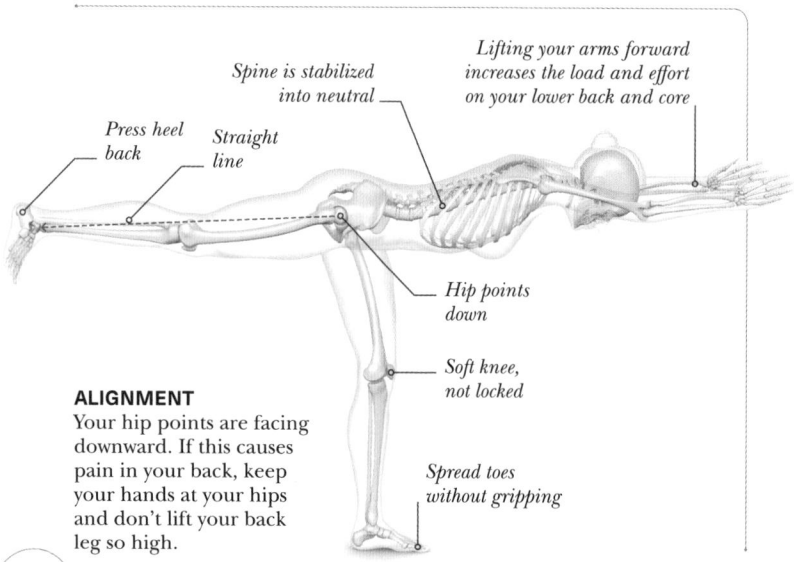

ALIGNMENT
Your hip points are facing downward. If this causes pain in your back, keep your hands at your hips and don't lift your back leg so high.

Lifted **leg**

Your **hip extensors** engage while your **hip flexors** stretch. Your **quadriceps** engage to extend your knee, while your **hamstrings** engage and lengthen. Push your heel back and feel your **ankle dorsiflexors** activate as though you are stepping firmly on a wall behind you. This helps your overall balance and stability.

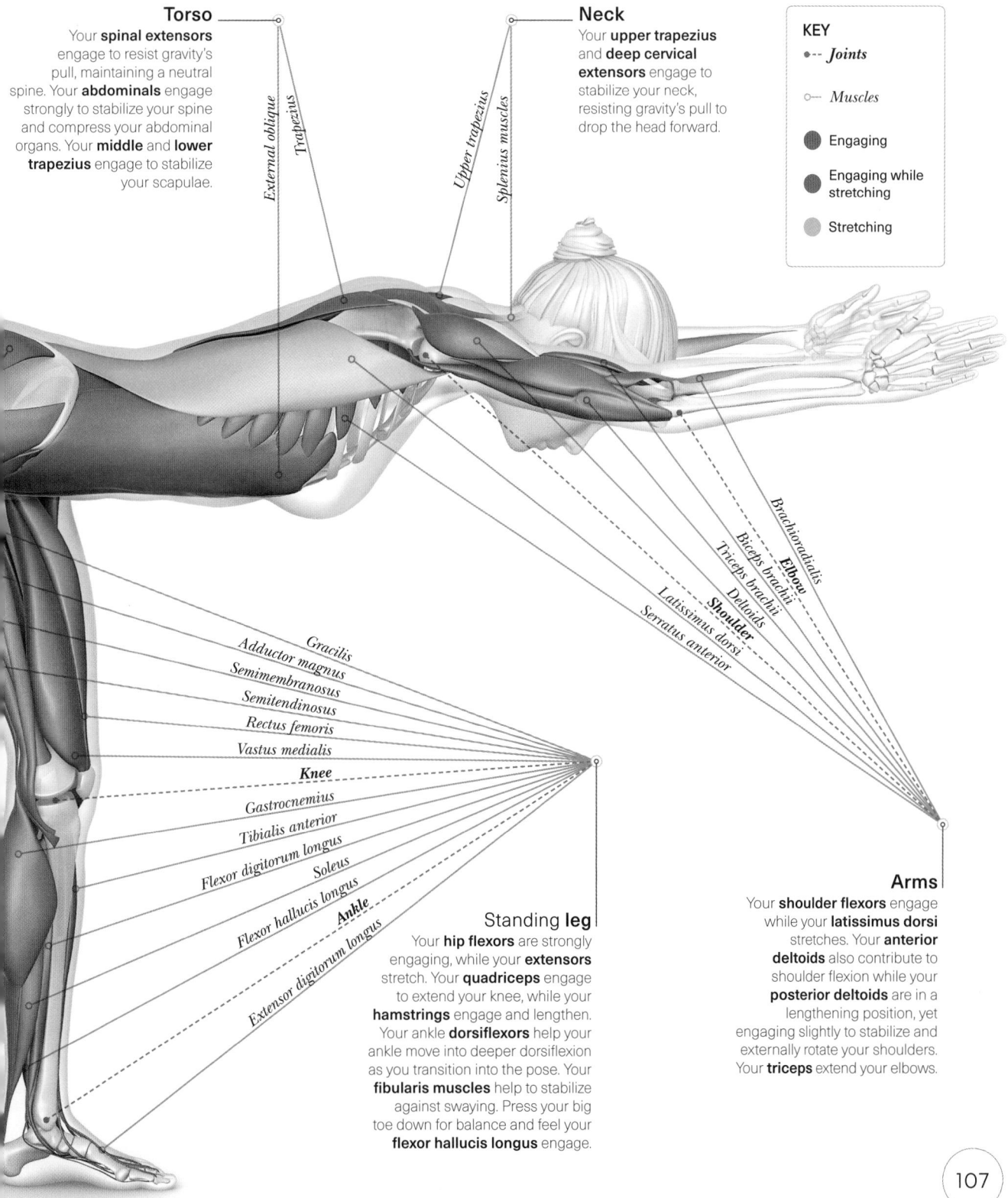
Torso
Your spinal extensors engage to resist gravity's pull, maintaining a neutral spine. Your abdominals engage strongly to stabilize your spine and compress your abdominal organs. Your middle and lower trapezius engage to stabilize your scapulae.
External oblique
Trapezius
Neck
Your upper trapezius and deep cervical extensors engage to stabilize your neck, resisting gravity's pull to drop the head forward.
Upper trapezius
Splenius muscles
KEY
Joints
Muscles
Engaging
Engaging while stretching
Stretching
Brachioradialis
Elbow
Biceps brachii
Triceps brachii
Deltoids
Shoulder
Latissimus dorsi
Serratus anterior
Gracilis
Adductor magnus
Semimembranosus
Semitendinosus
Rectus femoris
Vastus medialis
Knee
Gastrocnemius
Tibialis anterior
Flexor digitorum longus
Soleus
Flexor hallucis longus
Ankle
Extensor digitorum longus
Standing leg
Your hip flexors are strongly engaging, while your extensors stretch. Your quadriceps engage to extend your knee, while your hamstrings engage and lengthen. Your ankle dorsiflexors help your ankle move into deeper dorsiflexion as you transition into the pose. Your fibularis muscles help to stabilize against swaying. Press your big toe down for balance and feel your flexor hallucis longus engage.
Arms
Your shoulder flexors engage while your latissimus dorsi stretches. Your anterior deltoids also contribute to shoulder flexion while your posterior deltoids are in a lengthening position, yet engaging slightly to stabilize and externally rotate your shoulders. Your triceps extend your elbows.

» CLOSER LOOK

There are three mechanisms of balance: inner ear, visual, and proprioceptive input. Warrior III challenges each of these systems, improving your dynamic balance as you enter the pose and static balance while you hold it.

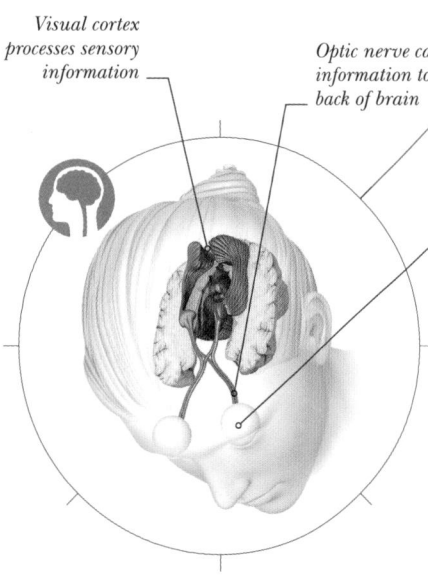

Visual input

"Drishti" is the yoga term for a focal point, which can help with both balance and concentration. Softly focus on a single stationary point ahead of you. You can also experiment with closing your eyes for a few moments—you will quickly realize how much your visual input contributes to your balance.

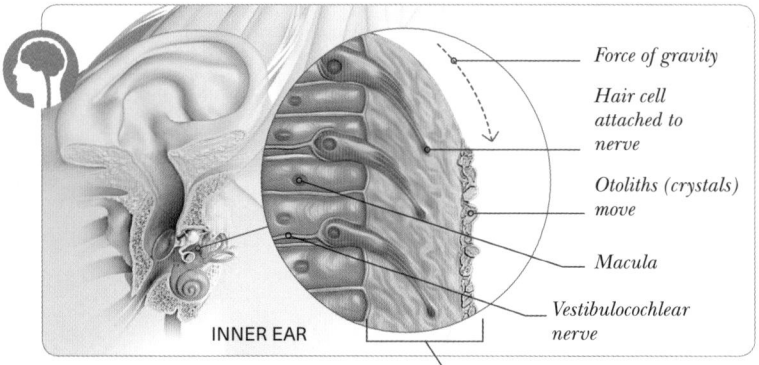

Inner ear input

Your inner ear has a bony labyrinth of tunnels filled with fluid to regulate your equilibrium or balance. When your head changes orientation, the fluid pushes on sensitive hair cells. Attached nerves tell your brain which direction your head is moving, to adjust for balance.

Torso comes as close toward parallel to the ground as comfortable

Hip extensors engage and lengthen

Quadriceps engage strongly

Kneecap faces forward

Tibialis anterior engages to maintain balance

Toes spread out and relaxed

ANTERIOR VIEW

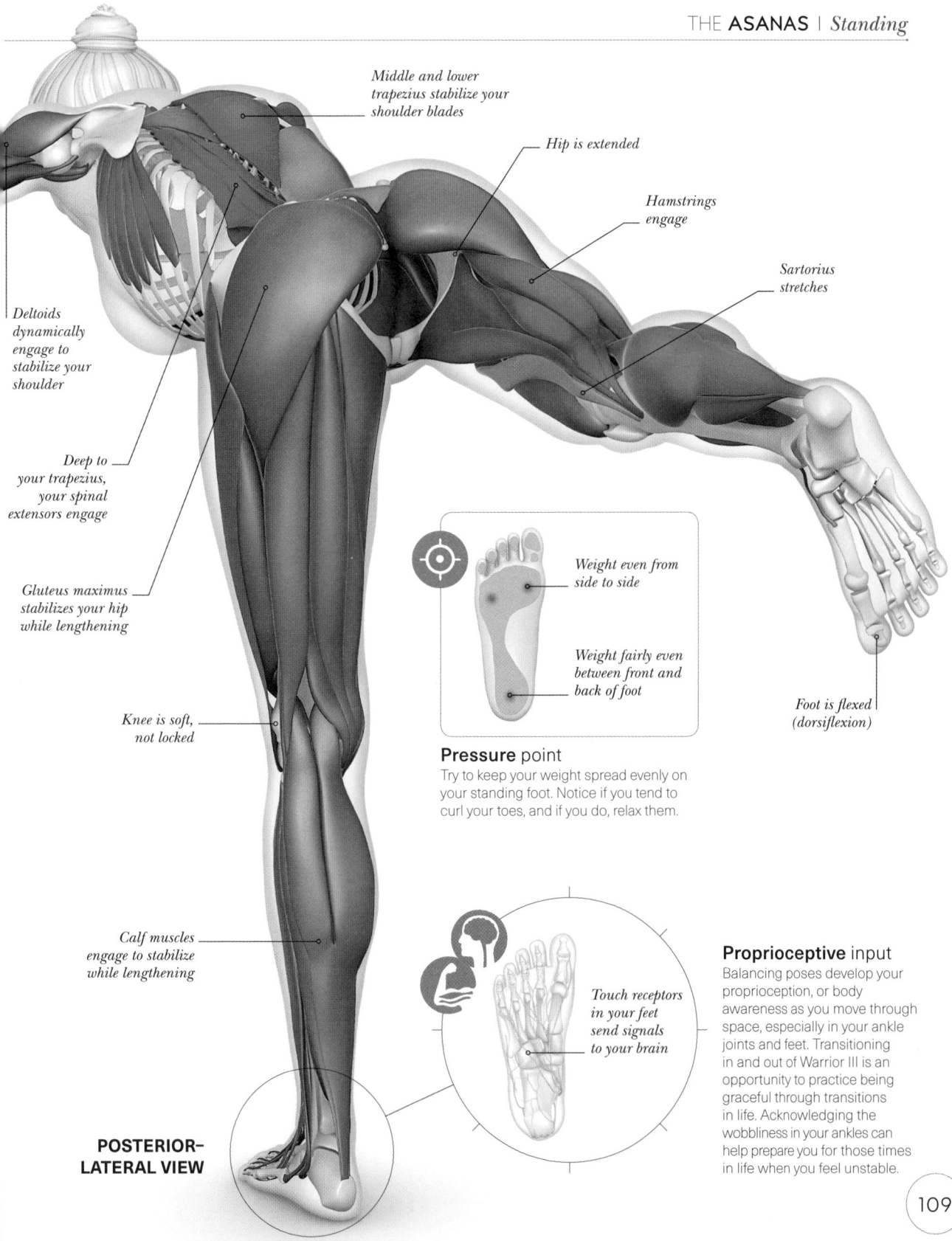

Pressure point

Try to keep your weight spread evenly on your standing foot. Notice if you tend to curl your toes, and if you do, relax them.

Proprioceptive input

Balancing poses develop your proprioception, or body awareness as you move through space, especially in your ankle joints and feet. Transitioning in and out of Warrior III is an opportunity to practice being graceful through transitions in life. Acknowledging the wobbliness in your ankles can help prepare you for those times in life when you feel unstable.

TREE

Vrksasana

Tree pose builds static balance, which can be facilitated by allowing a smooth and steady breath and focused mind. In this iconic yoga pose, unsteadiness is completely natural. Wobbling means you are strengthening muscles key for joint stabilization.

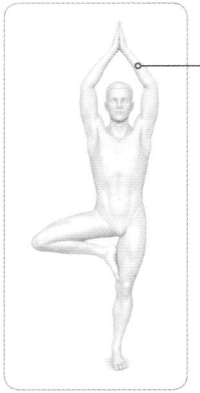

VARIATION
Raising your arms overhead shifts your center of gravity higher. Challenge your balance further by lifting your gaze. You may also hold your ams in a wide V.

THE BIG PICTURE

Large muscles in your standing thigh and lower leg engage to give your body a stable base. Muscles in your torso and on your raised thigh work to keep your leg lifted and rotated outward. Your upper body remains neutral and stable.

Rhomboids
Spinal extensors
Spine
Rectus abdominis
Transversus abdominis
Shoulder
Pectoralis major
Triceps brachii
Wrist
Biceps brachii
Brachialis
Elbow

Torso

Your **spinal extensors** and **transversus abdominis** engage to elongate and stabilize your spine into its neutral curves. Your **rhomboids** and **middle** and **lower trapezius** engage to retract your scapulae.

Arms

Your **brachialis**, **biceps**, and **brachioradialis** flex your elbows while your **pectoralis major** helps to adduct your shoulders. Your **wrist flexors** stretch, while your **wrist extensors** engage as you firmly press your palms together at your sternum.

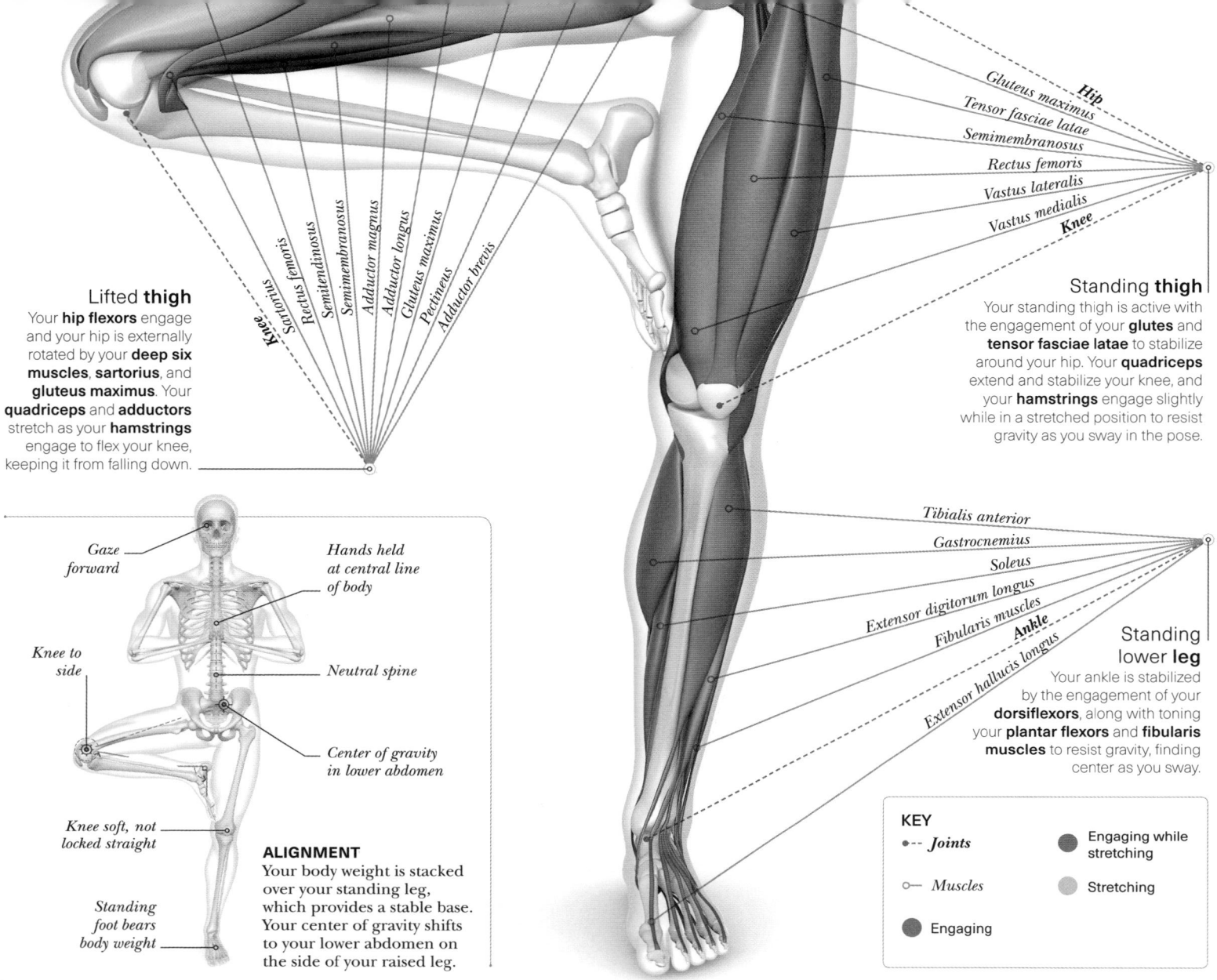

Lifted **thigh**

Your **hip flexors** engage and your hip is externally rotated by your **deep six muscles**, **sartorius**, and **gluteus maximus**. Your **quadriceps** and **adductors** stretch as your **hamstrings** engage to flex your knee, keeping it from falling down.

Standing **thigh**

Your standing thigh is active with the engagement of your **glutes** and **tensor fasciae latae** to stabilize around your hip. Your **quadriceps** extend and stabilize your knee, and your **hamstrings** engage slightly while in a stretched position to resist gravity as you sway in the pose.

Standing lower **leg**

Your ankle is stabilized by the engagement of your **dorsiflexors**, along with toning your **plantar flexors** and **fibularis muscles** to resist gravity, finding center as you sway.

ALIGNMENT

Your body weight is stacked over your standing leg, which provides a stable base. Your center of gravity shifts to your lower abdomen on the side of your raised leg.

» CLOSER LOOK

Tree pose stabilizes your hips in a unique position. Holding the pose increases body awareness, particularly in the sole of your standing foot. Breathe steadily and focus.

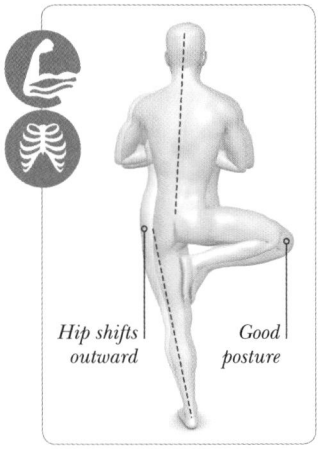

Hip abductors

If you are not engaging your hip abductors, particularly your gluteus medius, on your standing thigh, your hip will hike out. This is tough for balance and a common bad postural habit you may mindlessly do. To counteract this, press your standing hip in, bringing your pelvis to neutral.

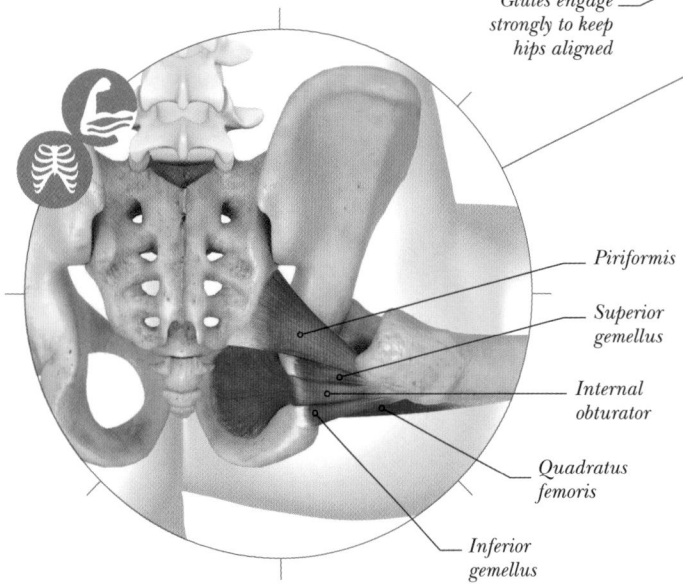

Deep six

To rotate your hip out to the side you engage a set of six small muscles deep within your hip joint. Strong standing poses, such as Tree, dynamically stretch and strengthen the deep six external rotators. To get a deeper stretch in these muscles, try stretches like King Pigeon (see pp.80–83).

Upper trapezius relaxes

Biceps brachi engage as hands are he at the heart

Spinal extensors engage to maintain posture

Quadriceps lengthen as knee flexes

Glutes engage strongly to keep hips aligned

Foot and thigh press into each other with equal and opposite force

Hamstrings engage to maintain balance

Ligaments around your ankles help to stabilize

POSTERIOR VIEW

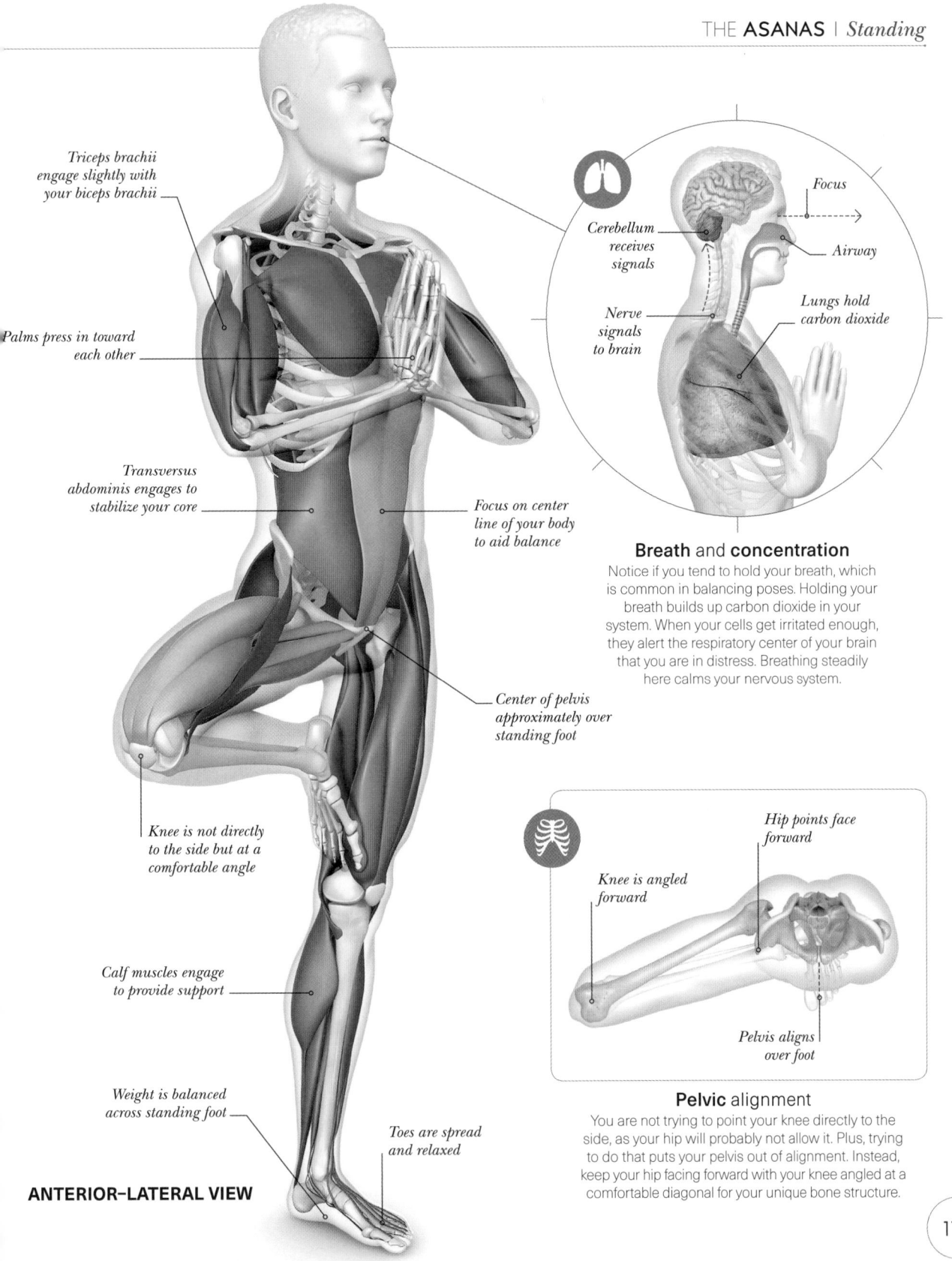

ANTERIOR–LATERAL VIEW

Breath and **concentration**

Notice if you tend to hold your breath, which is common in balancing poses. Holding your breath builds up carbon dioxide in your system. When your cells get irritated enough, they alert the respiratory center of your brain that you are in distress. Breathing steadily here calms your nervous system.

Pelvic alignment

You are not trying to point your knee directly to the side, as your hip will probably not allow it. Plus, trying to do that puts your pelvis out of alignment. Instead, keep your hip facing forward with your knee angled at a comfortable diagonal for your unique bone structure.

DANCER

Natarajasana

Dancer pose is a challenging static balancing pose, which also develops strength, flexibility, and agility. Dynamic balance skills are required to transition in and out of the pose with grace, though you can always hold onto a wall or chair for steadiness.

THE BIG PICTURE

Large muscles of your standing hip, thigh, and leg dramatically engage to help you balance on one leg. The front of your lifted hip and thigh stretch, as you kick back as a counterbalance. Your back muscles engage to come into a backbend, while your chest and abdomen stretch. Your neck is extended out long, and your shoulders are relaxed.

KEY

- *Joints*
- *Muscles*
- Engaging
- Engaging while stretching
- Stretching

Arms

In your front arm, your **anterior deltoid**, **pectoralis major**, and **coracobrachialis** flex your shoulder, while your **triceps** extend your elbow. In your back arm, your **posterior deltoid**, **latissimus dorsi**, and **teres major** engage to extend your shoulder, while your **triceps** extend your elbow. Your **elbow flexors** also engage in a stretched position to isometrically pull your leg inward.

Brachioradialis
Elbow
Biceps brachii
Triceps brachii
Deltoids
Shoulder
Pectoralis major

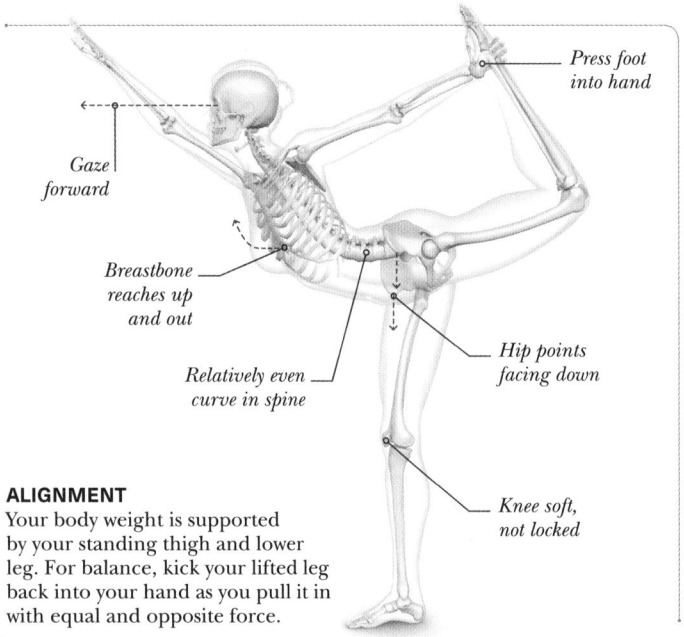

ALIGNMENT
Your body weight is supported by your standing thigh and lower leg. For balance, kick your lifted leg back into your hand as you pull it in with equal and opposite force.

VARIATION
For a challenge, reach both arms up and back to grasp your big toe. If you feel pinching in your lower back, don't go as deeply into the bend. Also try using a strap looped around your ankle.

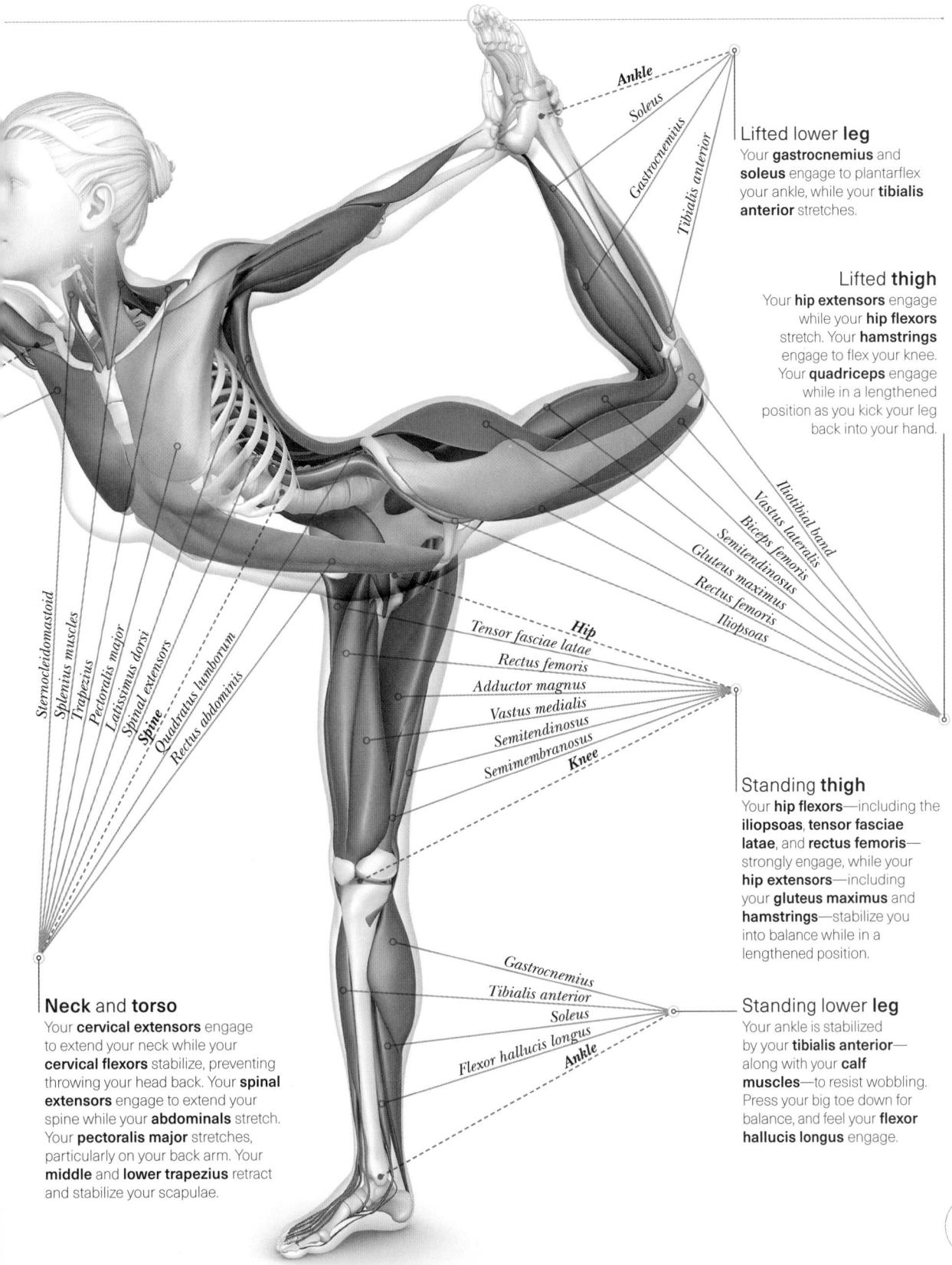

Lifted lower **leg**

Your **gastrocnemius** and **soleus** engage to plantarflex your ankle, while your **tibialis anterior** stretches.

Lifted **thigh**

Your **hip extensors** engage while your **hip flexors** stretch. Your **hamstrings** engage to flex your knee. Your **quadriceps** engage while in a lengthened position as you kick your leg back into your hand.

Standing **thigh**

Your **hip flexors**—including the **iliopsoas**, **tensor fasciae latae**, and **rectus femoris**—strongly engage, while your **hip extensors**—including your **gluteus maximus** and **hamstrings**—stabilize you into balance while in a lengthened position.

Standing lower **leg**

Your ankle is stabilized by your **tibialis anterior**—along with your **calf muscles**—to resist wobbling. Press your big toe down for balance, and feel your **flexor hallucis longus** engage.

Neck and **torso**

Your **cervical extensors** engage to extend your neck while your **cervical flexors** stabilize, preventing throwing your head back. Your **spinal extensors** engage to extend your spine while your **abdominals** stretch. Your **pectoralis major** stretches, particularly on your back arm. Your **middle** and **lower trapezius** retract and stabilize your scapulae.

» CLOSER LOOK

Dancer strikes a balance between stability and mobility, along with effort and ease. Muscles build in strengthening poses like this when microtears heal.

Elbow soft, not locked

Neck muscles stabilize chin, lifting slightly

Adductors engage and lengthen

ANTERIOR VIEW

Fiber bundle

Muscle fibers

Microtears on muscle tissue after exercise

How **muscles build**

Throughout your life, you can't grow more skeletal muscle cells. Instead, the cells can grow larger in diameter. After exercising, your muscle cells are left damaged—covered in microtears—as a response to the beneficial stress on your tissue. Your body sends nutrients to the area, helping it heal and become even stronger than before.

Scapula

Glenoid cavity

Humerus

SHOULDER

Acetabulum is deep, securing head of femur

Neck of femur

Femur

HIP

Ball and **socket** joints

Both your shoulder and hip joints are ball and socket joints (see pp.16–17). Your shoulder joint is shallow with a lot of mobility; it is only limited by ligaments and muscles. Your hip joint, in contrast, is deeper with more joint structures to help hold it securely in place.

First metatarsal

Your big toe is key for balance

Toe joints

In poses like Dancer, spreading your toes helps you to balance. It can also counteract hallux valgus (bunions) in which the big toe joint turns inward, leading to bone deformity and inflammation around the joint.

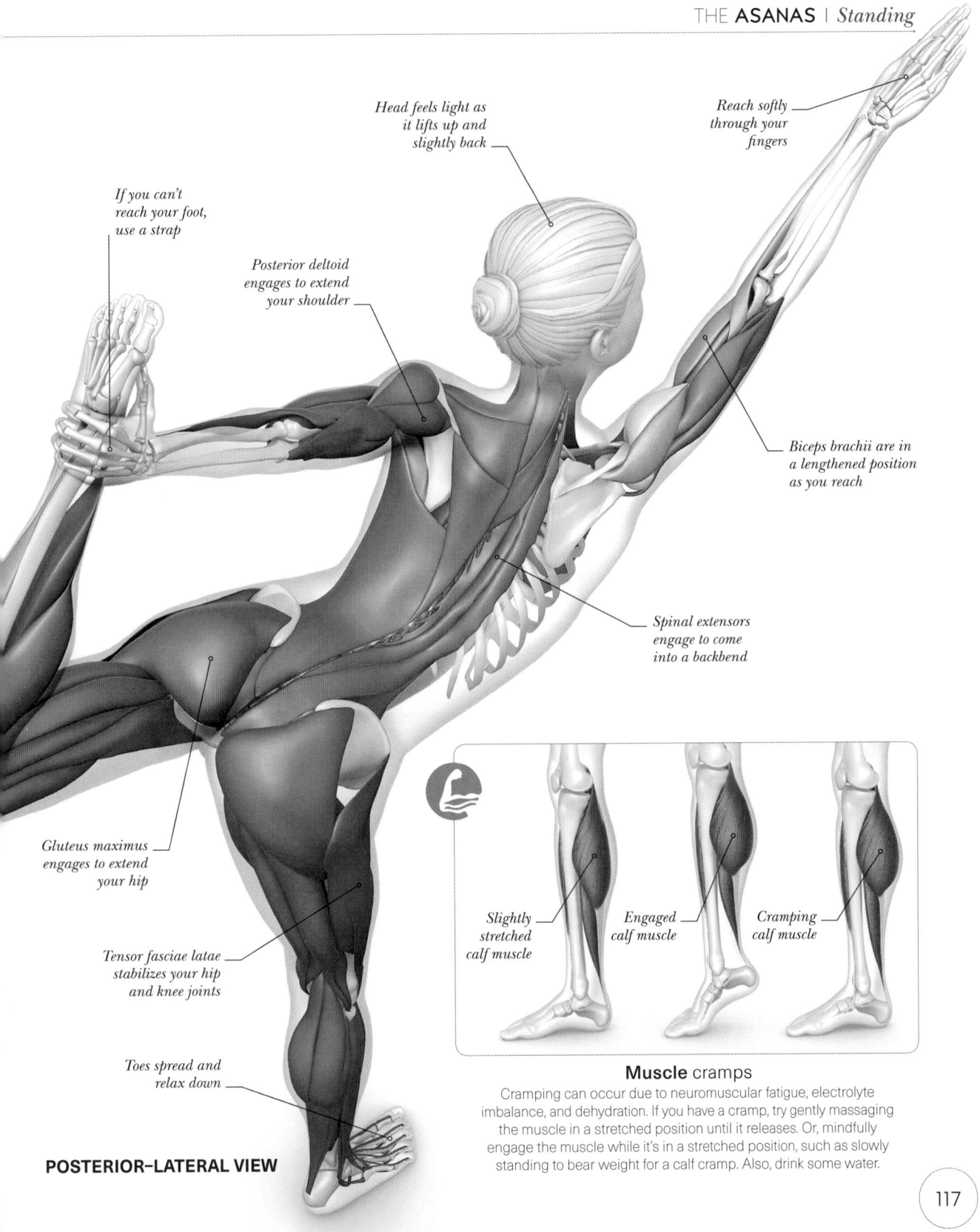

POSTERIOR–LATERAL VIEW

Muscle cramps

Cramping can occur due to neuromuscular fatigue, electrolyte imbalance, and dehydration. If you have a cramp, try gently massaging the muscle in a stretched position until it releases. Or, mindfully engage the muscle while it's in a stretched position, such as slowly standing to bear weight for a calf cramp. Also, drink some water.

TRIANGLE

Trikonasana

Triangle is a strengthening and grounding standing pose. It involves twisting your spine and ribcage to move against gravity and the tendency to round forward and down. Strong poses like this can strengthen both your muscles and bones.

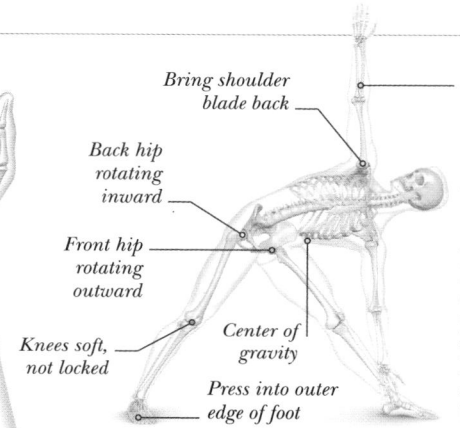

ALIGNMENT
Rotate your back hip inward by turning your toes inward. Rotate your front hip outward by turning your toes toward the front of your mat. Rotate your spine to stack your shoulder blades vertically.

THE BIG PICTURE

This pose particularly strengthens your core, thighs, and legs. Deep muscles close to your spine engage to stabilize your spine and give your brain feedback, enhancing your mind–body connection.

Neck and **torso**
To rotate your neck, on the side nearer the ground (model's left), your **sternocleidomastoid**, **rotatores**, **multifidus**, and **semispinalis cervicis** engage, while stretching on the upward-facing side (model's right). On the upward side, your **splenius capitis** and **splenius cervicis** engage, while stretching on the downward side. Your **transversus abdominis** engages to stabilize your spine. On the upward-facing side, your **external obliques** stretch, while your **internal obliques** engage to rotate your spine. On the downward side, your **external obliques** engage to rotate your spine.

Arms
Your shoulders are abducted by your **middle deltoids** and **supraspinatus** and stabilized by your **rotator cuff muscles**. Your **posterior deltoids** externally rotate your shoulders. Your elbows are extended by your **triceps** and forearms turned palms forward by your **supinators**.

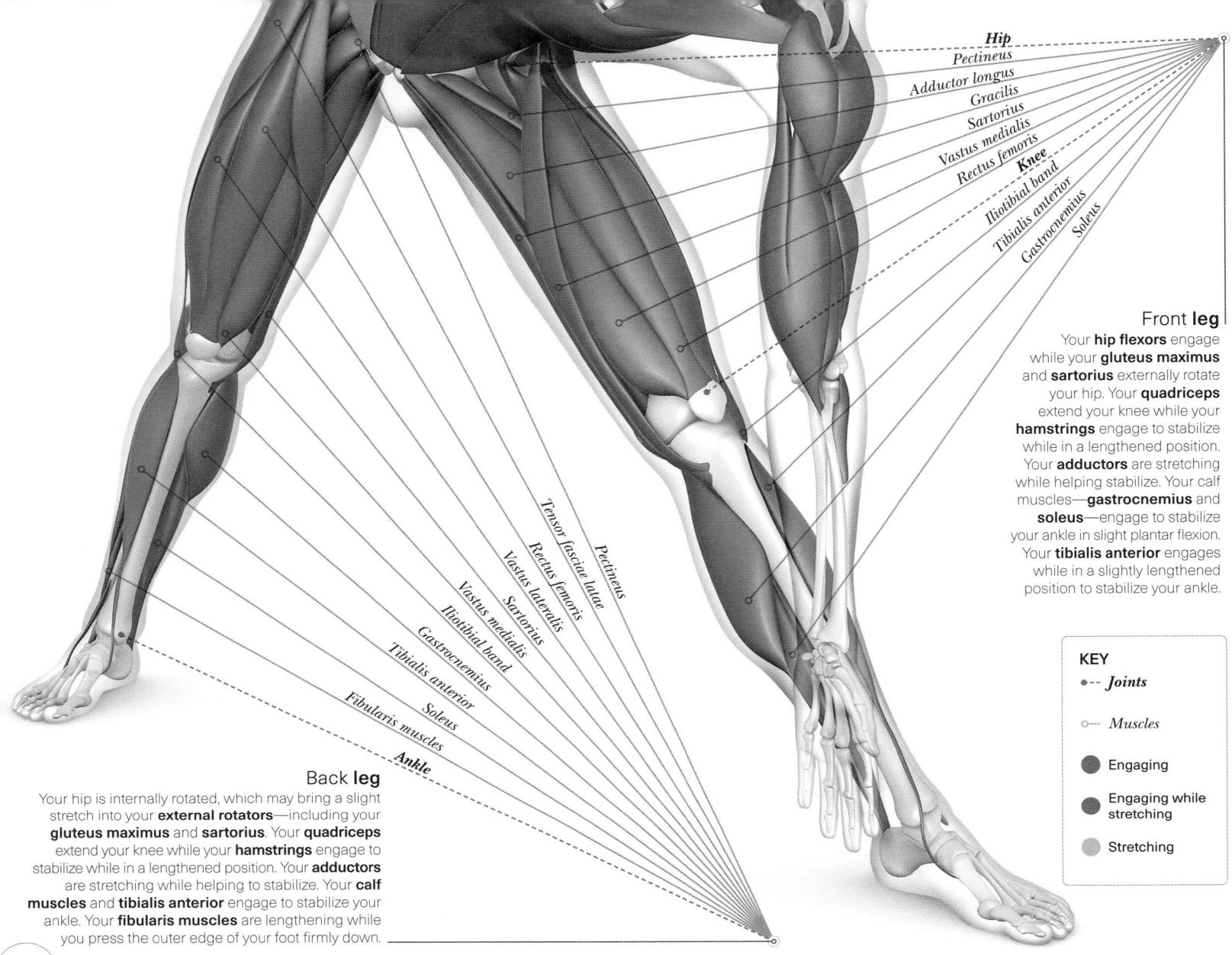

Front **leg**

Your **hip flexors** engage while your **gluteus maximus** and **sartorius** externally rotate your hip. Your **quadriceps** extend your knee while your **hamstrings** engage to stabilize while in a lengthened position. Your **adductors** are stretching while helping stabilize. Your calf muscles—**gastrocnemius** and **soleus**—engage to stabilize your ankle in slight plantar flexion. Your **tibialis anterior** engages while in a slightly lengthened position to stabilize your ankle.

Back **leg**

Your hip is internally rotated, which may bring a slight stretch into your **external rotators**—including your **gluteus maximus** and **sartorius**. Your **quadriceps** extend your knee while your **hamstrings** engage to stabilize while in a lengthened position. Your **adductors** are stretching while helping to stabilize. Your **calf muscles** and **tibialis anterior** engage to stabilize your ankle. Your **fibularis muscles** are lengthening while you press the outer edge of your foot firmly down.

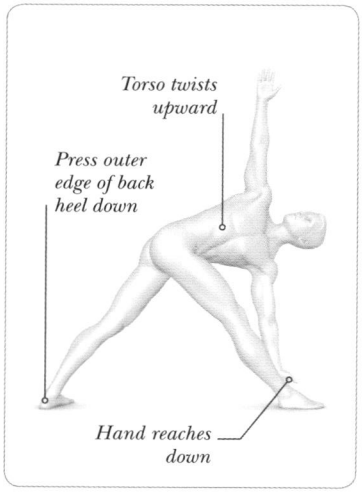

VARIATION
Twisted Triangle adds a torso twist to the pose, which challenges your stability. With your right foot forward, reach over your front leg and rotate your torso to the right. Avoid this pose if you have back issues. Feel free to place your left hand on your leg, a block, or the floor.

» CLOSER LOOK

Strengthening the muscles of your thighs, hips, and back in poses like Triangle may have the added benefit of boosting bone density. This pose should be practiced with care—listen to your body and ease out of the pose if you experience any pain or tingling, and be mindful of your knee joints.

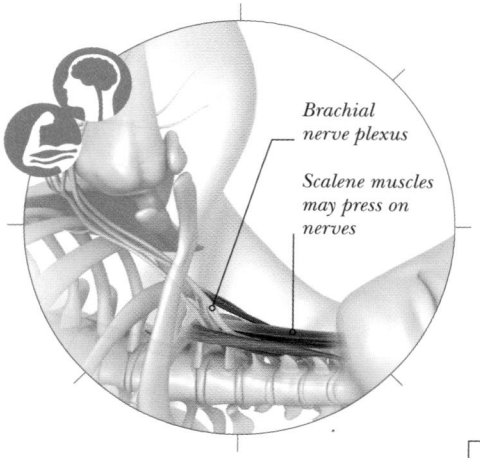

Pressure points

Ease off or come out of any pose that causes numbness, sharpness, or shooting pain. This may be due to pressure or impingement on nerves. Likewise, stop if you experience any tingling, coolness, or a dull, lifeless feeling like when you fall asleep on your arm. This can be caused by pressure occluding blood vessels.

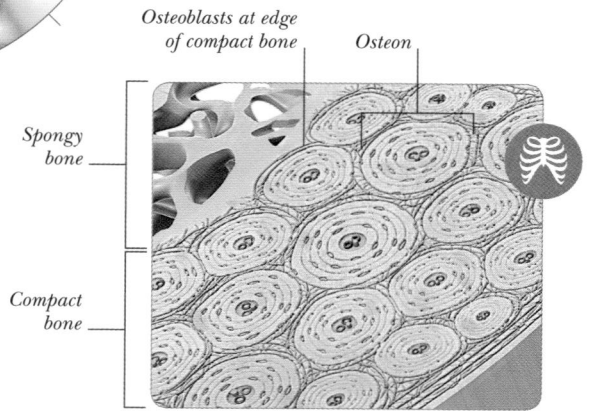

Bone growth

Large muscles in your thighs engage firmly, beneficially stressing your bones. This may wake up cells in the bone called osteoblasts, which triggers bone building. A 10-year trial concluded that yoga appears to raise bone mineral density in the spine and the femur.

Rotator cuff muscles stabilize the shoulders

Transversospinales (including the rotatores and multifidus) engage to rotate with awareness

Neck muscles dynamically engage and stretch

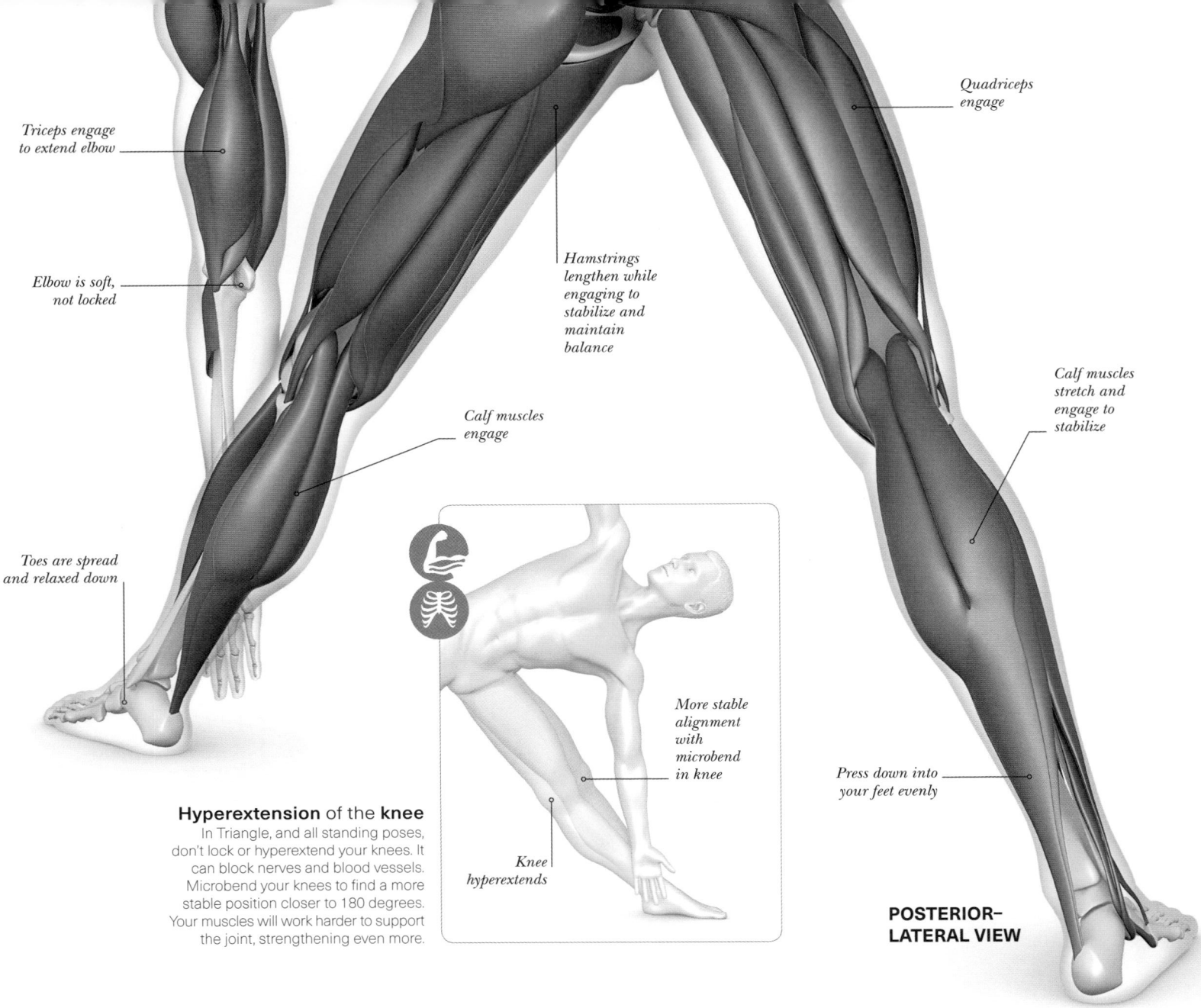

Hyperextension of the **knee**

In Triangle, and all standing poses, don't lock or hyperextend your knees. It can block nerves and blood vessels. Microbend your knees to find a more stable position closer to 180 degrees. Your muscles will work harder to support the joint, strengthening even more.

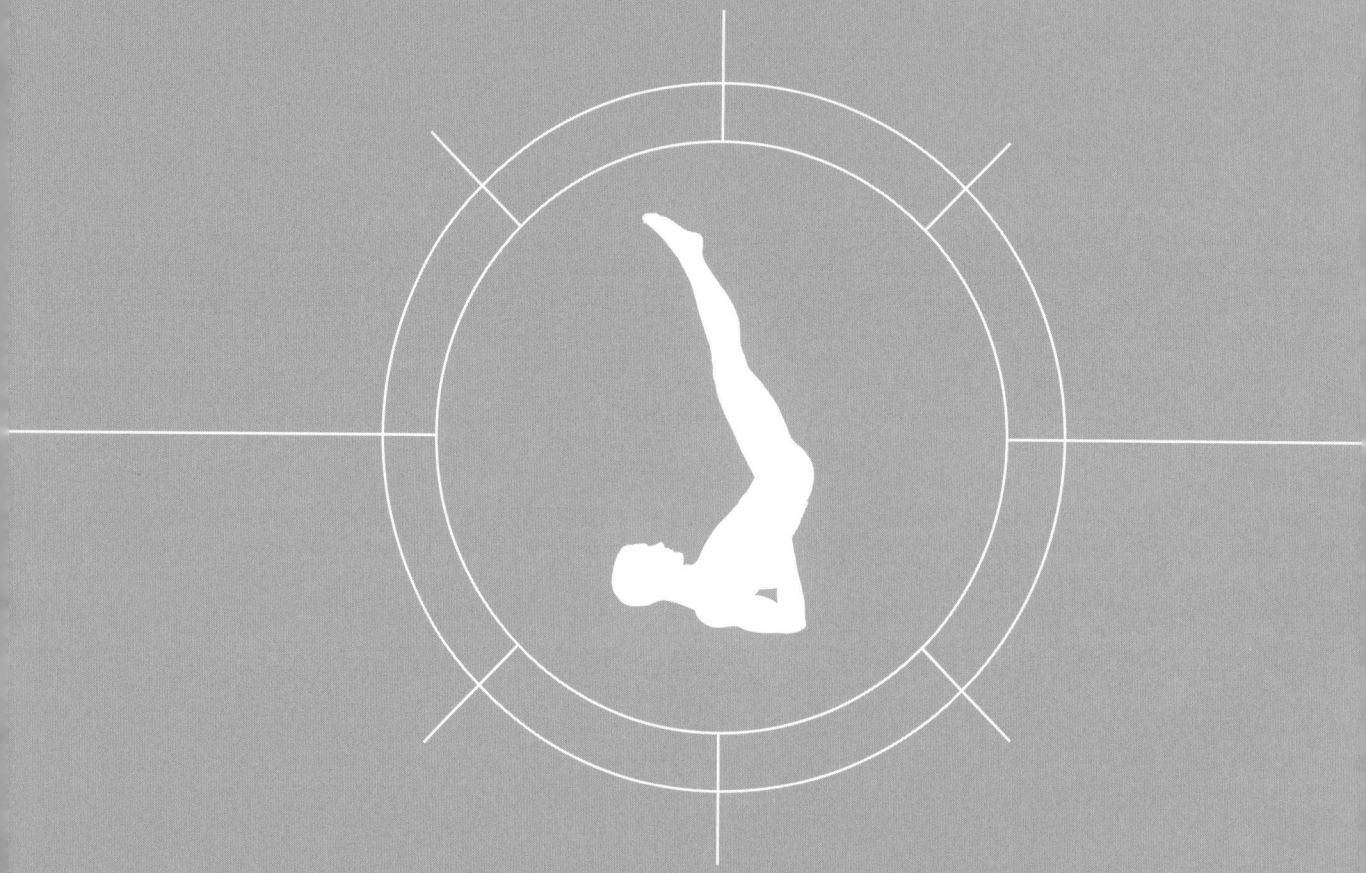

INVERSION ASANAS

Inversions are defined here as poses that turn your body upside down. Having your head below your heart has certain physiological effects and benefits, such as boosting circulation and aiding lymphatic drainage. Full inversions can be a great exploration of getting a new perspective—both literally and figuratively.

DOWNWARD-FACING DOG

Adho mukha svanasana

Also known as "Downward dog," this is a common pose in modern yoga classes, particularly as an integral part of sun salutations or flow sequences. This arm balance is a forward fold and partial inversion, stretching the back of your legs and strengthening your shoulders.

THE BIG PICTURE

In this pose, the back of your body—including your buttocks, thighs, and calf muscles—is stretching. Your shoulders are strengthening as you press into the floor.

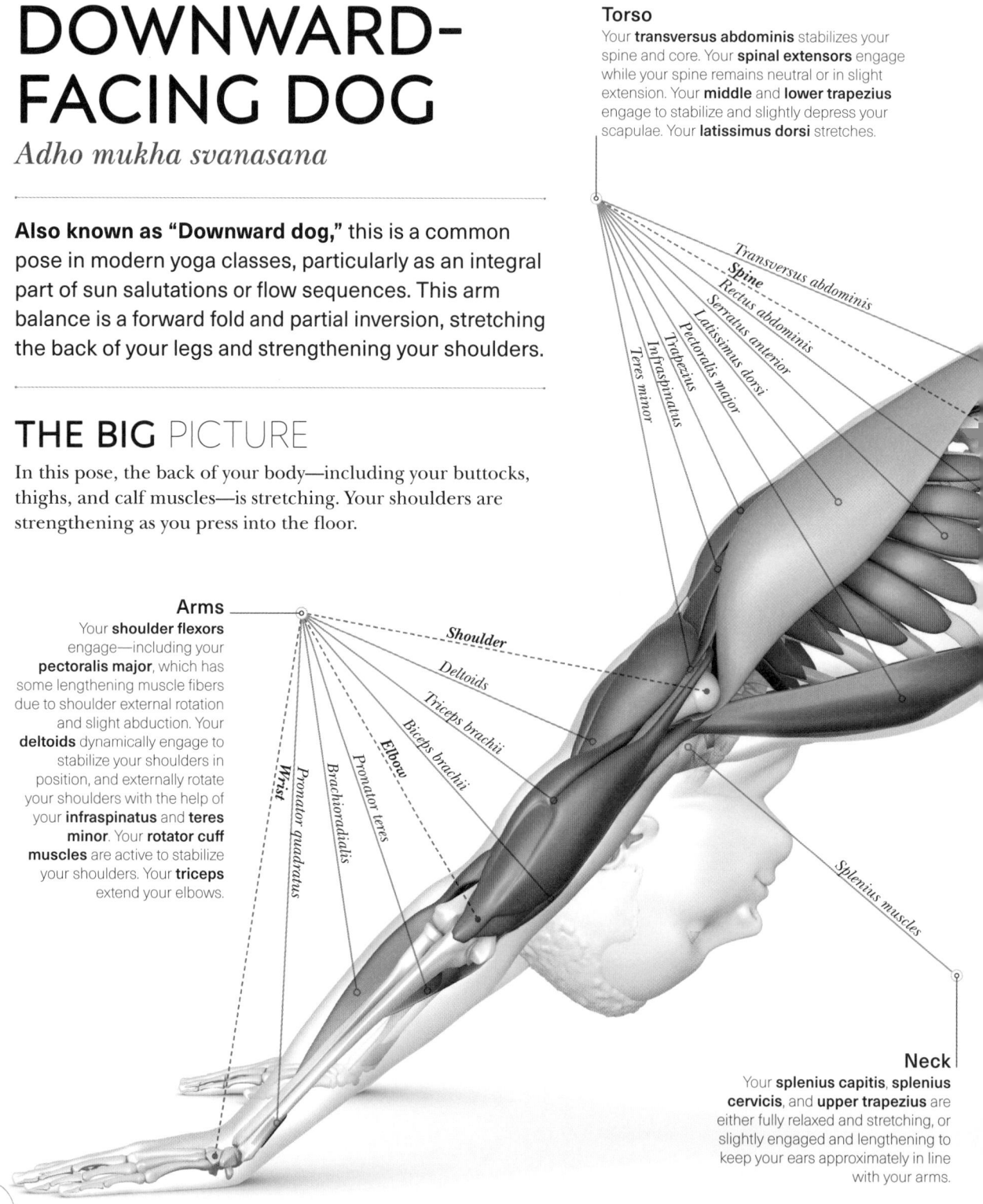

Torso

Your **transversus abdominis** stabilizes your spine and core. Your **spinal extensors** engage while your spine remains neutral or in slight extension. Your **middle** and **lower trapezius** engage to stabilize and slightly depress your scapulae. Your **latissimus dorsi** stretches.

Arms

Your **shoulder flexors** engage—including your **pectoralis major**, which has some lengthening muscle fibers due to shoulder external rotation and slight abduction. Your **deltoids** dynamically engage to stabilize your shoulders in position, and externally rotate your shoulders with the help of your **infraspinatus** and **teres minor**. Your **rotator cuff muscles** are active to stabilize your shoulders. Your **triceps** extend your elbows.

Neck

Your **splenius capitis**, **splenius cervicis**, and **upper trapezius** are either fully relaxed and stretching, or slightly engaged and lengthening to keep your ears approximately in line with your arms.

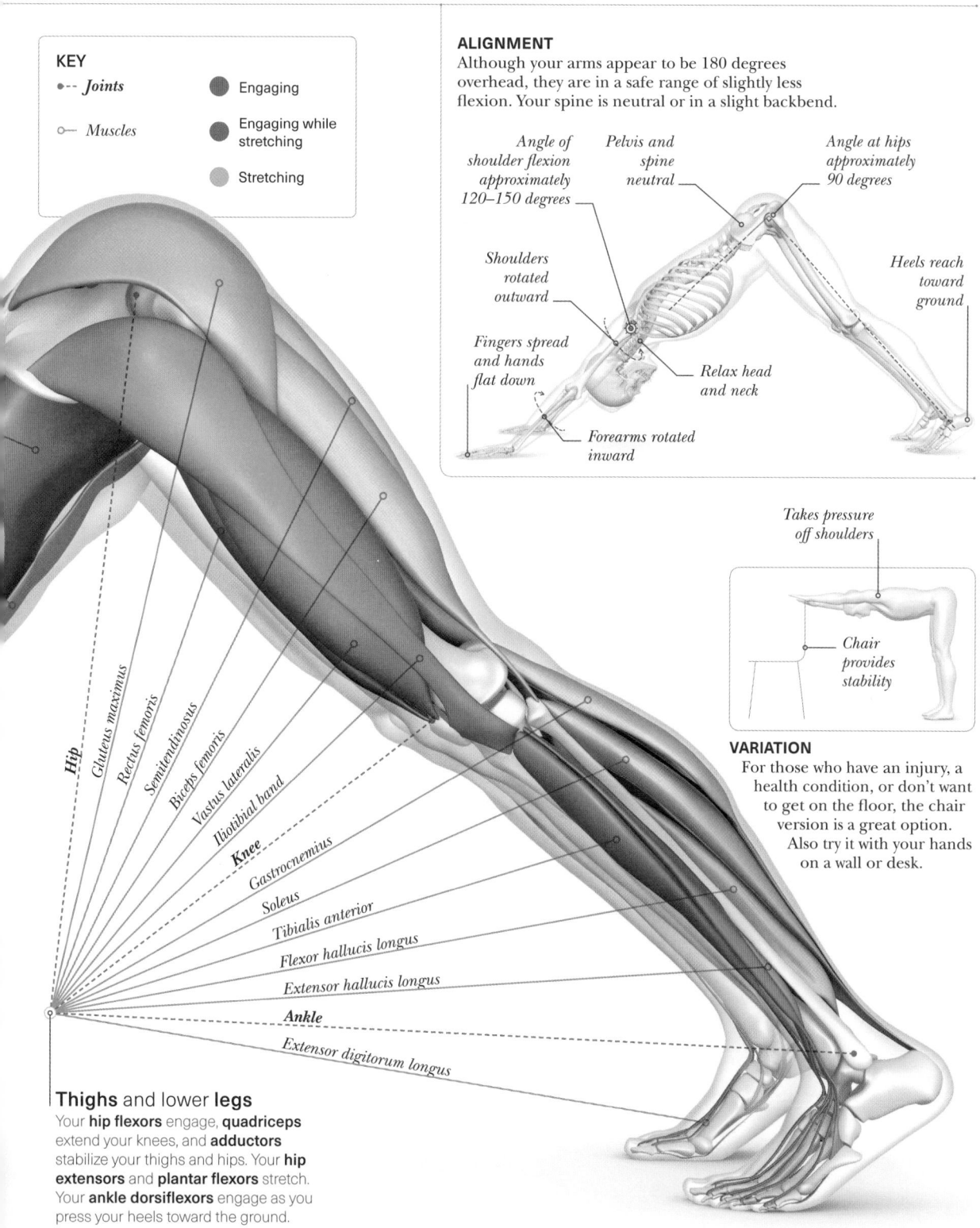

ALIGNMENT

Although your arms appear to be 180 degrees overhead, they are in a safe range of slightly less flexion. Your spine is neutral or in a slight backbend.

VARIATION

For those who have an injury, a health condition, or don't want to get on the floor, the chair version is a great option. Also try it with your hands on a wall or desk.

Thighs and lower **legs**

Your **hip flexors** engage, **quadriceps** extend your knees, and **adductors** stabilize your thighs and hips. Your **hip extensors** and **plantar flexors** stretch. Your **ankle dorsiflexors** engage as you press your heels toward the ground.

» CLOSER LOOK

Being too tight or too flexible can both present challenges when finding effective alignment in Downward-facing Dog. However, modifications can make it accessible for everyone.

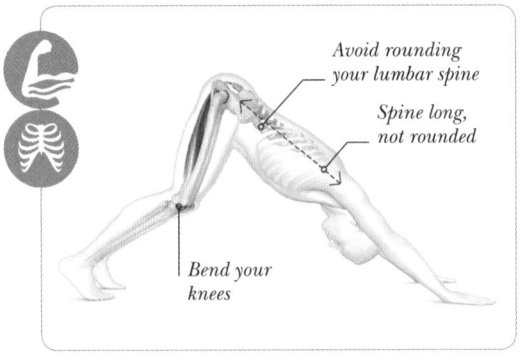

Tight **hamstrings**

When the hamstrings are tight, the pelvis is pulled and the back rounded. The integrity of your spine is more important than having your legs straight in this pose, so flex your knees and press into the floor—this will help lengthen your spine and bring the pelvis closer to neutral.

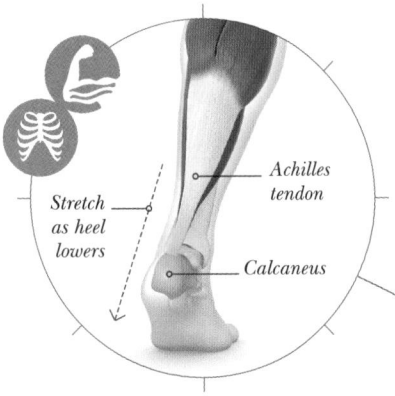

Achilles tendon

This tendon was named after the Greek mythological figure, Achilles, who only had one weakness: his calcaneal tendon. For many, it is very tight, preventing touching the heel to the ground in this pose. It can stretch with practice; however, there is some functional benefit to it maintaining tension as it stores potential energy.

Gluteus maximus stretches

Adductor magnus engages while in a neutral position

Pressing your heels actively down strengthens your tibialis anterior

POSTERIOR-LATERAL VIEW

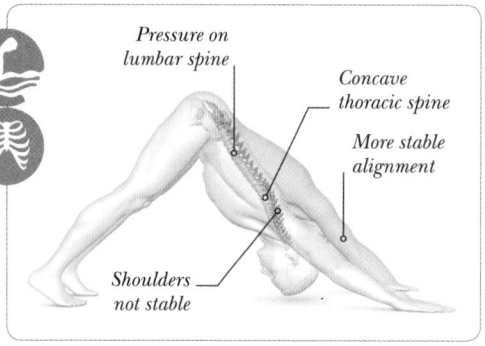

Spinal flexibility

If you are very flexible or hypermobile, your Downward Dog might look something like this. However, this is not stable on your shoulders or spine. You can squeeze your belly button in to engage your transversus abdominis. This will bring in your low ribs, and your spine and pelvis closer to neutral.

Spinal extensors engage—including your erector spinae

Latissimus dorsi stretches

Lower and middle trapezius engage to stabilize upper body

All deltoid heads may engage as you press into the floor

Biceps lengthen as elbow extends but engage to stabilize

Head releases down to comfort level

Fingers face forward

Pronator quadratus engages as you press into your thumb

ANTERIOR–LATERAL VIEW

HEADSTAND

Sirsasana

This full inversion turns you physically upside down. A multitude of benefits can be ascribed to this pose: from helping you breathe more efficiently to strengthening your upper body—especially the muscles around your shoulder joints—and your core.

THE BIG PICTURE

This pose strengthens your arms and shoulders. Your core and thighs activate to stabilize your body at its center, preventing you from falling to either side. Despite the name, it is your arms that are supporting your weight in this pose, not your head.

KEY

- *Joints*
- *Muscles*
- Engaging
- Engaging while stretching
- Stretching

VARIATION
This version of the pose has a reduced risk of falling and takes weight off your upper body. Push your forearms into the floor, lower your heels, and lift hips up and back. Allow your head to drop.

Lower legs
Your **ankle dorsiflexors** engage, to dorsiflex your feet and extend your toes. Your **calf muscles** are in a stretched position.

Ankle
Extensor hallucis longus
Soleus
Extensor digitorum longus
Tibialis anterior
Gastrocnemius

Knee
Gracilis
Vastus medialis
Rectus femoris
Adductor magnus
Adductor longus
Pectineus

Hip
Gluteus minimus
Gluteus maximus
Gluteus medius

Thighs
Your **quadriceps** engage to extend your knees. Your **hip adductors** engage to adduct your hips, while your **abductors** stabilize, keeping you from falling over to either side.

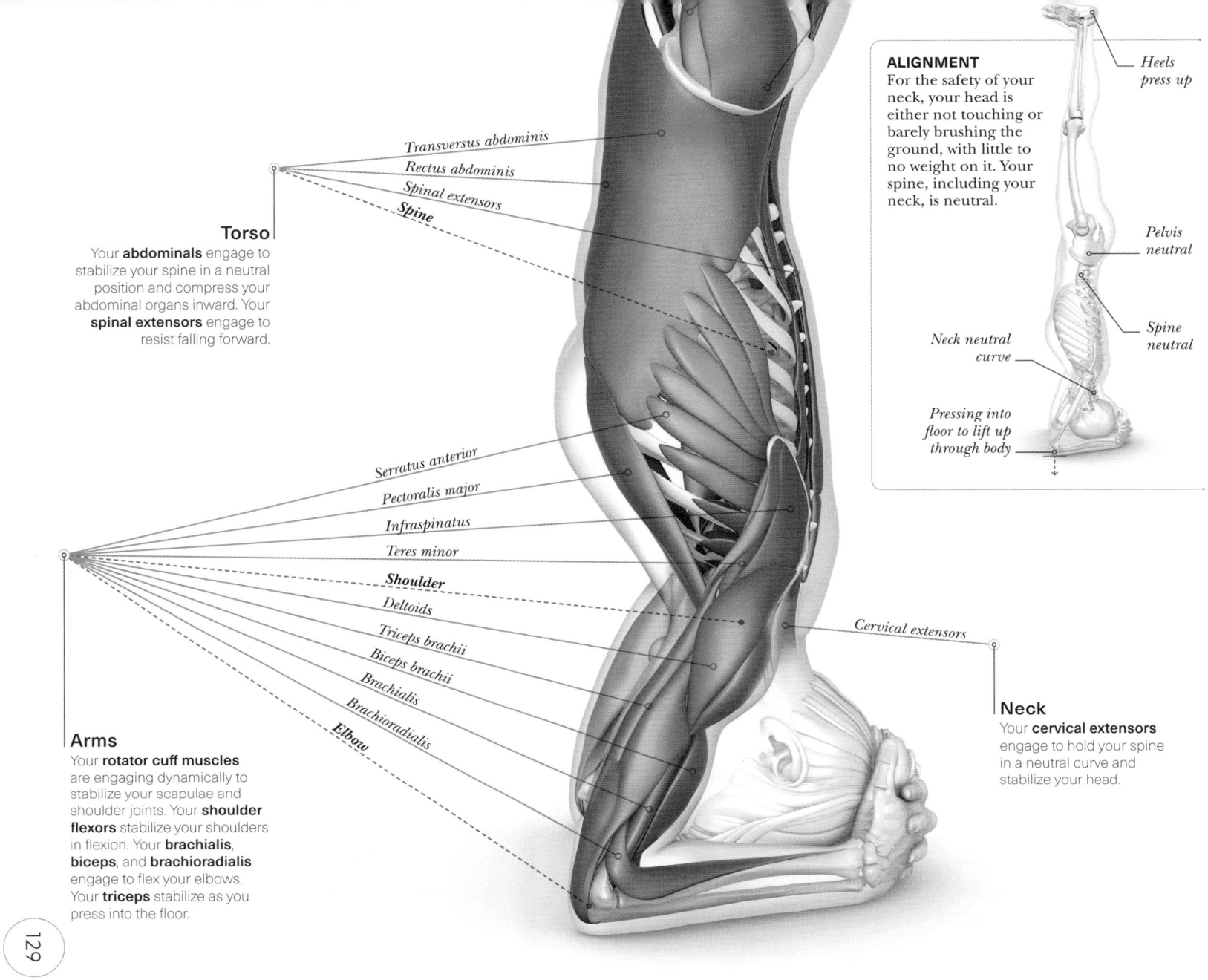
Transversus abdominis
Rectus abdominis
Spinal extensors
Spine
Torso
Your abdominals engage to stabilize your spine in a neutral position and compress your abdominal organs inward. Your spinal extensors engage to resist falling forward.
ALIGNMENT
For the safety of your neck, your head is either not touching or barely brushing the ground, with little to no weight on it. Your spine, including your neck, is neutral.
Heels press up
Pelvis neutral
Spine neutral
Neck neutral curve
Pressing into floor to lift up through body
Serratus anterior
Pectoralis major
Infraspinatus
Teres minor
Shoulder
Deltoids
Triceps brachii
Biceps brachii
Brachialis
Brachioradialis
Elbow
Cervical extensors
Neck
Your cervical extensors engage to hold your spine in a neutral curve and stabilize your head.
Arms
Your rotator cuff muscles are engaging dynamically to stabilize your scapulae and shoulder joints. Your shoulder flexors stabilize your shoulders in flexion. Your brachialis, biceps, and brachioradialis engage to flex your elbows. Your triceps stabilize as you press into the floor.

» CLOSER LOOK

Headstand can be safely practiced with little to no pressure on the head and neck. It has many health benefits, from improving respiratory and shoulder function to helping you better regulate your blood pressure.

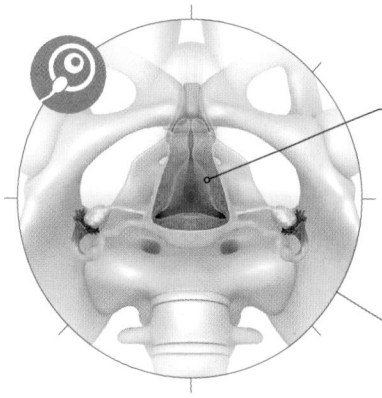

Menstruation

From a yoga perspective, inverting during menstruation could interfere with the natural downward flow of energy (*apana vayu*), which could be reason enough to avoid inversions during menstruation. However, from a medical perspective, there is no scientific evidence yet to support these claims. Choose what feels right for you.

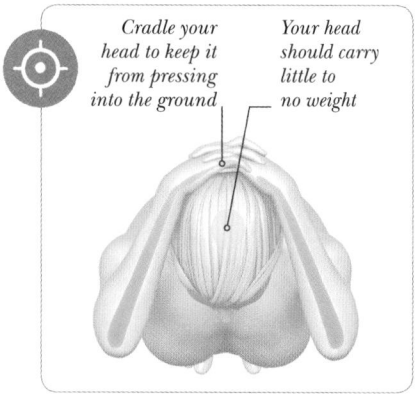

Pressure and balance

A modern way of practicing Headstand, and perhaps the safest, involves only allowing 0–10 percent of your weight to rest on your head. Cradle your head in your forearms so it barely touches the ground. Press firmly into your forearms, reaching your feet upward energetically.

Heels are lifted toward ceiling

Kneecaps face forward

Quadriceps tighten to keep legs lifted

Hip adductors engage to draw legs together

Tensor fasciae latae stabilizes your hips

Abdominal muscles engage to maintain balance

Pectoralis major engages and stretches as the chest expands

Slowly transitioning in and out of the pose reduces neck pressure

Gaze is forward

ANTERIOR VIEW

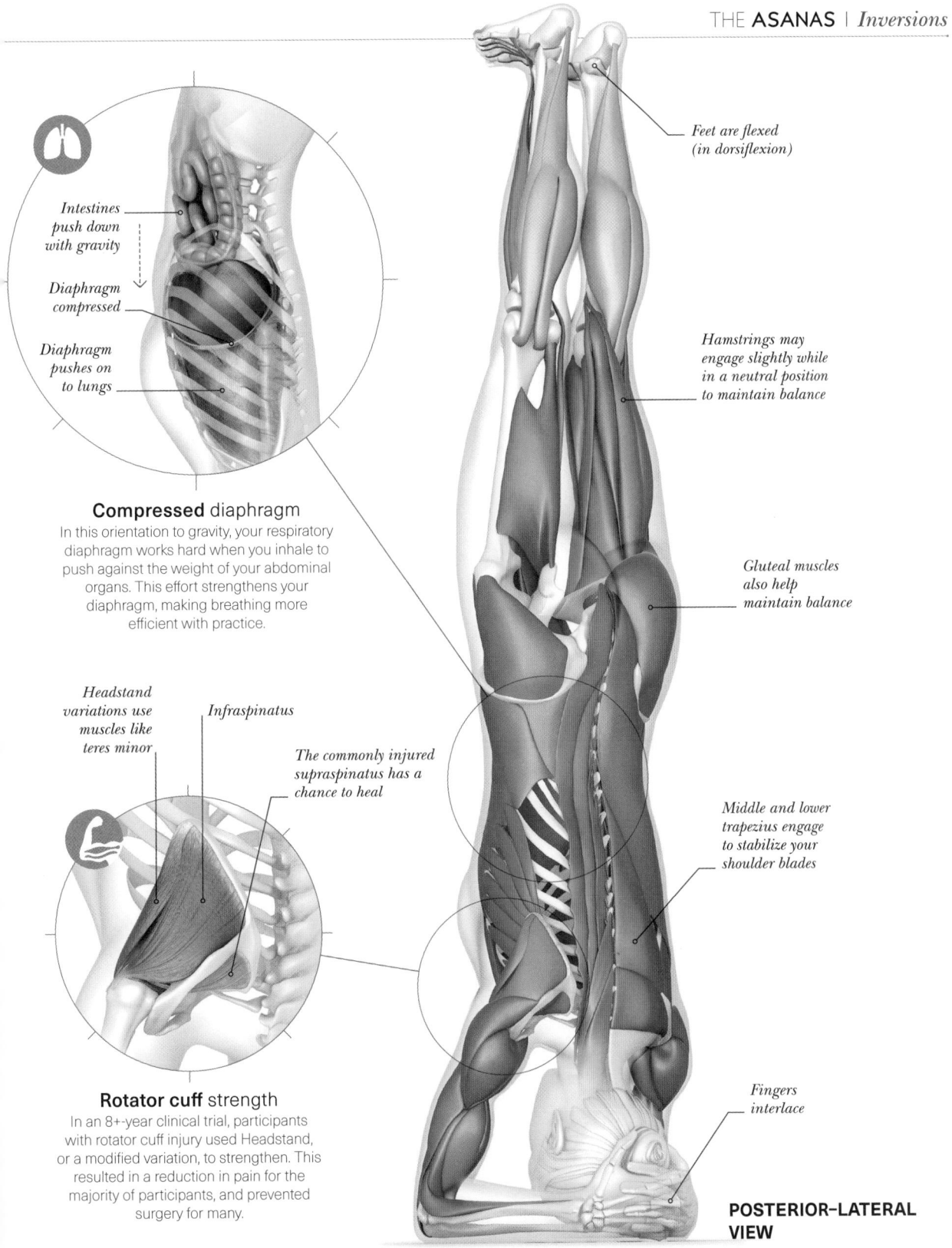

Compressed diaphragm

In this orientation to gravity, your respiratory diaphragm works hard when you inhale to push against the weight of your abdominal organs. This effort strengthens your diaphragm, making breathing more efficient with practice.

Rotator cuff strength

In an 8+-year clinical trial, participants with rotator cuff injury used Headstand, or a modified variation, to strengthen. This resulted in a reduction in pain for the majority of participants, and prevented surgery for many.

POSTERIOR-LATERAL VIEW

HALF SHOULDERSTAND

Ardha Sarvangasana

Shoulderstand is a classic inversion, often done at the end of an asana class to relax. It can help lower your blood pressure and activate the rest, digest, and rejuvenate part of your nervous system. The version shown here reduces pressure on the neck.

THE BIG PICTURE

This pose gently strengthens the muscles at the front of your neck, while your upper back and neck muscles stretch. The muscles of your core and thighs engage to stabilize you and hold your body in an inverted position.

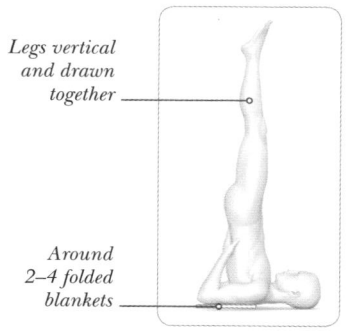

VARIATION
Supported shoulderstand allows you to bring your legs vertical to perform the traditional pose safely. Folded blankets under the shoulders take pressure and the sharp angle off your neck. Reducing the degree of neck flexion in this way lessens the risk of injury, particularly if you have neck issues.

Lower **legs**
With your feet in plantar flexion, your **calf muscles** are engaged with some stretch in your **dorsiflexors**, particularly your **tibialis anterior muscles**. Try dorsiflexing your ankles, with your heels to the ceiling, and feel the stretch shift to your **calf muscles**.

Ankle
Soleus
Tibialis anterior
Gastrocnemius

ALIGNMENT
By flexing your hips, you distribute more weight toward your hands and off your upper body. This is shoulderstand not neckstand. Avoid anything that causes pain or intense pressure in your neck.

Legs drawn together
Weight of legs slightly forward
Hips flexed, to create a counterbalance
Weight on shoulders and upper arms
Neck flexion less than or equal to 50 degrees

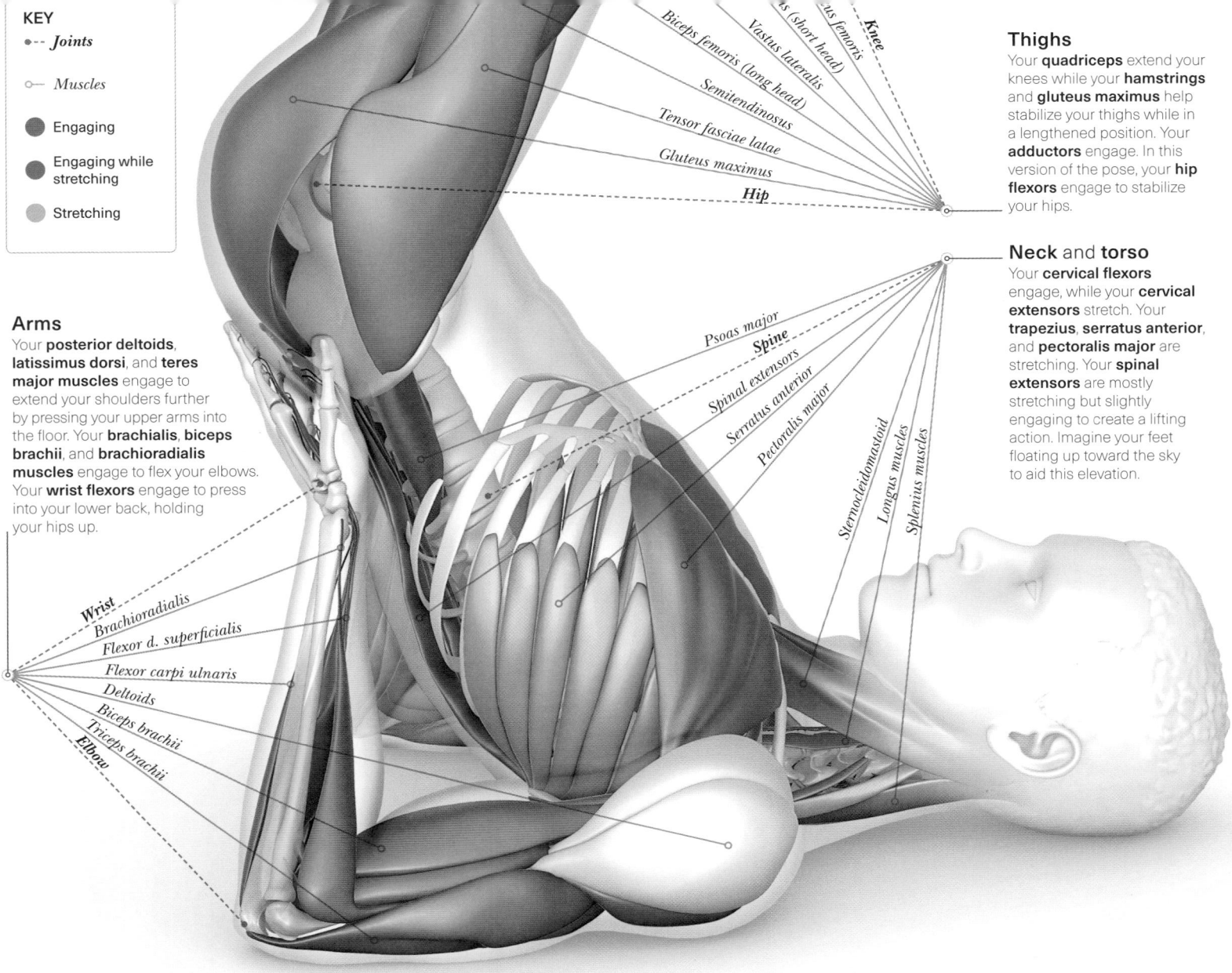

Thighs

Your **quadriceps** extend your knees while your **hamstrings** and **gluteus maximus** help stabilize your thighs while in a lengthened position. Your **adductors** engage. In this version of the pose, your **hip flexors** engage to stabilize your hips.

Neck and torso

Your **cervical flexors** engage, while your **cervical extensors** stretch. Your **trapezius**, **serratus anterior**, and **pectoralis major** are stretching. Your **spinal extensors** are mostly stretching but slightly engaging to create a lifting action. Imagine your feet floating up toward the sky to aid this elevation.

Arms

Your **posterior deltoids**, **latissimus dorsi**, and **teres major muscles** engage to extend your shoulders further by pressing your upper arms into the floor. Your **brachialis**, **biceps brachii**, and **brachioradialis muscles** engage to flex your elbows. Your **wrist flexors** engage to press into your lower back, holding your hips up.

» CLOSER LOOK

Shoulderstand is particularly effective at encouraging lymphatic drainage and improving overall circulation. Although it may not stimulate your thyroid, it can stimulate baroreceptors to lower your blood pressure.

Lymphatic drainage

Lymph vessels rely on movement to pump lymphatic fluid around your body. Like veins, they have one-way valves that prevent backflow (see opposite). Inverting encourages these valves to open, preventing or alleviating edema (a buildup of fluids) in your ankles.

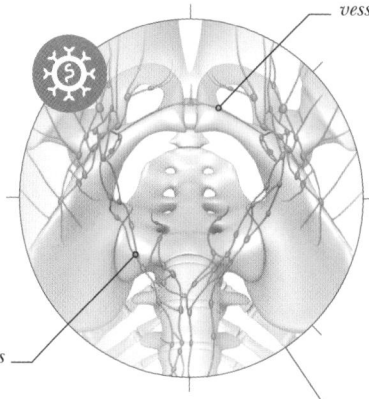

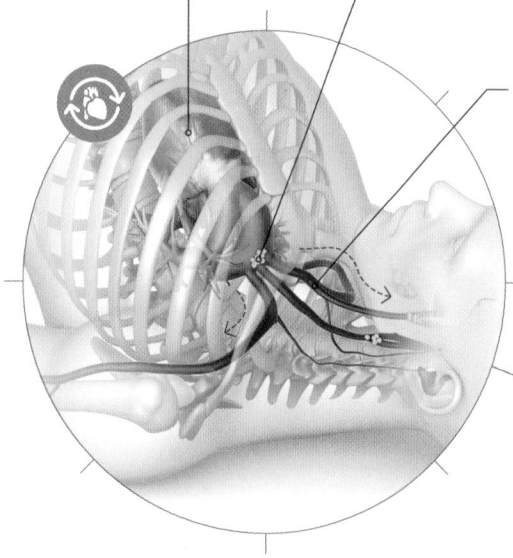

Baroreceptors

When inverting, blood pressure initially goes up. Then, signals to your brain should set in motion a cascade of events to lower your blood pressure and maintain homeostasis. This drop may be enhanced by the physical pressure placed on carotid baroreceptors in neck flexion. Regular practice of inversions could lower your blood pressure over time.

ANTERIOR VIEW

Tibialis anterior lengthens

Knees can be straight or flexed slightly

Hip flexors are engaged

Psoas major particularly engages

Shoulders rest on the floor

Avoid putting pressure on your head or neck

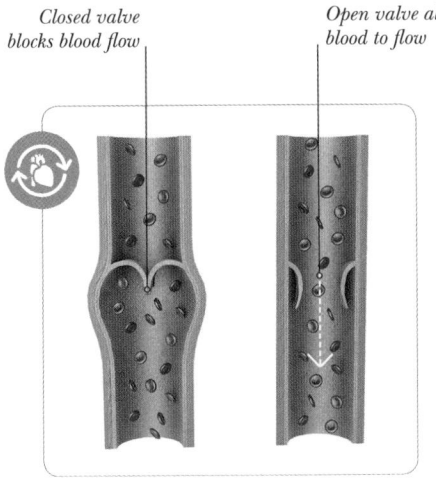

Venous blood flow

Unlike arteries, veins do not have muscular walls to transport blood around your body. Instead they have one-way valves that prevent backflow and help carry deoxygenated blood back to your heart. Inversions allow gravity to open the valves for you, encouraging venous return to your heart and improving circulation.

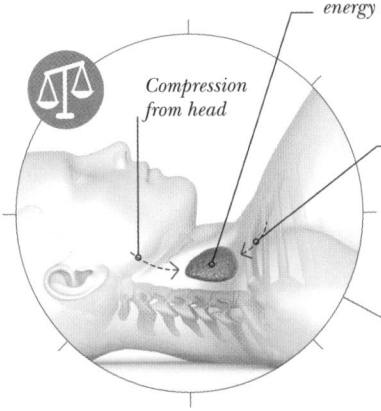

Thyroid stimulation

Yogis claim that this pose stimulates and regulates your thyroid gland. However, unlike your digestive tract, the thyroid doesn't function through mechanical pressure and movement. Although it is possible that the increased blood flow could affect its function, scientific evidence does not seem to support these claims.

In this version, feet are pointed (in plantar flexion)

Hamstrings may engage slightly to maintain balance

Gluteal muscles may also engage minimally

Arms support your lower back

POSTERIOR–LATERAL VIEW

BRIDGE

Setu Bandhasana

Bridge is a gentle and accessible backbend that can help relieve back pain, particularly discomfort caused by sitting down too much. It is a calming pose, used by many to wind down at the end of a practice or at the end of the day in preparation for sleep.

THE BIG PICTURE

Bridge pose stretches the muscles along the front of your body—including your thighs, hips, abdomen, and chest. The back of your body strengthens—including your thighs, buttocks, back, and shoulders—as the muscles here work to support and hold you in an elevated backbend.

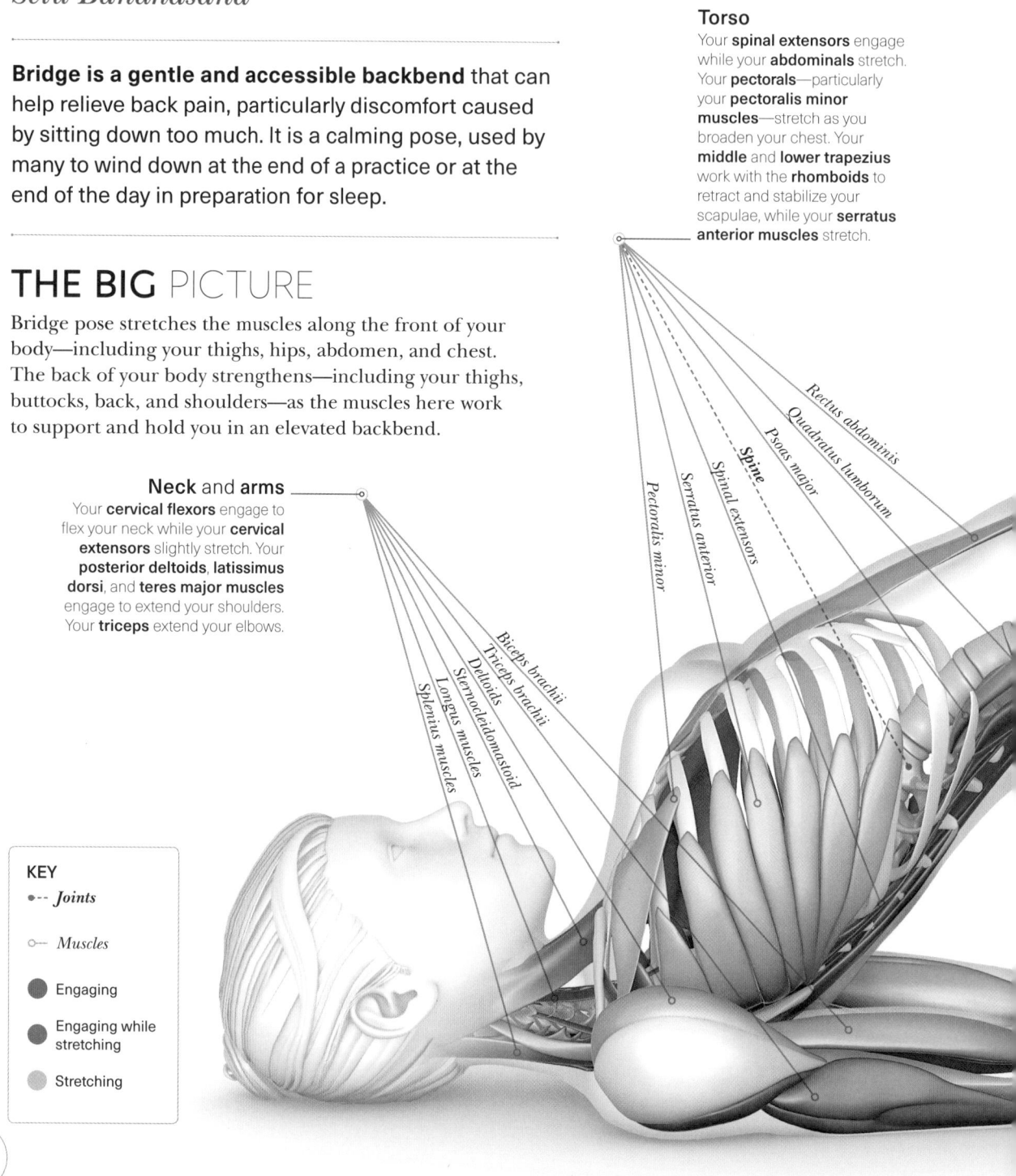

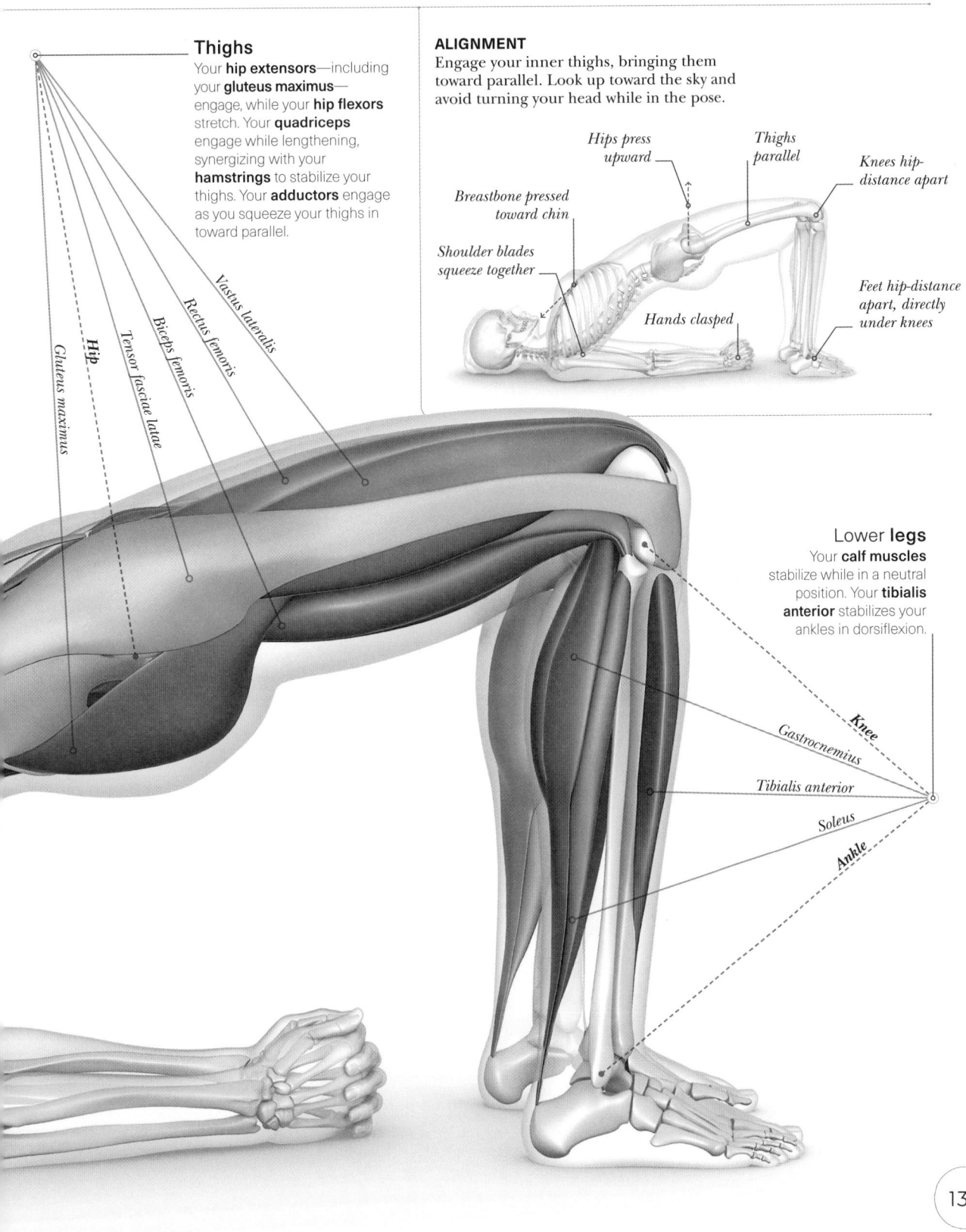
Thighs
Your hip extensors—including your gluteus maximus—engage, while your hip flexors stretch. Your quadriceps engage while lengthening, synergizing with your hamstrings to stabilize your thighs. Your adductors engage as you squeeze your thighs in toward parallel.
ALIGNMENT
Engage your inner thighs, bringing them toward parallel. Look up toward the sky and avoid turning your head while in the pose.
Hips press upward
Thighs parallel
Knees hip-distance apart
Breastbone pressed toward chin
Shoulder blades squeeze together
Hands clasped
Feet hip-distance apart, directly under knees
Gluteus maximus
Hip
Tensor fasciae latae
Biceps femoris
Rectus femoris
Vastus lateralis
Lower legs
Your calf muscles stabilize while in a neutral position. Your tibialis anterior stabilizes your ankles in dorsiflexion.
Knee
Gastrocnemius
Tibialis anterior
Soleus
Ankle

» CLOSER LOOK

Backbends like this could also be considered "heart openers," because broadening your chest area may leave you feeling openhearted. Your glutes strengthen and tone.

VARIATION
To challenge your pelvic stability, try raising one leg while in Bridge pose. Engage your core muscles to support your back as you raise one leg upward. Focus on keeping your hips parallel. Press your standing foot into the floor to find support.

Hips are extended

Rectus abdominis stretches

Send knees away from torso to find length through the hips

Hips press directly upward

Heart pumps oxygenated blood around body

Lungs fill chest cavity as you inhale

Feel your ribcage expand three-dimensionally

Oxygenated blood flows to brain

ANTERIOR VIEW

Blood rush

Some claim that inversions send a rush of oxygenated blood to your head. This may happen briefly but your brain regulates its blood flow (see p.134). If you do get a head rush, come down. Backbends are called "heart openers" because they create space in your chest. Be aware of this sensation and feel your ribcage expand as you inhale.

Glute engagement

You must engage your glutes and hamstrings to extend your hips into Bridge. However, you should not do this by forcefully clenching together your buttocks. Instead, isometrically engage by gliding your heels back toward your head (without moving them on the mat) while imagining driving your knees forward.

Closed chain movement

A kinetic chain is a line of joints and muscles that affect each other. Bridge is a closed chain because both ends of the chain (shoulders and feet) are fixed in place. In an open chain, one side of the chain is not fixed, like the arms in Warrior poses. Closed chains tend to be more stable.

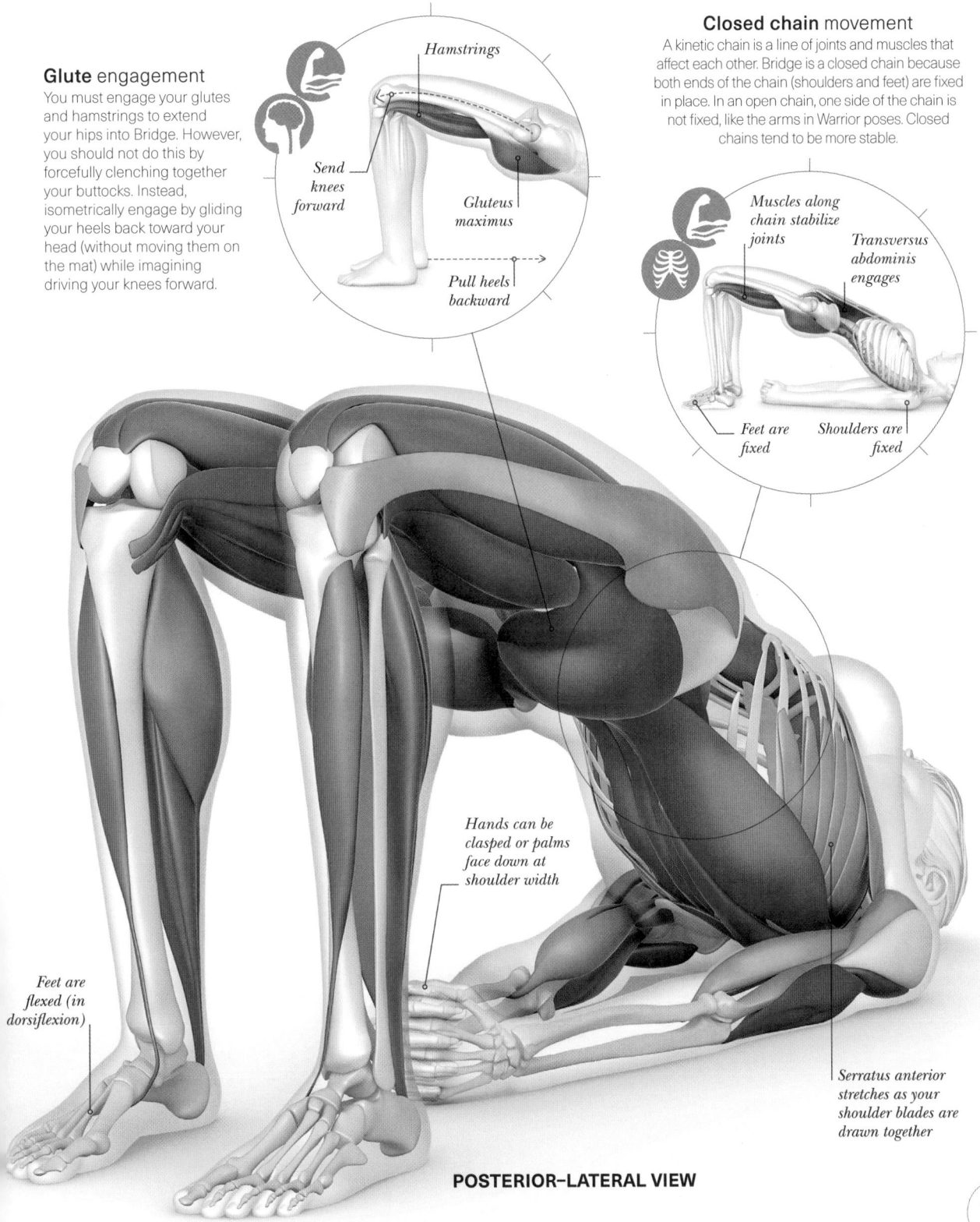

POSTERIOR–LATERAL VIEW

WHEEL

Urdhva Dhanurasana

Wheel is a full backbend and inversion, bringing your head below the level of your heart. This pose is often done toward the end of a class as it requires warming up to be safe for most people. With practice, Wheel can improve the strength and flexibility of your back.

THE BIG PICTURE

This pose strongly stretches the muscles at the front of your body—including your thighs, hips, abdomen, and chest. It strengthens your shoulders and the back of your body—particularly your back muscles, buttocks, and thighs—as they support you in this deep backbend and elevation.

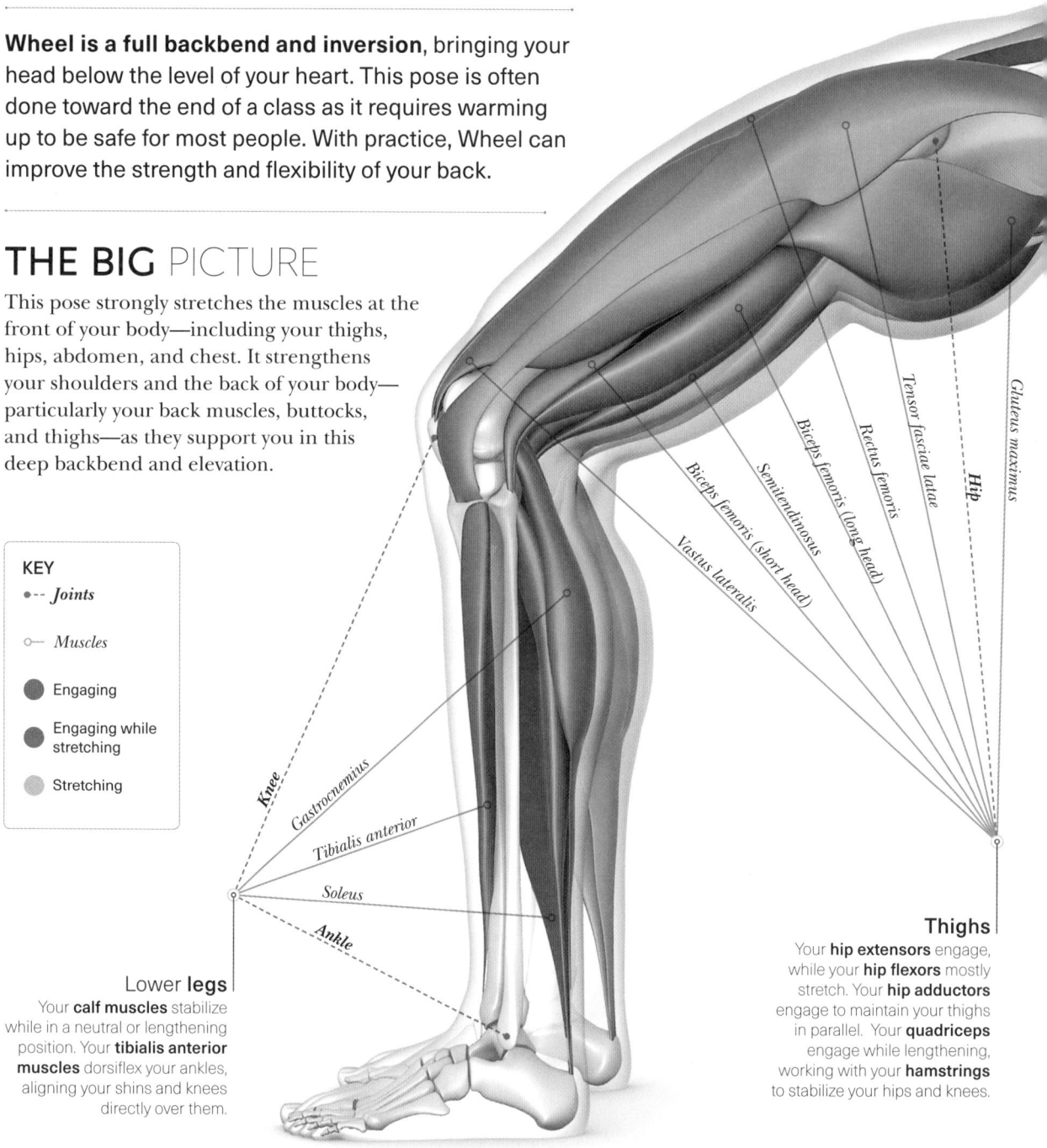

Lower legs

Your **calf muscles** stabilize while in a neutral or lengthening position. Your **tibialis anterior muscles** dorsiflex your ankles, aligning your shins and knees directly over them.

Thighs

Your **hip extensors** engage, while your **hip flexors** mostly stretch. Your **hip adductors** engage to maintain your thighs in parallel. Your **quadriceps** engage while lengthening, working with your **hamstrings** to stabilize your hips and knees.

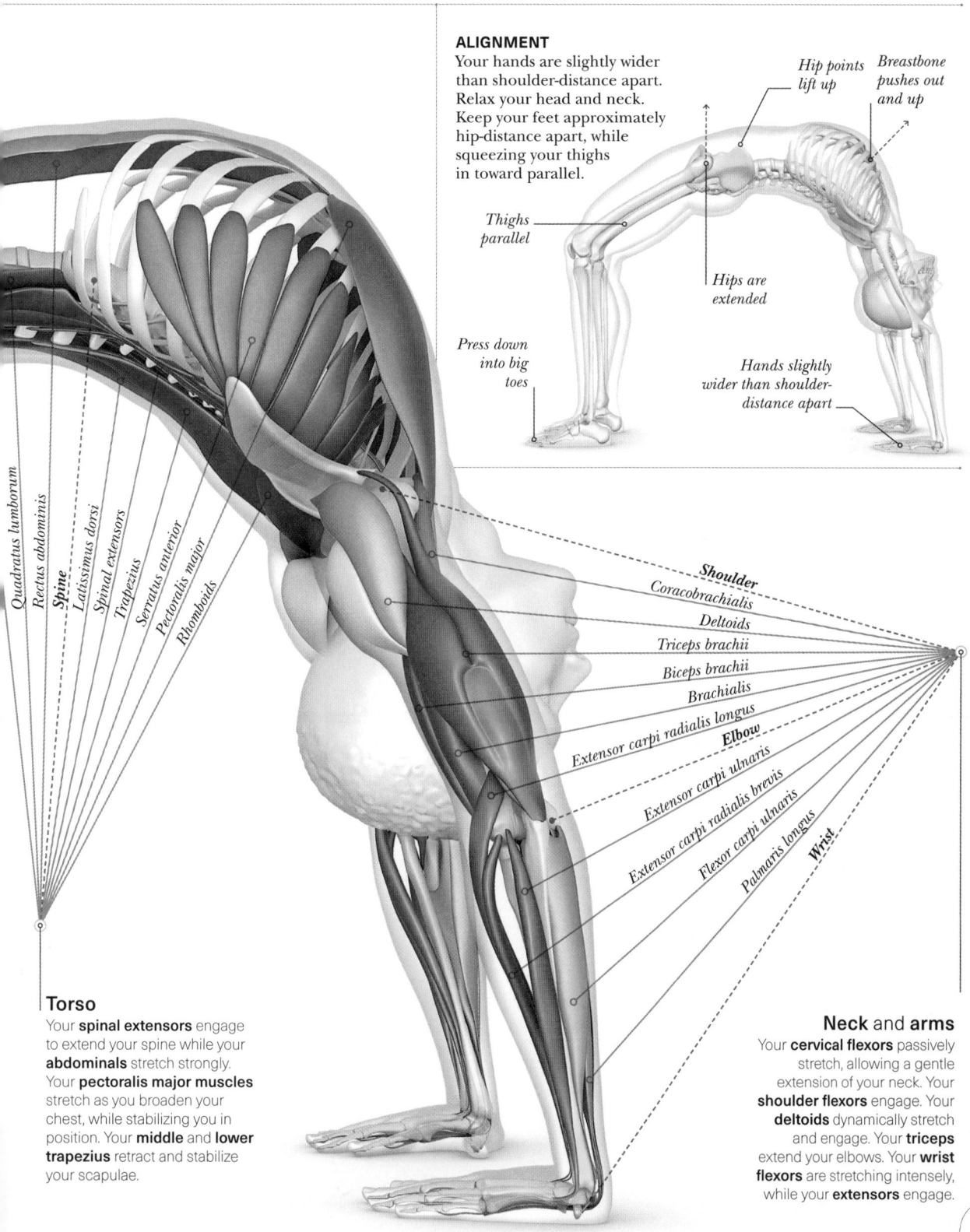
ALIGNMENT
Your hands are slightly wider than shoulder-distance apart. Relax your head and neck. Keep your feet approximately hip-distance apart, while squeezing your thighs in toward parallel.
Hip points lift up
Breastbone pushes out and up
Thighs parallel
Hips are extended
Press down into big toes
Hands slightly wider than shoulder-distance apart
Quadratus lumborum
Rectus abdominis
Spine
Latissimus dorsi
Spinal extensors
Trapezius
Serratus anterior
Pectoralis major
Rhomboids
Shoulder
Coracobrachialis
Deltoids
Triceps brachii
Biceps brachii
Brachialis
Extensor carpi radialis longus
Elbow
Extensor carpi ulnaris
Extensor carpi radialis brevis
Flexor carpi ulnaris
Palmaris longus
Wrist
Torso
Your spinal extensors engage to extend your spine while your abdominals stretch strongly. Your pectoralis major muscles stretch as you broaden your chest, while stabilizing you in position. Your middle and lower trapezius retract and stabilize your scapulae.
Neck and arms
Your cervical flexors passively stretch, allowing a gentle extension of your neck. Your shoulder flexors engage. Your deltoids dynamically stretch and engage. Your triceps extend your elbows. Your wrist flexors are stretching intensely, while your extensors engage.

» CLOSER LOOK

Wheel puts your shoulder joints and spine in a unique position that can be challenging both for people who are tight and those who are very flexible. It can be quite demanding, yet energizing and uplifting.

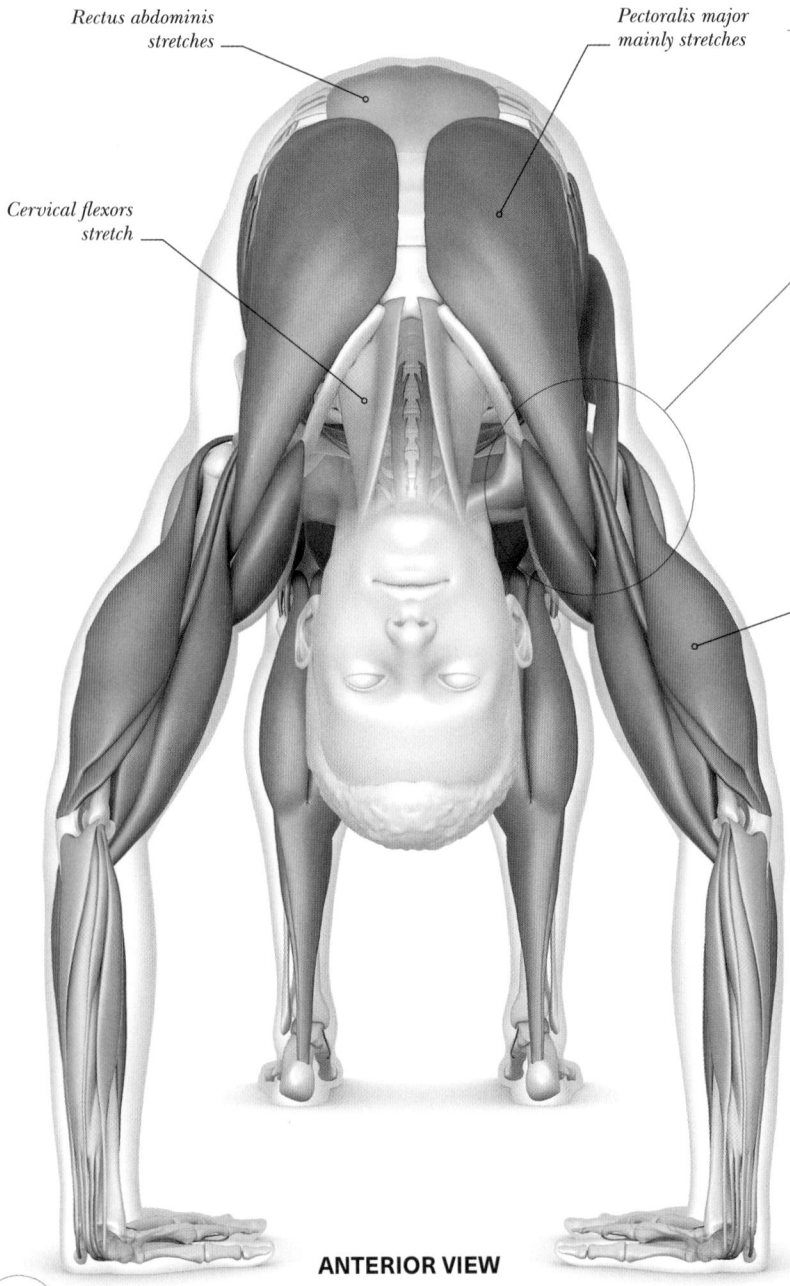

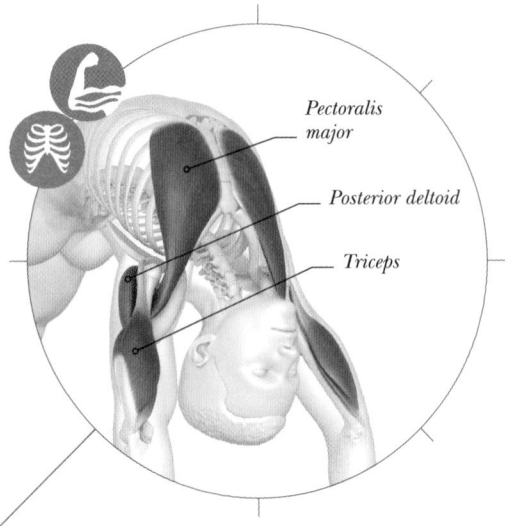

Tight **shoulders**

Tight shoulders are a common limiting factor for this pose. Many people lack the range of motion to go into full shoulder flexion, bringing the arms directly overhead. Make sure you thoroughly warm up your shoulders before doing Wheel. Stretch your shoulders over time with poses like Cow Face (see pp.60–63).

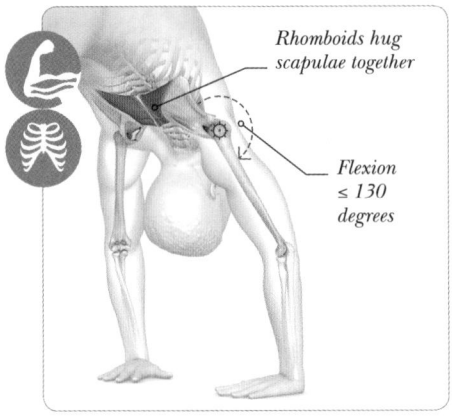

Shoulder flexion

In shoulder flexion there is little stability, particularly when weight bearing as in Wheel. If you are very flexible, especially with a tendency toward dislocation, be mindful when doing this pose, or try Bridge pose for a more stable shoulder position in extension instead (see pp.136–39).

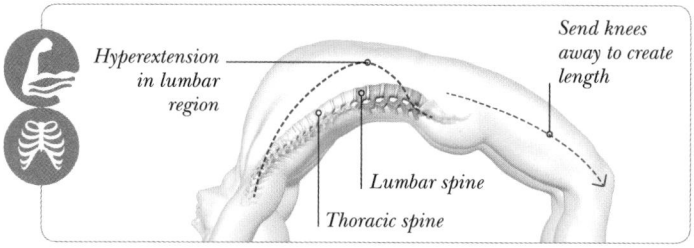

Spinal flexibility

Many yogis allow too much bending or hyperextension in the lower back, like this. If this is you, focus on lengthening your lower back instead of crunching and sinking into it. Although your lumbar spine has a greater capacity for extension than your thoracic, try to make the extension more even.

Knees stacked approximately over your ankles

Sternum lifts upward

Even curve in lumbar spine

Gluteus maximus engages to extend your hips

Heart pumps to deliver blood

Engaging muscles need more blood

Feet are fixed

Wrists are extended

Heart pounding

In this intense backbend, large muscles are active, your heart is pounding, and breaths are full. Wheel often leaves people feeling energized and confident. You may feel empowered to know that you can consciously choose certain yoga practices to lift your energy when feeling sluggish (or calm yourself when agitated).

ANTERIOR–LATERAL VIEW

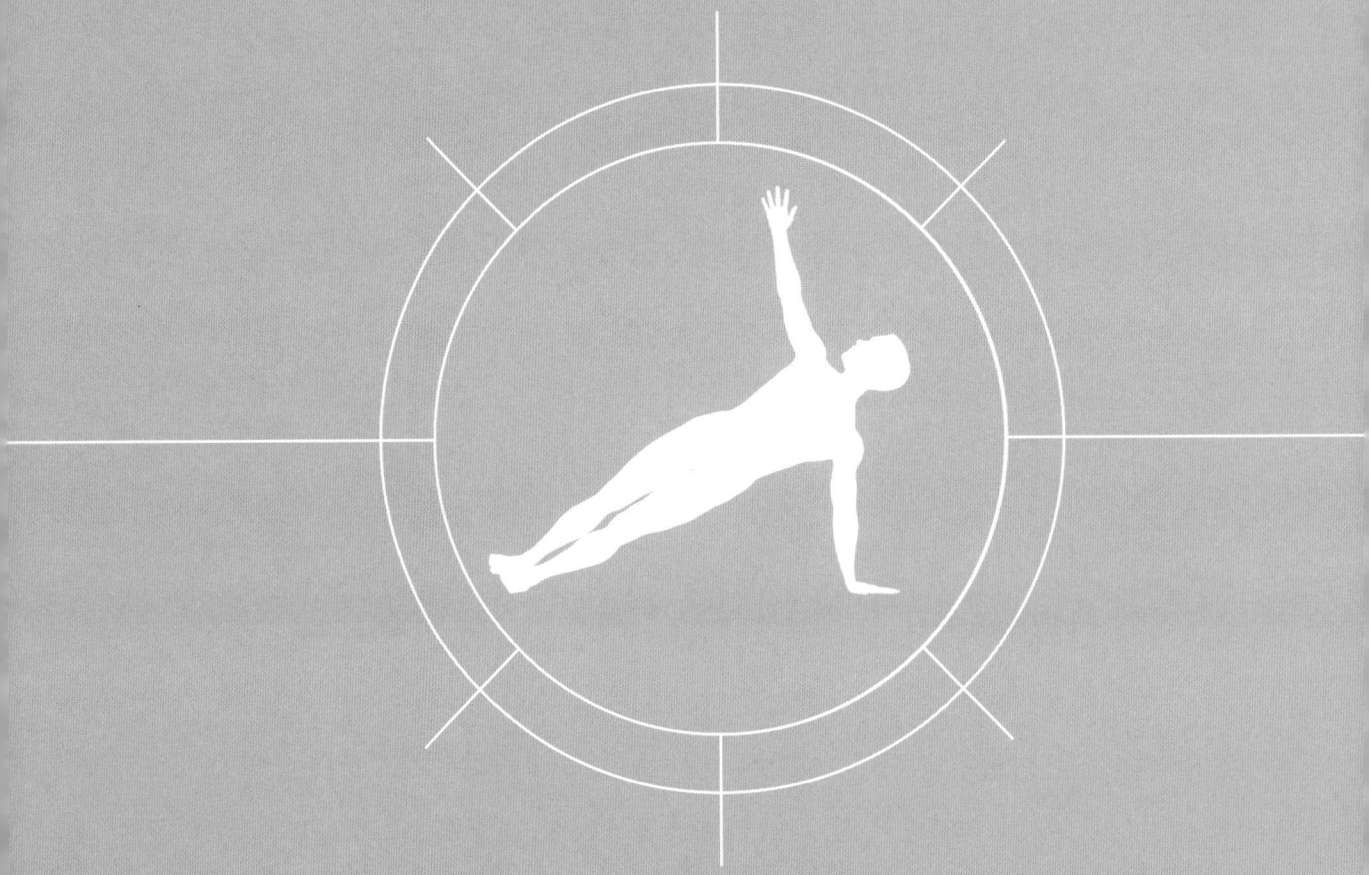

FLOOR ASANAS

These floor asanas include arm balance poses, prone (on your front) poses, and supine (on your back) poses. The asanas in this section range from intense and strong, like Plank, to soft and subtle, like the Supine Twist. No matter what the intensity, they all provide a rich opportunity to inquire within yourself.

CROW

Bakasana

Crow pose is an arm balance that uniquely develops your strength, flexibility, balance, and agility. Working your wrist muscles is a great antidote to typing on a computer all day. Plus, this challenging pose offers an opportunity to face your fears and be playful in your practice.

THE BIG PICTURE

Practicing Crow strengthens the muscles of your wrists, shoulders, arms, hips, and abdomen. In this pose, you are fully weight-bearing on your hands, with your upper body working to support you and keep you balanced.

KEY

- *Joints*
- *Muscles*
- Engaging
- Engaging while stretching
- Stretching

ALIGNMENT

Your knees rest on a shelf created by your upper arms. Gaze forward with your chin slightly lifted. Press down into the floor and be prepared to fall backward with grace.

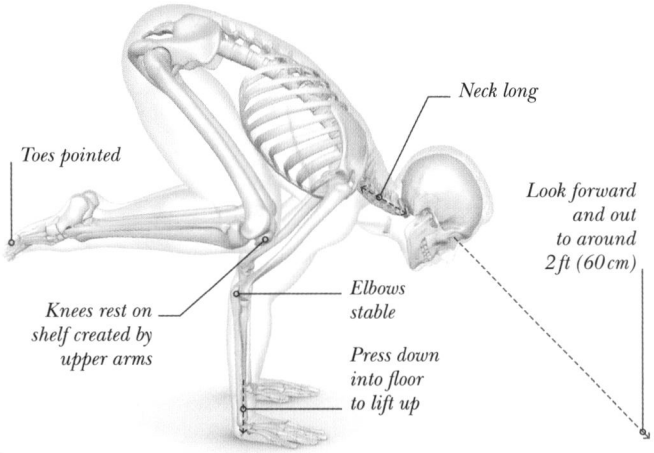

Thighs

Your **hip flexors** engage to flex your hips. Your **hamstrings** flex your knees and your **quadriceps** stretch. Your **adductors** are recruited to adduct and stabilize your hips and thighs.

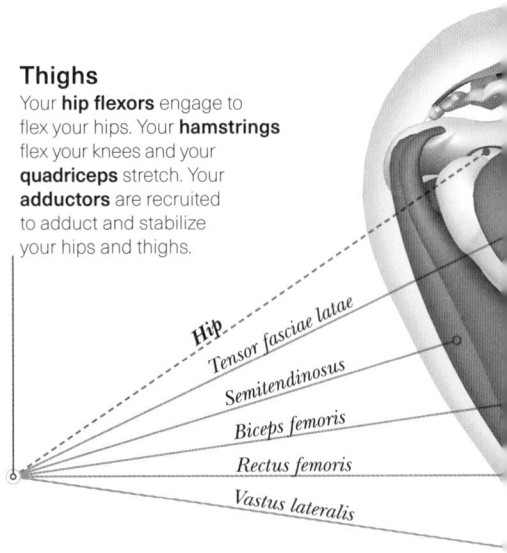

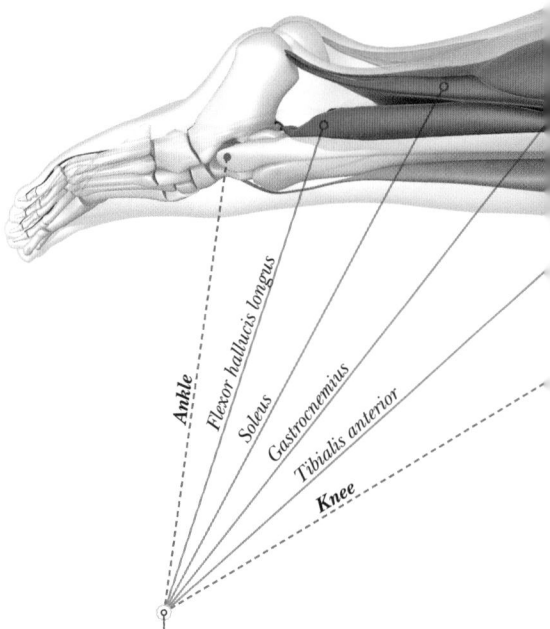

Lower legs

Your **plantar flexors** engage to point your toes, while your **dorsiflexors** slightly stretch—particularly your **tibialis anterior**.

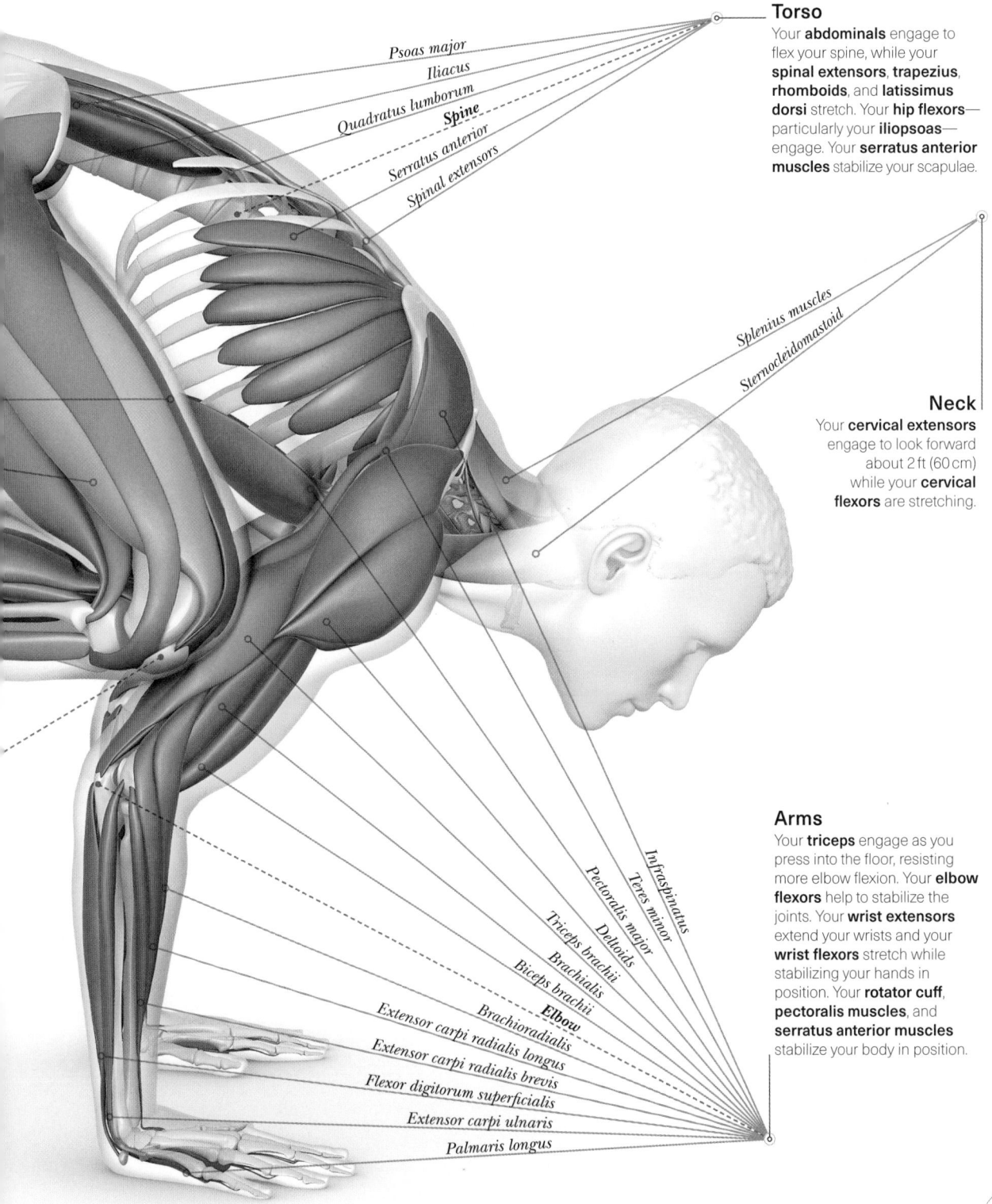

Torso

Your **abdominals** engage to flex your spine, while your **spinal extensors**, **trapezius**, **rhomboids**, and **latissimus dorsi** stretch. Your **hip flexors**—particularly your **iliopsoas**—engage. Your **serratus anterior muscles** stabilize your scapulae.

Neck

Your **cervical extensors** engage to look forward about 2 ft (60 cm) while your **cervical flexors** are stretching.

Arms

Your **triceps** engage as you press into the floor, resisting more elbow flexion. Your **elbow flexors** help to stabilize the joints. Your **wrist extensors** extend your wrists and your **wrist flexors** stretch while stabilizing your hands in position. Your **rotator cuff**, **pectoralis muscles**, and **serratus anterior muscles** stabilize your body in position.

» CLOSER LOOK

Crow is a challenging balancing pose that strengthens your wrists. Finding playfulness in the pose can help you reveal a sense of bravery and resilience.

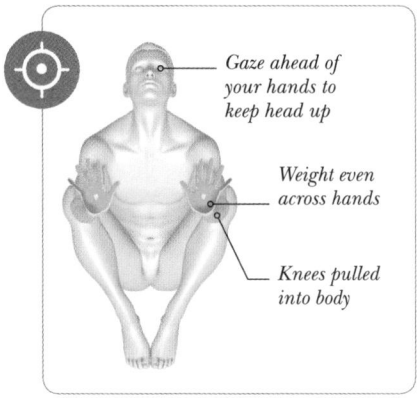

Pressure and **balance**

Your forearms and hands bear your weight in Crow. Spread the weight evenly across both hands and across the knuckles of each hand. Find a sense of play in this pose and laugh when you fall back. Research suggests that play is not only vital for children but enhances life satisfaction and well-being in adults.

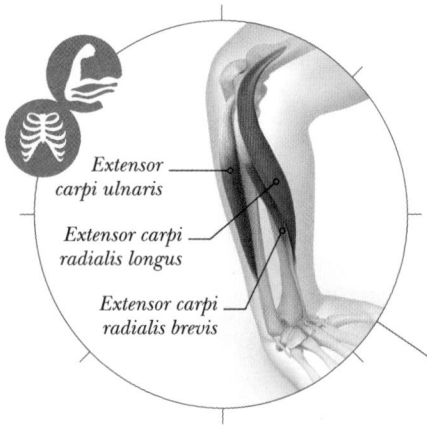

Wrist strengthening

Arm balances strengthen your wrist extensors, which can often be weak due to lack of use. Likewise, your wrist flexors may be tight from typing, texting, and grasping things. This action stretches them to help prevent carpal tunnel syndrome. However, if you currently have wrist issues, this amount of weight bearing is probably too much.

Spinal extensors, such as the erector spinae, slightly stretch

Underneath the erector spinae, your quadratus lumborum stretches

Quadriceps stretch as knee is flexed

Chin is lifted for balance

Wrists are extended

ANTERIOR VIEW

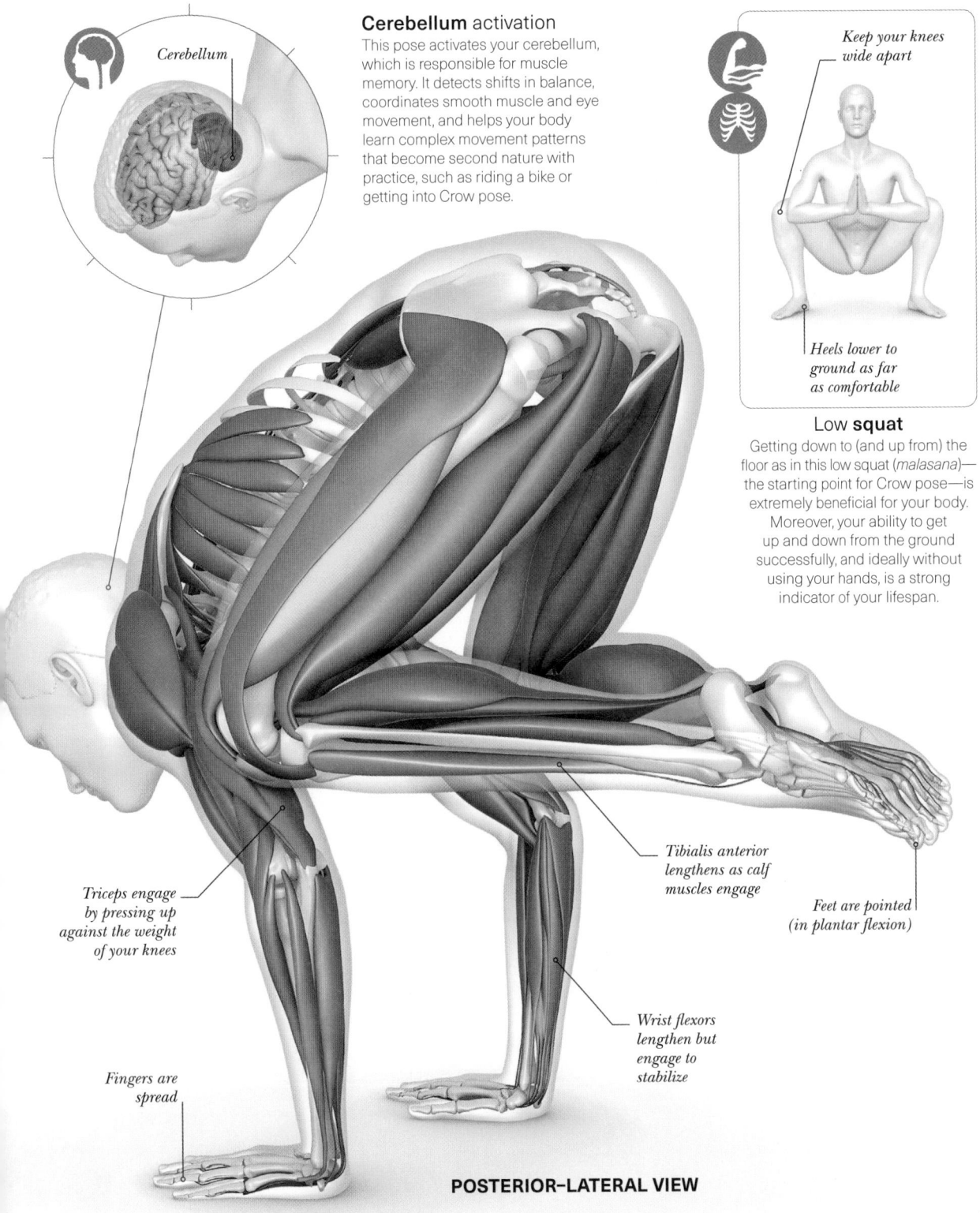

Cerebellum activation

This pose activates your cerebellum, which is responsible for muscle memory. It detects shifts in balance, coordinates smooth muscle and eye movement, and helps your body learn complex movement patterns that become second nature with practice, such as riding a bike or getting into Crow pose.

Low **squat**

Getting down to (and up from) the floor as in this low squat (*malasana*)—the starting point for Crow pose—is extremely beneficial for your body. Moreover, your ability to get up and down from the ground successfully, and ideally without using your hands, is a strong indicator of your lifespan.

PLANK

Kumbhakasana

Plank pose is the top point of a press up. It is a strong, stabilizing pose that works muscles from the deepest layers inside you to the most superficial. When holding Plank you are giving your body a thorough, strengthening workout.

THE BIG PICTURE

Plank pose particularly strengthens your shoulders and entire core—including your abdominals, back muscles, and pelvic floor muscles. It builds heat and energy throughout your body when held for several breaths or more.

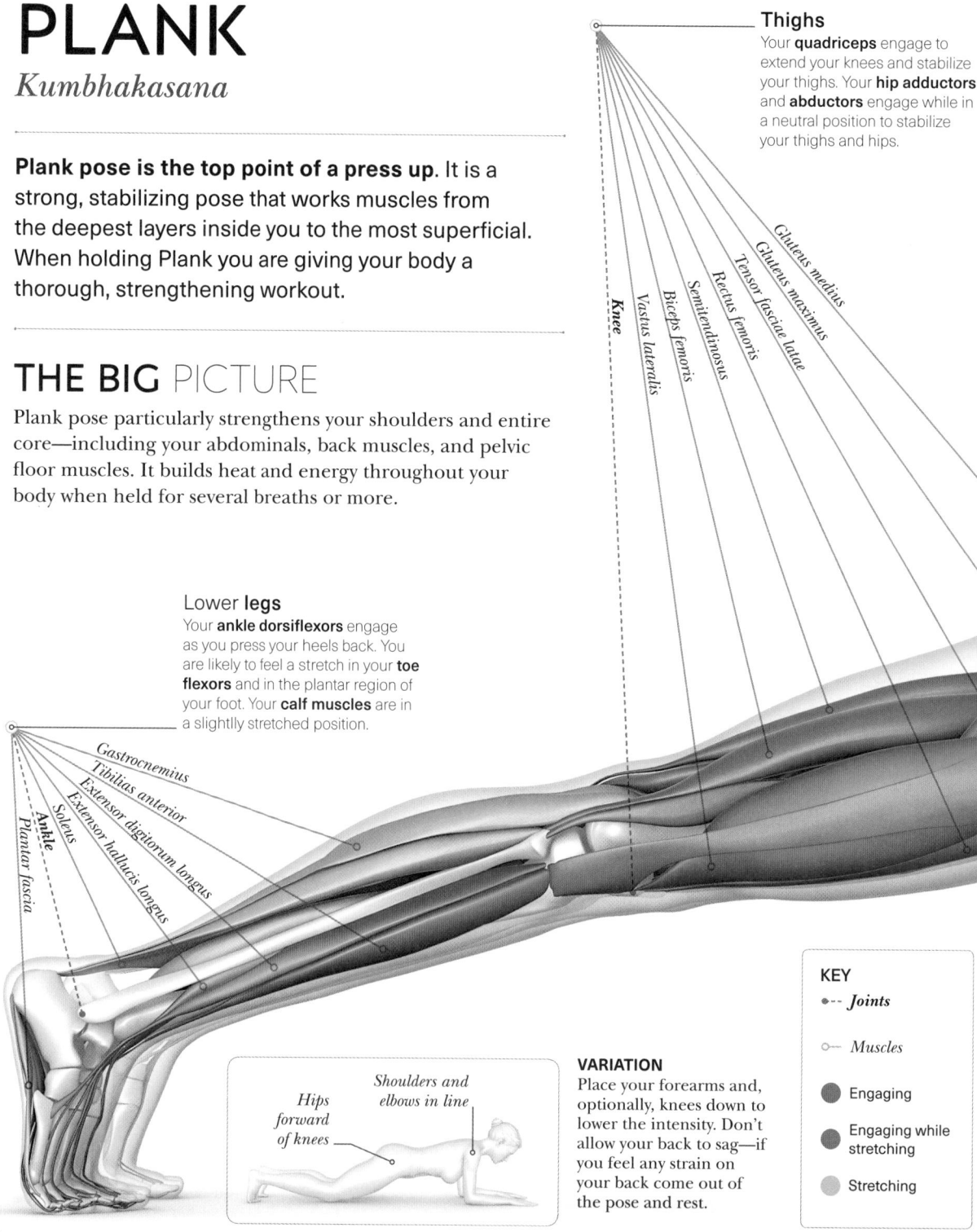

VARIATION
Place your forearms and, optionally, knees down to lower the intensity. Don't allow your back to sag—if you feel any strain on your back come out of the pose and rest.

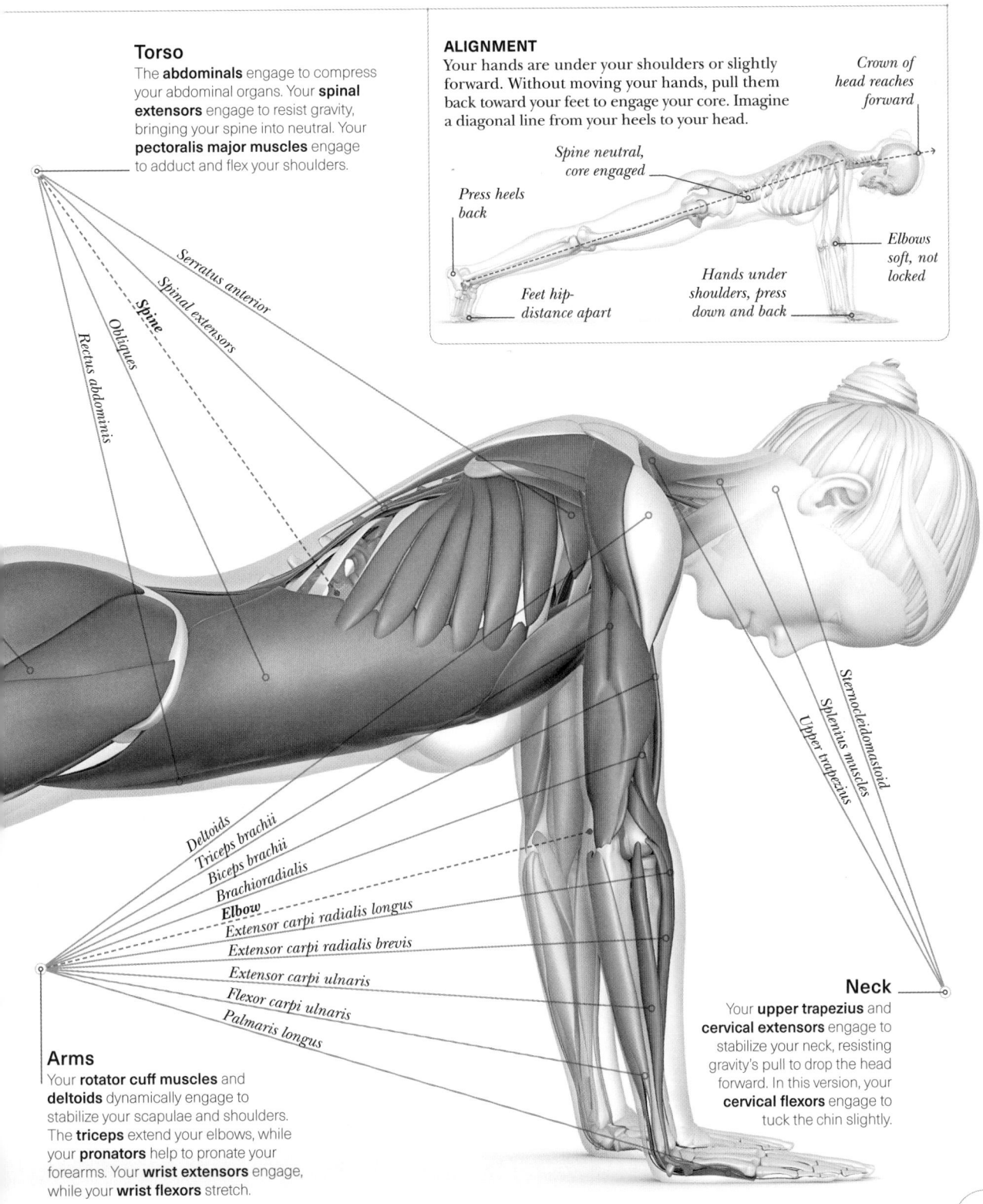
Torso
The abdominals engage to compress your abdominal organs. Your spinal extensors engage to resist gravity, bringing your spine into neutral. Your pectoralis major muscles engage to adduct and flex your shoulders.
ALIGNMENT
Your hands are under your shoulders or slightly forward. Without moving your hands, pull them back toward your feet to engage your core. Imagine a diagonal line from your heels to your head.
Crown of head reaches forward
Spine neutral, core engaged
Press heels back
Elbows soft, not locked
Hands under shoulders, press down and back
Feet hip-distance apart
Serratus anterior
Spinal extensors
Spine
Obliques
Rectus abdominis
Sternocleidomastoid
Splenius muscles
Upper trapezius
Deltoids
Triceps brachii
Biceps brachii
Brachioradialis
Elbow
Extensor carpi radialis longus
Extensor carpi radialis brevis
Extensor carpi ulnaris
Flexor carpi ulnaris
Palmaris longus
Neck
Your upper trapezius and cervical extensors engage to stabilize your neck, resisting gravity's pull to drop the head forward. In this version, your cervical flexors engage to tuck the chin slightly.
Arms
Your rotator cuff muscles and deltoids dynamically engage to stabilize your scapulae and shoulders. The triceps extend your elbows, while your pronators help to pronate your forearms. Your wrist extensors engage, while your wrist flexors stretch.

» CLOSER LOOK

Plank can be used to explore yogic energetic locks, called *bandhas*. In this version, notice subtle activation in the areas of the *bandhas* while breathing. You can practice engaging the *bandhas* under instruction from a qualified teacher.

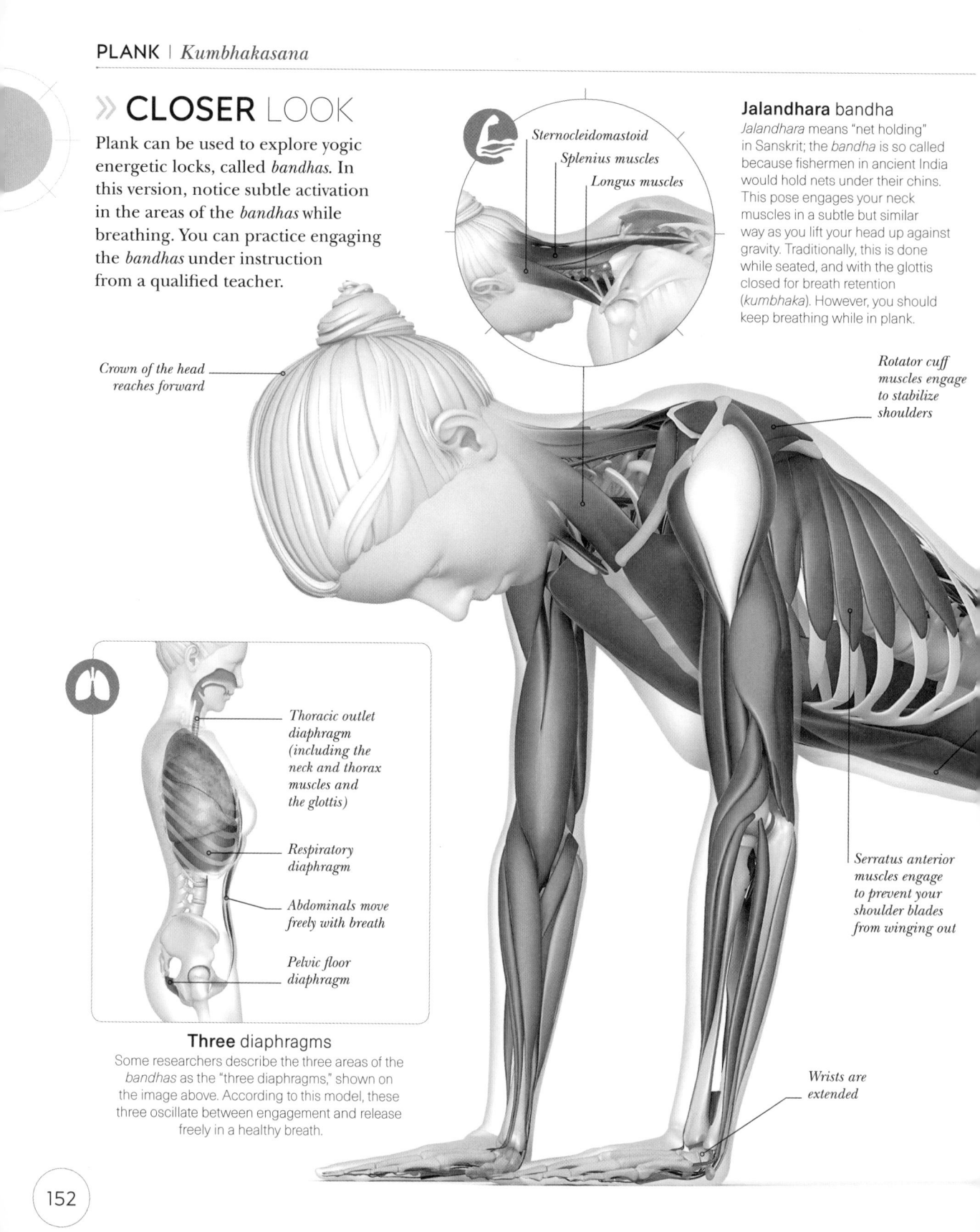

Jalandhara bandha

Jalandhara means "net holding" in Sanskrit; the *bandha* is so called because fishermen in ancient India would hold nets under their chins. This pose engages your neck muscles in a subtle but similar way as you lift your head up against gravity. Traditionally, this is done while seated, and with the glottis closed for breath retention (*kumbhaka*). However, you should keep breathing while in plank.

Three diaphragms

Some researchers describe the three areas of the *bandhas* as the "three diaphragms," shown on the image above. According to this model, these three oscillate between engagement and release freely in a healthy breath.

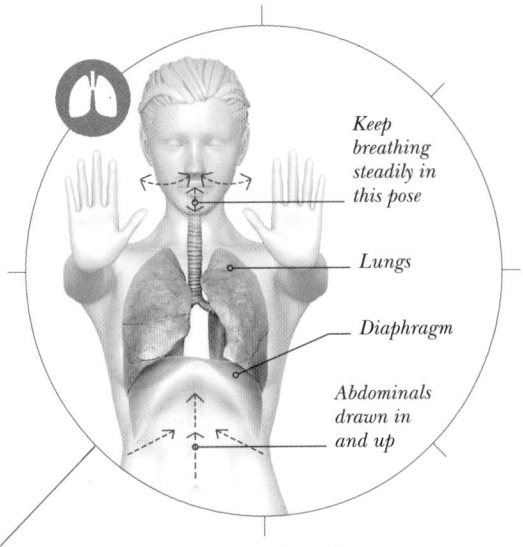

Uddiyana bandha

Uddiyana bandha involves engagement of the abdominal muscles to pull the belly in, and diaphragm up, with the exhale, and often incorporates a breath hold. In this pose, you are only doing a subtle version of this, but drawing your abdominals in and up while still breathing.

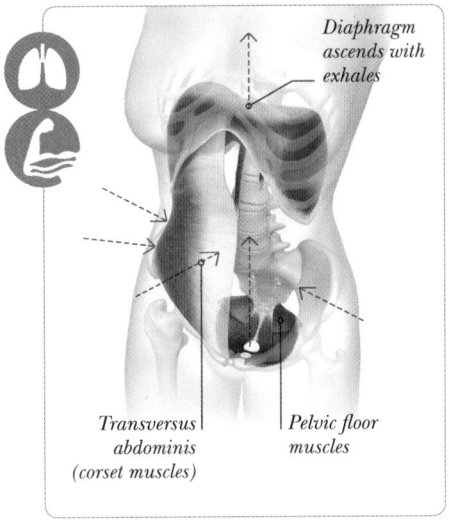

Mula bandha

Mula means "root" in Sanskrit; this *bandha* involves a pelvic floor engagement. When your transversus abdominis or "corset muscles" engage, your pelvic floor also naturally engages. Notice if you feel a lift (in this case, horizontally) in your pelvic floor, particularly with your exhales. This is soft and subtle; don't squeeze.

Deep core muscles engage to support torso

Hamstrings lengthen and engage slightly to stabilize

Calf muscles stretch as heels press back

Heels press back

Quadriceps work to hold legs in extended position

Knees extended but not locked

Toes are curled under

ANTERIOR–LATERAL VIEW

SIDE PLANK

Vasisthasana

Side Plank is a challenging arm balance that may get you sweating and your heart pounding. This pose is particularly beneficial for anyone looking to improve their focus and endurance. Holding Side Plank takes concentration to keep your hips from sagging.

THE BIG PICTURE

This pose strengthens your core—including your abdominals and back muscles. Your supporting arm and shoulder muscles are also engaging strongly to maintain balance. Even your leg muscles are working to support you and keep you aligned and balanced.

KEY

- *Joints*
- *Muscles*
- Engaging
- Engaging while stretching
- Stretching

ALIGNMENT

Try to stack your hips and shoulders on top of each other. If comfortable, reach your top arm up and gaze skyward. Alternatively, you may find looking down at your supporting hand helps you to stay balanced.

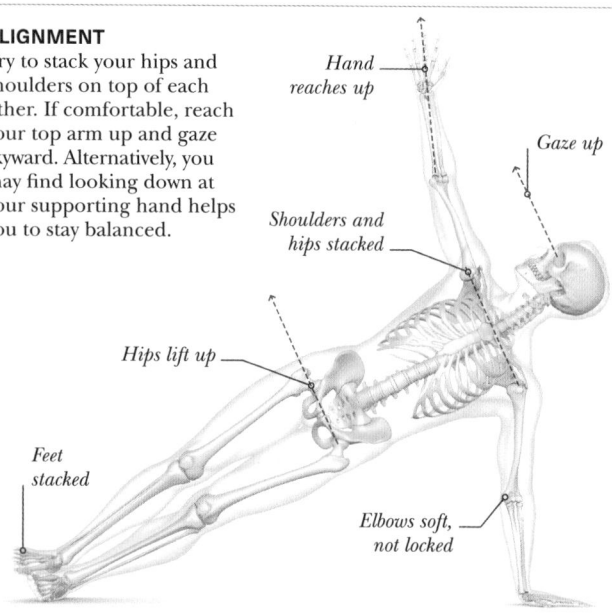

Top **thigh**

Your **hip adductors** engage on both sides to stabilize your thighs.

Lower **legs**

Your **ankle dorsiflexors** engage to dorsiflex your ankles and extend your toes. Your **calf muscles** are in a stretched position. Press the side of your foot into the floor to activate your **fibularis muscles** in your bottom leg, preventing your ankle from rolling downward.

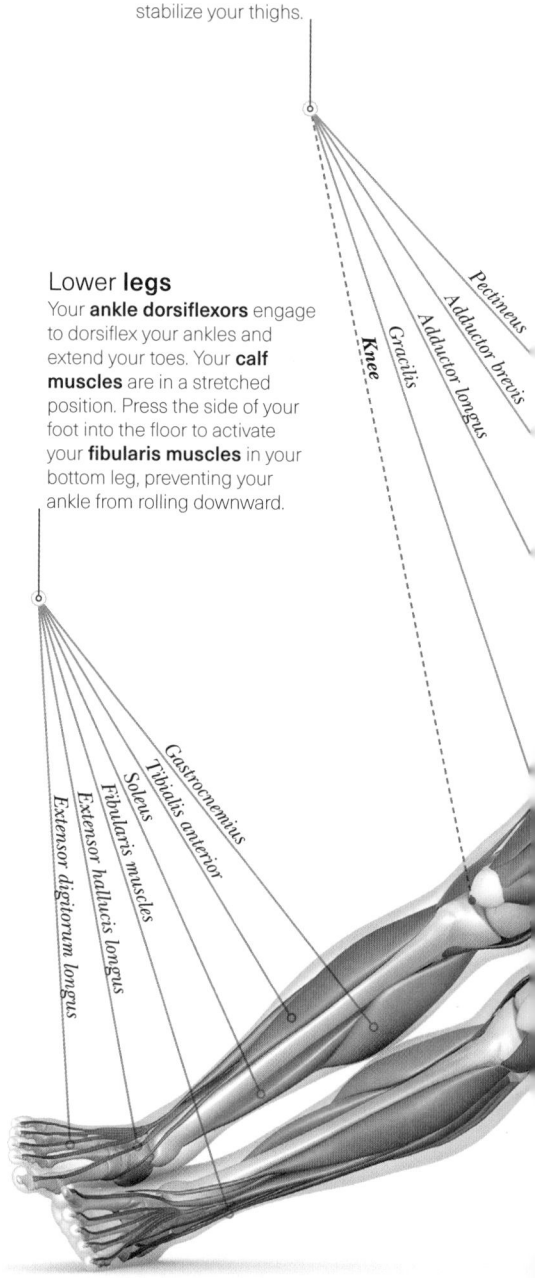

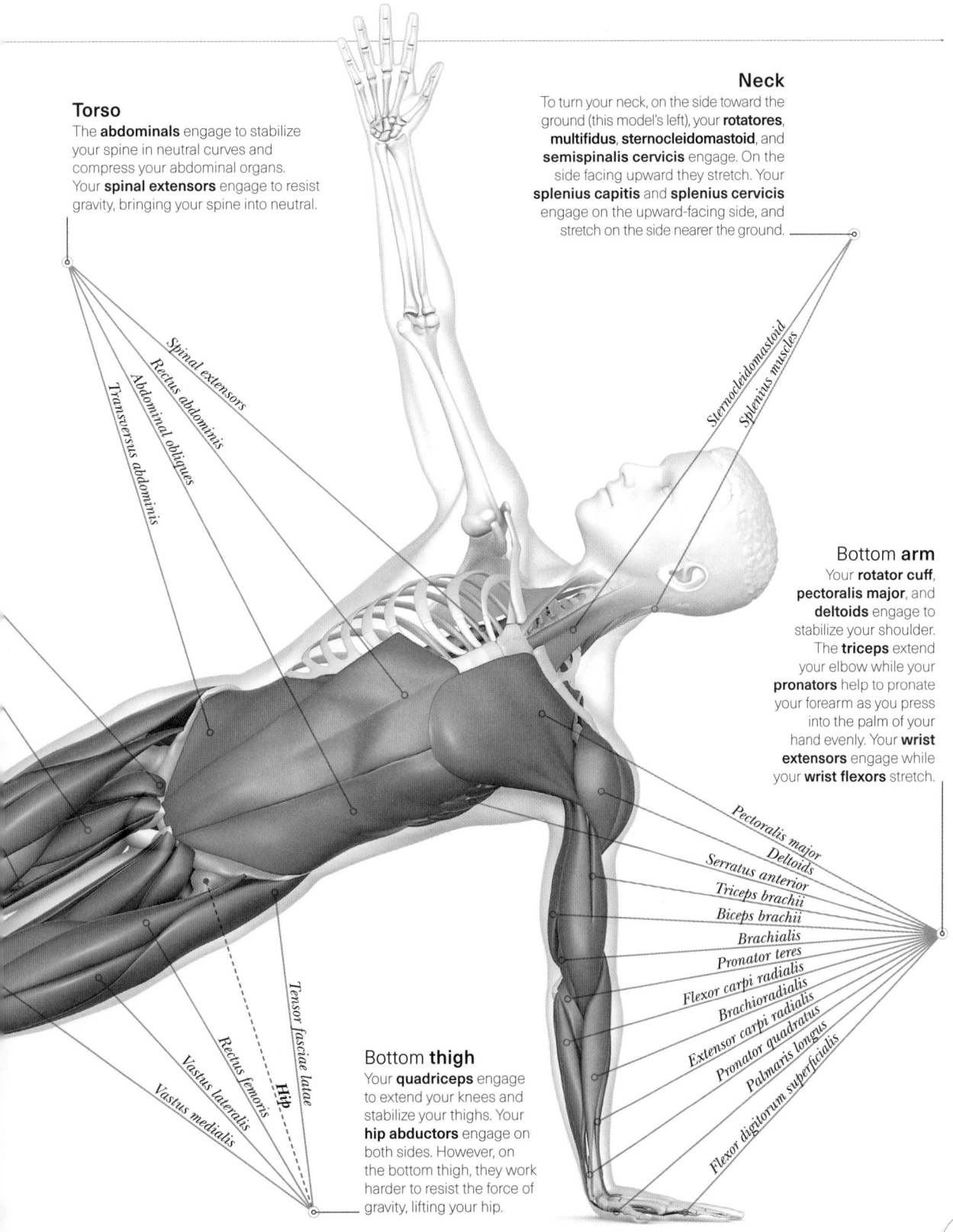
Torso
The abdominals engage to stabilize your spine in neutral curves and compress your abdominal organs. Your spinal extensors engage to resist gravity, bringing your spine into neutral.
Neck
To turn your neck, on the side toward the ground (this model's left), your rotatores, multifidus, sternocleidomastoid, and semispinalis cervicis engage. On the side facing upward they stretch. Your splenius capitis and splenius cervicis engage on the upward-facing side, and stretch on the side nearer the ground.
Spinal extensors
Rectus abdominis
Abdominal obliques
Transversus abdominis
Sternocleidomastoid
Splenius muscles
Bottom arm
Your rotator cuff, pectoralis major, and deltoids engage to stabilize your shoulder. The triceps extend your elbow while your pronators help to pronate your forearm as you press into the palm of your hand evenly. Your wrist extensors engage while your wrist flexors stretch.
Pectoralis major
Deltoids
Serratus anterior
Triceps brachii
Biceps brachii
Brachialis
Pronator teres
Flexor carpi radialis
Brachioradialis
Extensor carpi radialis
Pronator quadratus
Palmaris longus
Flexor digitorum superficialis
Tensor fasciae latae
Hip
Rectus femoris
Vastus lateralis
Vastus medialis
Bottom thigh
Your quadriceps engage to extend your knees and stabilize your thighs. Your hip abductors engage on both sides. However, on the bottom thigh, they work harder to resist the force of gravity, lifting your hip.

» CLOSER LOOK

Side Plank involves deep breaths, recruiting more respiratory muscles than usual. There is also significant core muscle engagement, which is good for scoliosis, but poses risks for pregnancy.

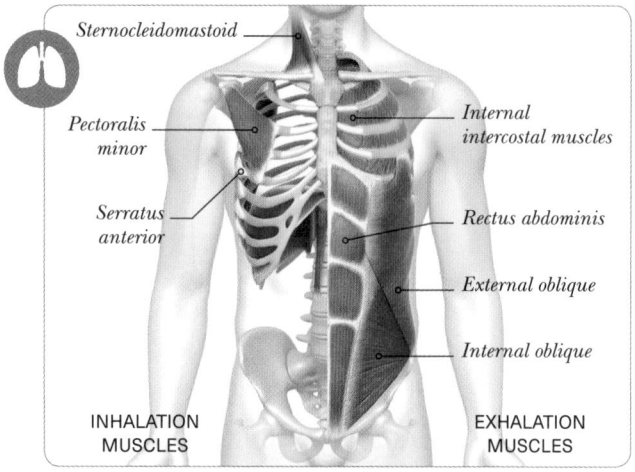

Respiratory muscles

In a natural breath, your diaphragm is the main player. When you breathe deeply, as in this pose, other accessory respiratory muscles can be recruited. The inhale involves the muscles above left, along with small muscles along your neck called the scalenes. The exhale also involves deep muscles along your ribs called transversus thoracis.

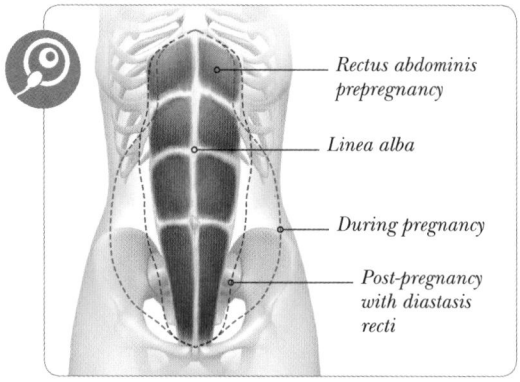

Pregnancy caution

The linea alba is the connective tissue joining the sections of the rectus abdominis. During pregnancy, pressure can separate this tissue, causing a condition called diastasis recti or abdominal separation. For this reason, be cautious of poses that involve abdominal engagement and pressure while pregnant, particularly later in pregnancy.

Stretch up to the tips of your fingers

This arm engages minimal compared to the supporting arm

Gaze up to your hand to challenge your balance

Pectoralis major has minimal engagement compared to the supporting side

Quadriceps engage to stabilize thighs

Elbow flexors lengthen and engage to stabilize

Feet stacked with weight on side of foot

To reduce pressure on wrist, come down to rest on your forearm

ANTERIOR–LATERAL VIEW

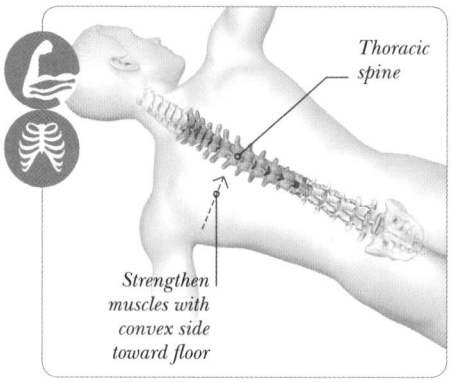

Scoliosis

In scoliosis the spine curves to the side making an S or, more commonly, a backward S shape. Some evidence suggests that strengthening the convex side of the primary curve with Side Plank —by practicing with the convex side toward the floor—can reduce sideways spinal curvature and symptoms. Consult a specialist if you are unsure which side needs strengthening more.

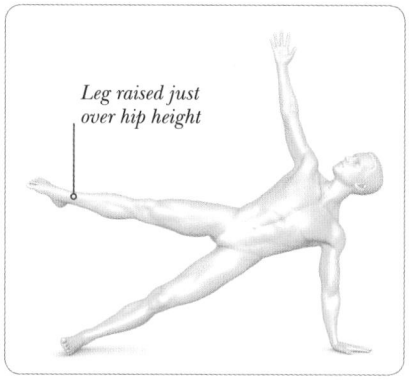

VARIATION

To further challenge your stability and strengthen your core muscles, slowly lift your top leg to a little over hip height. Keep your hips aligned and if you feel unstable lower your leg back down.

This side of your hip engages less than the supporting side

Back muscles engage more on this side of your spine

Knees extended but not locked

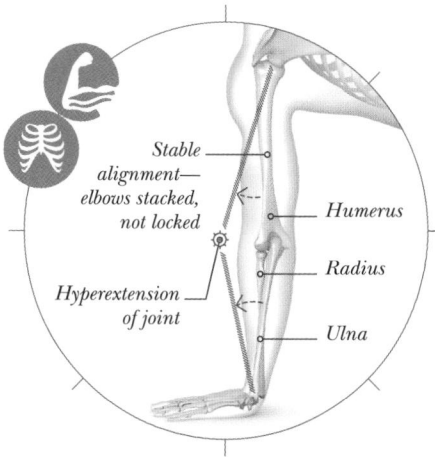

Locking **elbows**

Avoid locking your elbows into hyperextension (>180 degrees). This causes compression on the weight-bearing joint. When the bones aren't stacked optimally, the uneven load may lead to osteoarthritis and impaired function over time. Instead, microbend your elbow so it appears straight and stacked. This will require more muscular effort but will bring long-term stability.

Feet are flexed (in dorsiflexion)

Calf muscles are in a lengthened position

POSTERIOR–LATERAL VIEW

COBRA

Bhujangasana

Cobra pose is a key traditional yoga pose. This gentle backbend was believed to ignite a burning digestive fire and awaken the flow of dormant energy. It does seem to stimulate digestion and elimination, while helping to ease back pain for many.

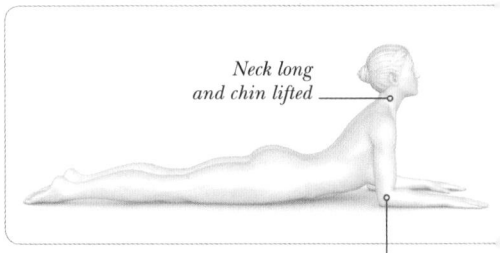

VARIATION
With your forearms on the floor, Sphinx can be a more accessible and more passive version of the pose.

THE BIG PICTURE

The front of your body—including your chest, abdominals, and hips—is stretching. Meanwhile, muscles in your back, shoulders, and arms are strengthening as you maintain the posture, creating an even curve along your neck and spine.

ALIGNMENT
Your pubic bone remains on the mat as you elongate your spine into an even backbend. If you feel pinching or pain in your lower back, come down lower.

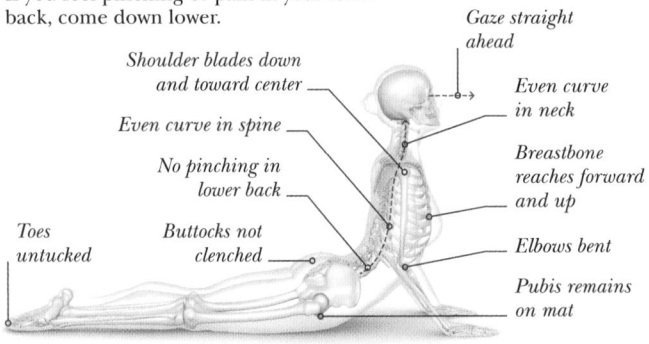

Thighs

Your **gluteus maximus**, **adductor magnus**, and **hamstrings** engage to hold your hips in extension, while your **tensor fasciae latae** and **iliotibial band** stabilize your hips.

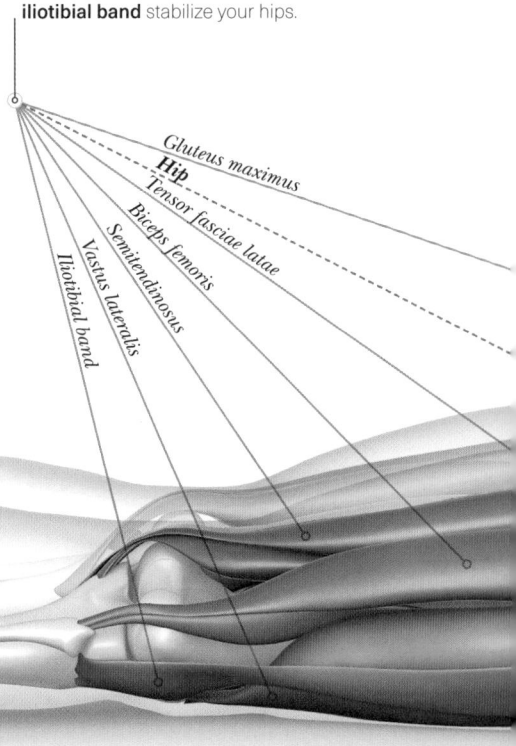

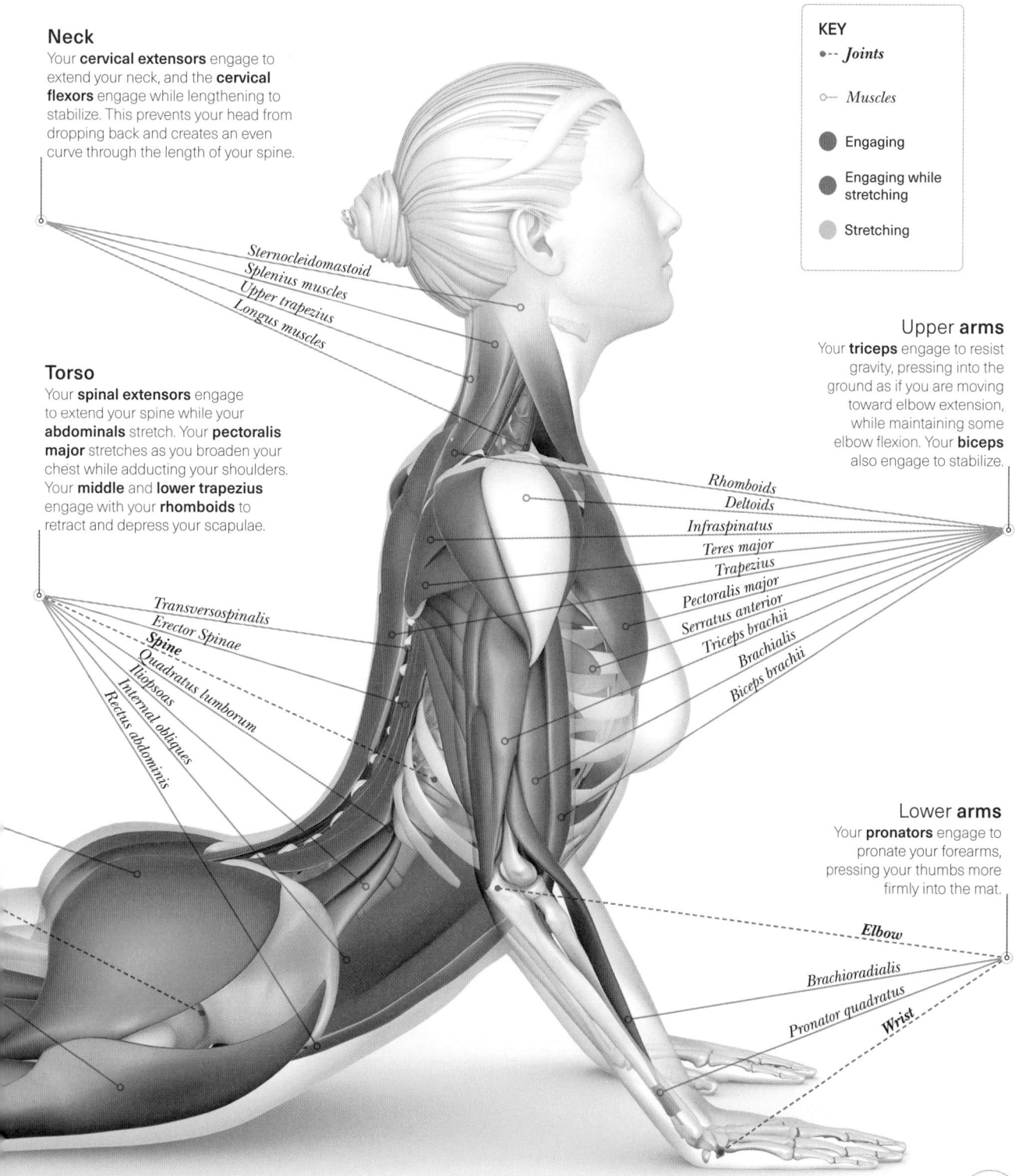
Neck
Your cervical extensors engage to extend your neck, and the cervical flexors engage while lengthening to stabilize. This prevents your head from dropping back and creates an even curve through the length of your spine.
KEY
Joints
Muscles
Engaging
Engaging while stretching
Stretching
Sternocleidomastoid
Splenius muscles
Upper trapezius
Longus muscles
Torso
Your spinal extensors engage to extend your spine while your abdominals stretch. Your pectoralis major stretches as you broaden your chest while adducting your shoulders. Your middle and lower trapezius engage with your rhomboids to retract and depress your scapulae.
Upper arms
Your triceps engage to resist gravity, pressing into the ground as if you are moving toward elbow extension, while maintaining some elbow flexion. Your biceps also engage to stabilize.
Rhomboids
Deltoids
Infraspinatus
Teres major
Trapezius
Pectoralis major
Serratus anterior
Triceps brachii
Brachialis
Biceps brachii
Transversospinalis
Erector Spinae
Spine
Quadratus lumborum
Iliopsoas
Internal obliques
Rectus abdominis
Lower arms
Your pronators engage to pronate your forearms, pressing your thumbs more firmly into the mat.
Elbow
Brachioradialis
Pronator quadratus
Wrist

» CLOSER LOOK

Cobra pose can be refined with activation of key muscles like the serratus anterior. It can be adapted to gentler versions such as Sphinx pose, or a deeper backbend such as Upward-facing Dog.

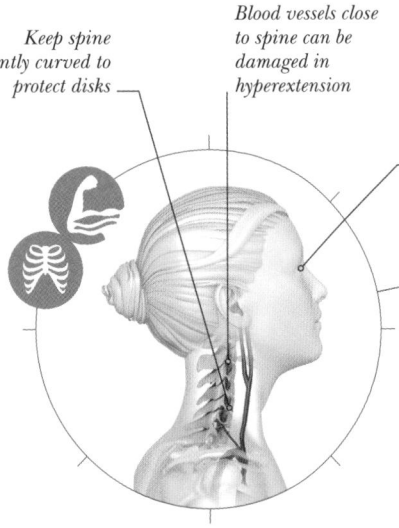

Neck problems

Traditional teachers have taught to throw the head back as far as possible. However, we now understand this has more risks than benefits. Based on case studies and anatomy knowledge, you can choose to respectfully adapt for safety and optimal function by avoiding hyperextension.

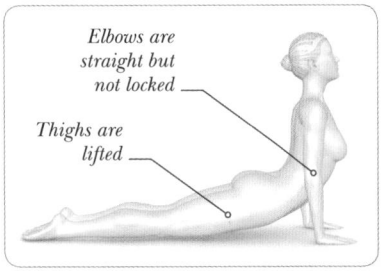

VARIATION

Upward-facing Dog is a similar pose to Cobra, used more in some styles of yoga. The thighs are lifted off the floor with the elbows straight to create a deeper backbend.

Eyes gaze up diagonally to where the ceiling meets the wall

Chin lifts slightly

Pectoralis minor mainly stretches

Iliopsoas stretches

ANTERIOR VIEW

Upper back strength

By strengthening your upper and middle back and stretching your torso, you counteract hyperkyphosis (see p.14). The more passive forearm version (Sphinx pose) can also be a great option for hyperkyphosis of the thoracic spine, preventing dysfunction or a "dowager's hump."

Serratus anterior activation

Try gliding your hands back without moving them on your mat, pressing your breastbone forward to reduce pressure in your spine. This also activates your serratus anterior muscles, which help with functional movements like reaching your arms forward. Strengthening them could relieve some neck and shoulder pain.

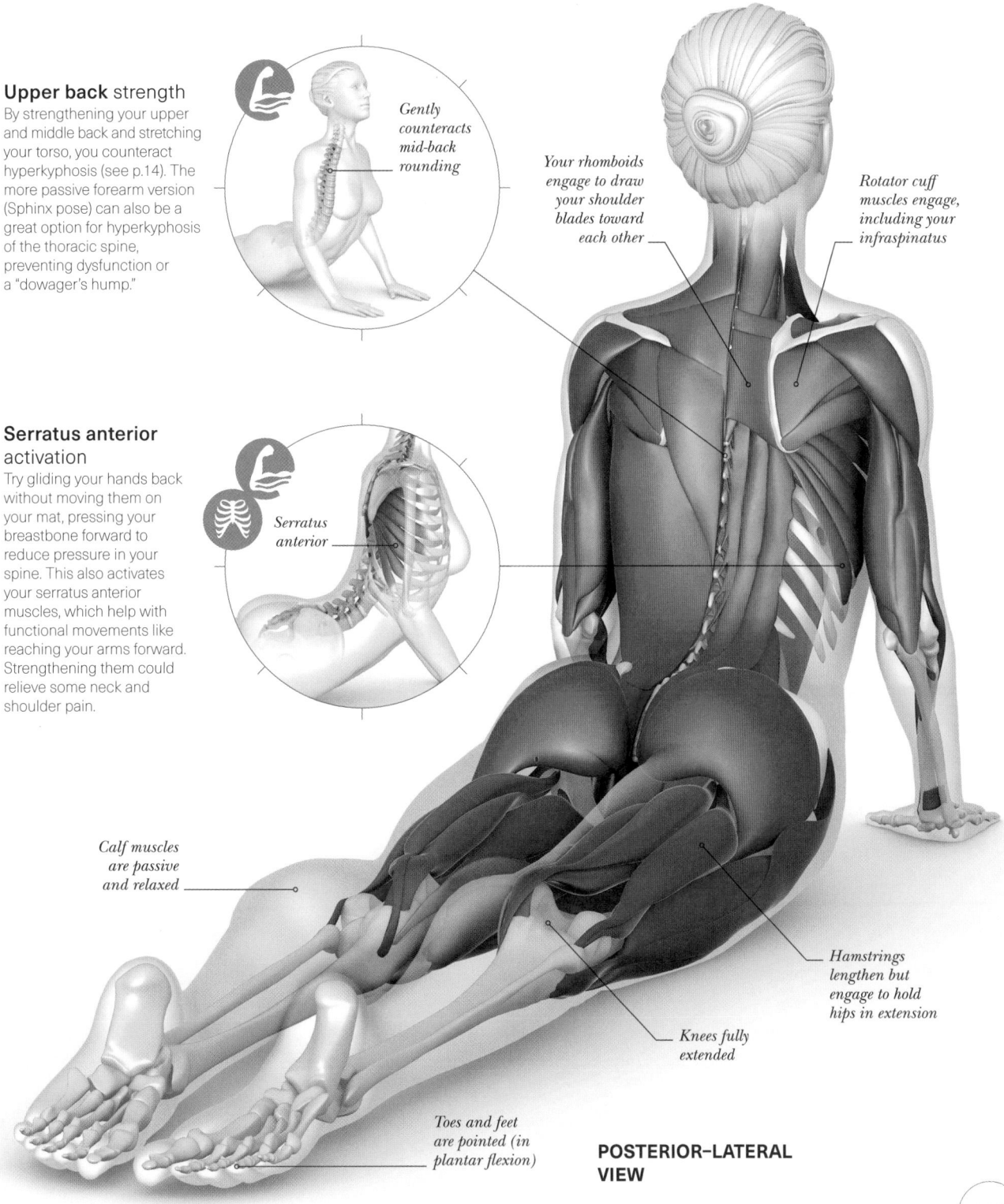

POSTERIOR–LATERAL VIEW

LOCUST

Salabhasana

Locust pose, also known as belly-down boat pose, can be helpful for relieving back pain. Elongating your spine in this way helps to counteract poor posture and related issues, as muscles along your back and legs engage to hold each end of your body off the ground.

THE BIG PICTURE

This pose particularly strengthens your back muscles, buttocks, and thighs as you lift your legs and shoulders from the ground. It can be challenging, but you don't have to lift very high to get the benefits of the pose.

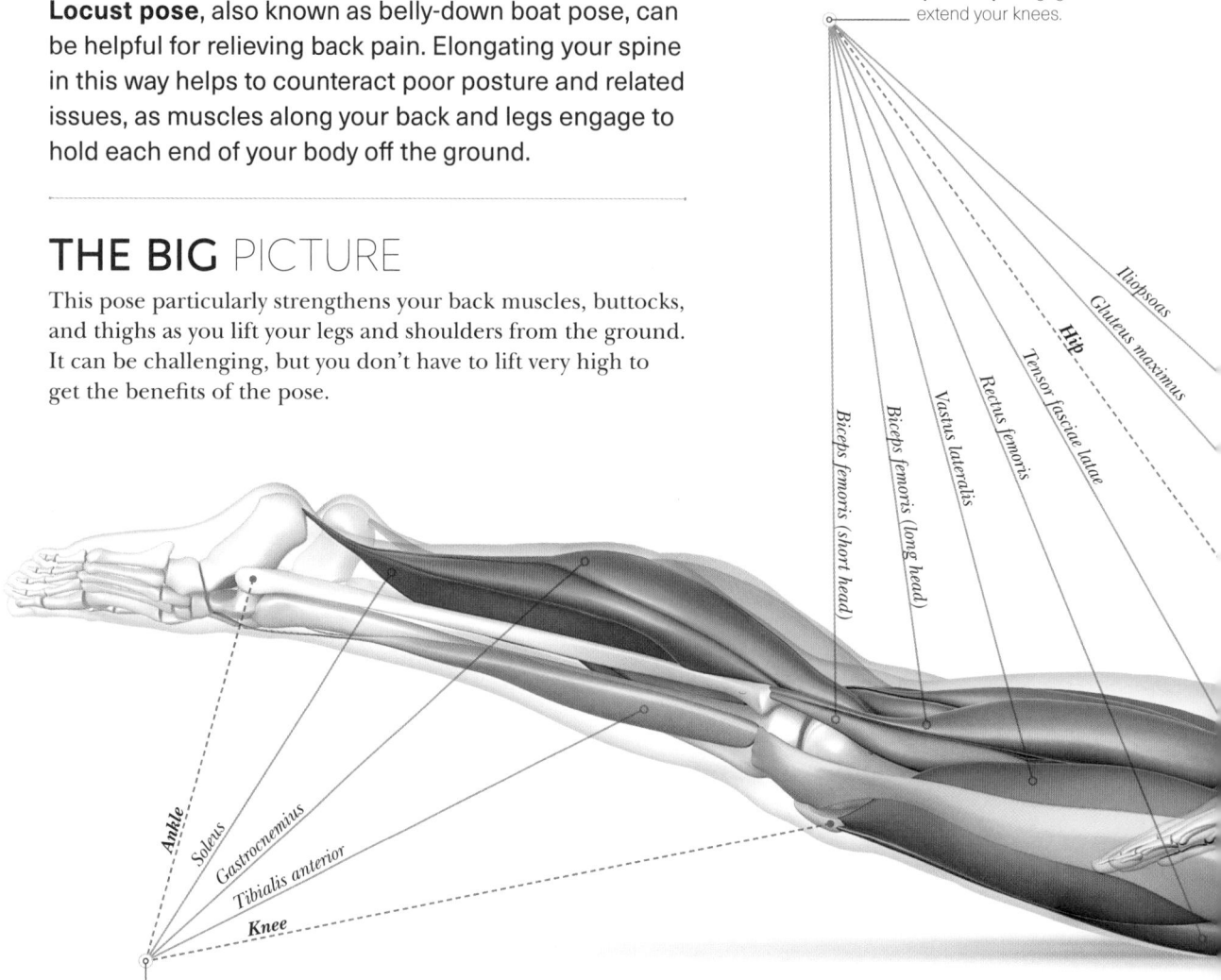

Lower **legs**
Your **gastrocnemius** and **soleus** engage to plantarflex your ankles while your **tibialis anterior muscles** and other **ankle dorsiflexors** are in a stretched position.

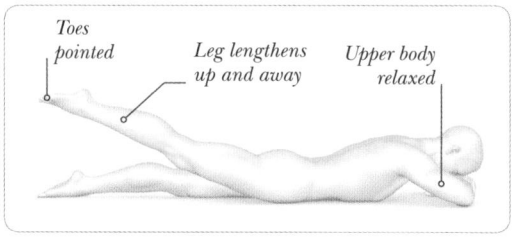

VARIATION
If you have neck issues, place your forehead on your hands and lift one leg at a time, trying to keep both of your front hip points toward the ground. Hold for several breaths, then switch legs.

ALIGNMENT
As you lift your shoulders and legs, focus on elongating your spine. Reach the crown of your head forward and upward, and aim for an even-feeling curve all the way along your spine, including in your neck.

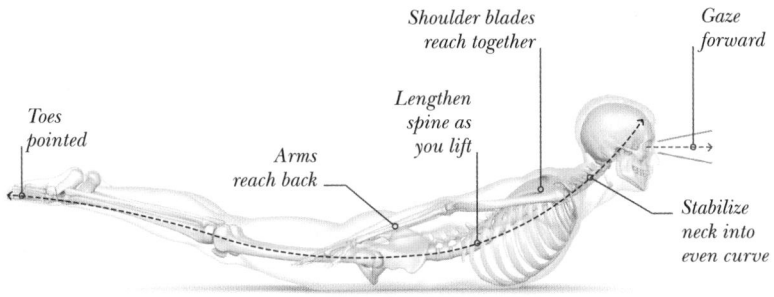

Neck and arms

Your **cervical extensors** engage to extend your neck, and your **cervical flexors** engage while lengthening to stabilize your neck, preventing throwing your head back and creating an even curve in your spine. Your **posterior deltoids**, **latissimus dorsi**, and **teres major muscles** engage to extend your shoulders, while your **triceps** extend your elbows.

Elbow
Triceps brachii
Biceps brachii
Deltoids
Longus muscles
Sternocleidomastoid
Splenius muscles
Pectoralis major
Serratus anterior
Rectus abdominis
Spine
Spinal extensors
Quadratus lumborum

KEY
- *Joints*
- *Muscles*
- Engaging
- Engaging while stretching
- Stretching

Torso

Your **spinal extensors** engage while your **abdominals** and **pectoralis major** stretch. Your **middle** and **lower trapezius** engage with your **rhomboids** to retract your scapulae, while your **serratus anterior muscles** engage to help stabilize them.

» CLOSER LOOK

Locust strengthens the entire back of your body, which can be particularly helpful for improving posture and core function. You do not have to lift yourself very far off the ground to gain the benefits of this pose.

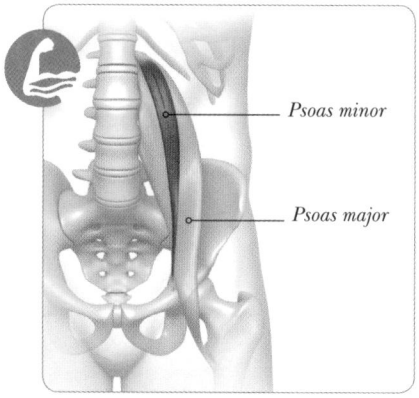

Psoas minor

You'll probably feel your psoas stretching in this pose. Approximately 40 percent of people have a psoas minor. This is further evidence of the variation between individuals; some people have more muscles or bones than others. Bodies are so different that, of course, each person's expression of a yoga pose will look unique.

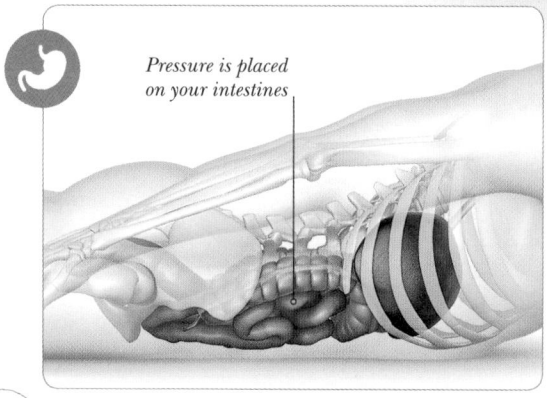

Boost digestion

Poses like this can help stimulate a bowel movement because of the floor's pressure on your digestive organs, and from the engagement of your core. This effect may be enhanced if you come in and out of the pose several times, simulating the rhythmic movements of your intestines.

Crown of your head reaches up at a diagonal

Anterior deltoid stretches

Pectoralis major stretches

Rectus abdominis stretches

ANTERIOR VIEW

Feet are pointed (in plantar flexion)

Back pain

The World Health Organization cites back pain, particularly lower back pain, as a major cause of disability worldwide. Research suggests that yoga is a safe, effective, nonpharmacological, noninvasive, and low-cost method of relieving back pain, and pain in general.

Degenerated disk

Disk degeneration is part of the natural process of aging. Over time, gravity wears down your disks. Cells, including those in intervertebral disks, also become drier with age, making them less resilient, and less resistant to gravity. Strengthening your back muscles may help prevent or slow these age-related changes by supporting your spinal posture.

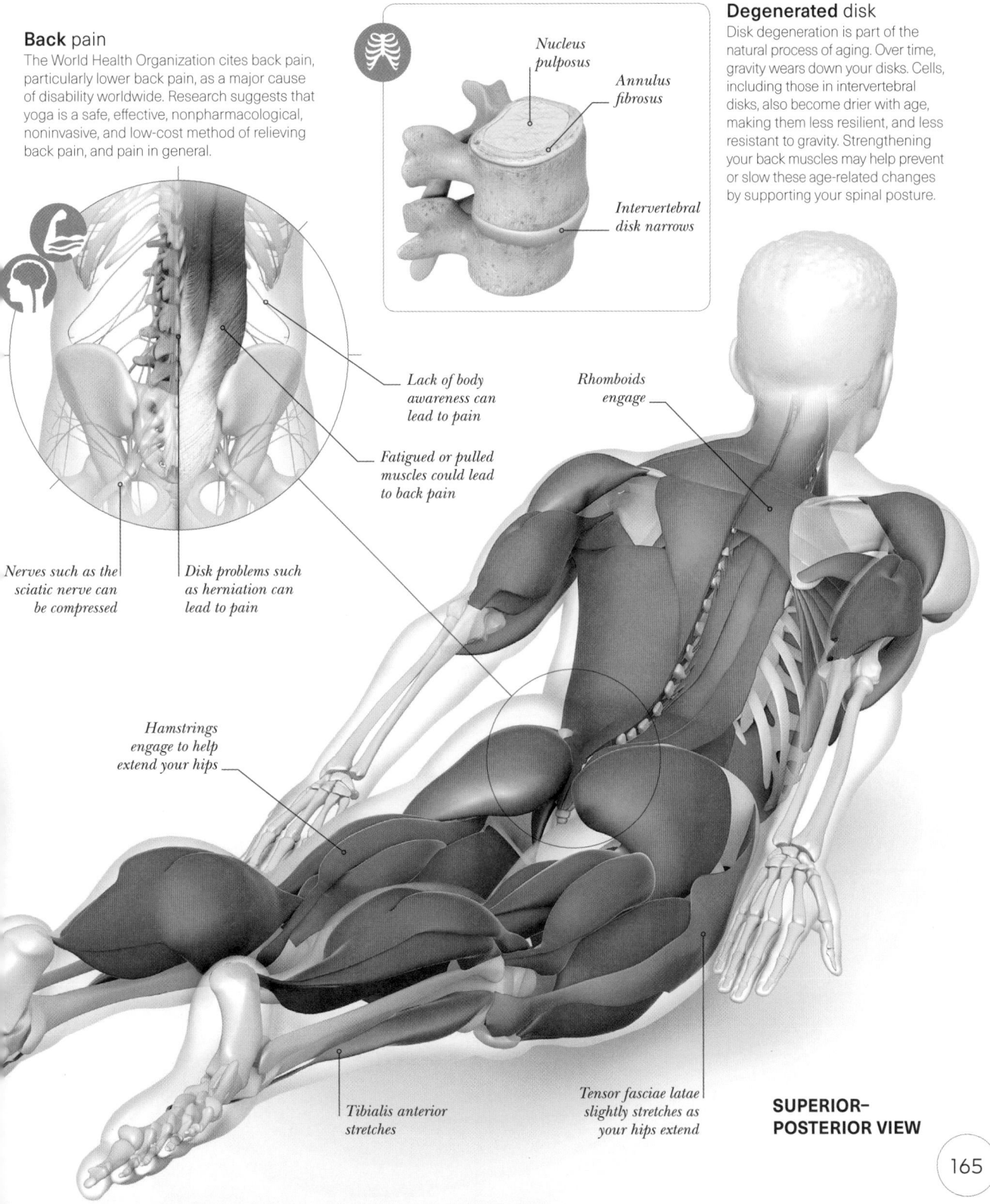

SUPERIOR–POSTERIOR VIEW

SUPINE LEG STRETCH

Supta Padangusthasana

This pose and its variations stretch your thighs in a way that is particularly safe for your lower back. This can be very relaxing and great for winding down after a long day. If you are unable to grasp your toes, try holding onto a strap around the bottom of your foot.

THE BIG PICTURE

The back of your lifted thigh and leg intensely stretch. Your arms gently pull your leg in, but you should try to relax any muscles that are not necessary for this action (like your jaw, neck, and shoulders).

ALIGNMENT
Your spine is neutral, or your lower back may be slightly flexed depending on how far into the pose you go. Pull your toe in until you feel a deep but comfortable stretch in your hamstrings.

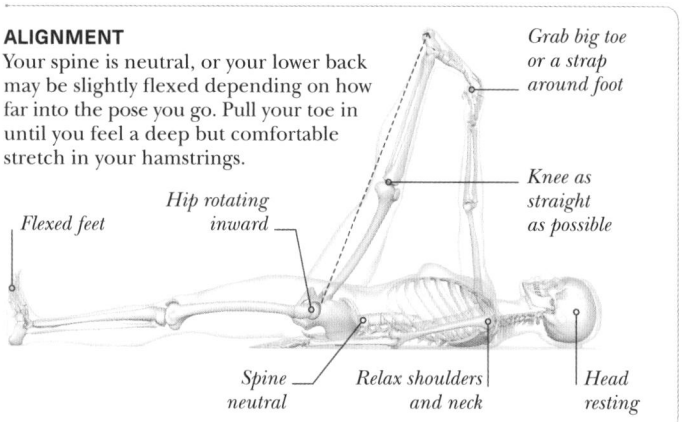

Lifted **thigh** and lower **leg**
Your **hip flexors** engage while your **quadriceps** extend your knee. Your **hip extensors**—particularly your **hamstrings** and **gluteus maximus**—stretch. As you grasp at your toes you'll likely feel your **ankle plantar flexors**—especially your **calf muscles**—stretching.

Lowered **thigh** and lower **leg**
In this version of the pose, your lowered thigh and leg are slightly engaged to stabilize. Your **hip flexors** are in a slightly lengthened position, your **quadriceps** extend your knees, and your **hamstrings** are slightly engaging. Your **ankle dorsiflexors** engage while your **plantar flexors** are in a neutral or lengthening position.

Tibialis anterior
Extensor digitorum longus
Gastrocnemius
Knee
Biceps femoris (short head)
Biceps femoris (long head)
Iliotibial band
Vastus lateralis
Rectus femoris

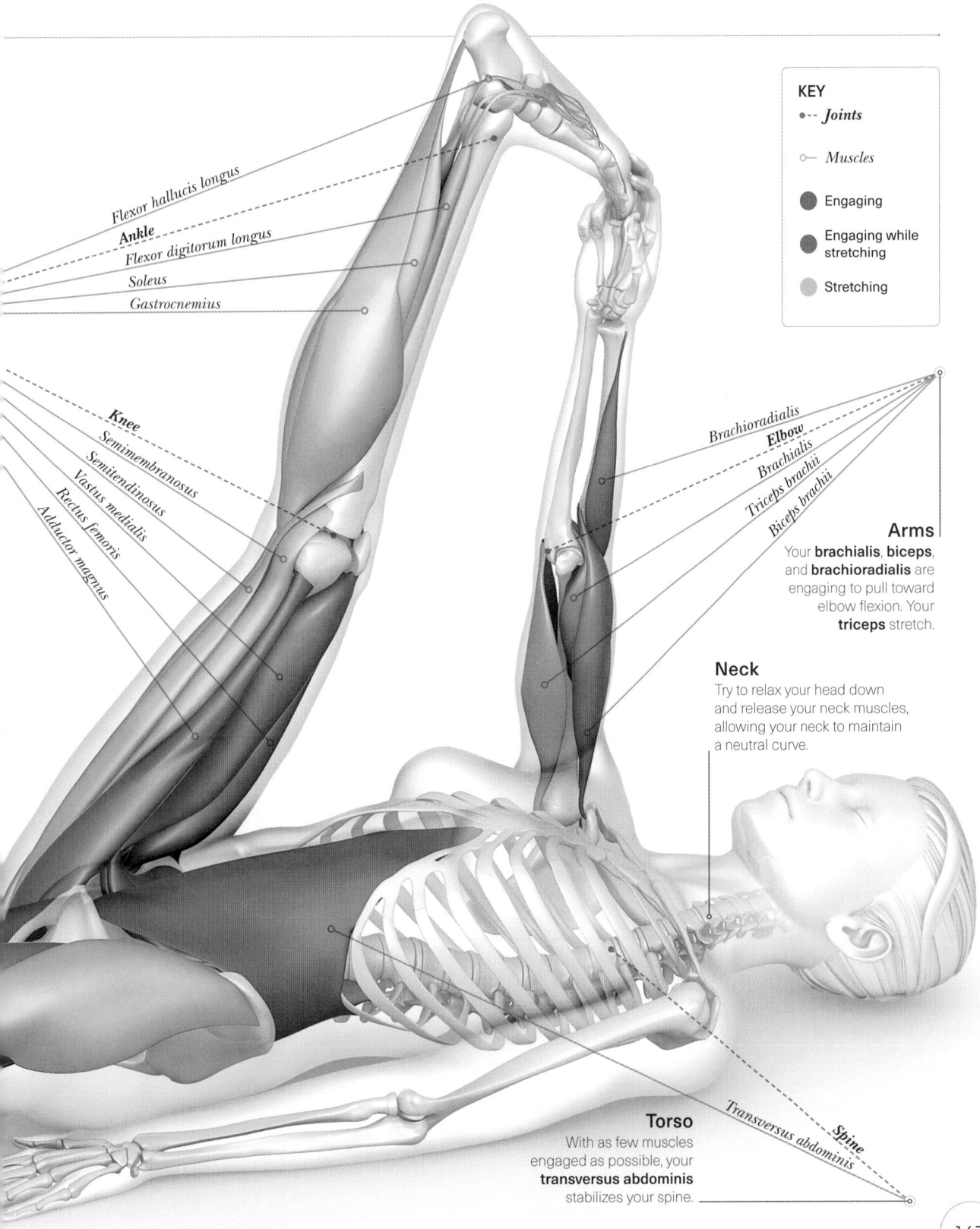
KEY
Joints
Muscles
Engaging
Engaging while stretching
Stretching
Flexor hallucis longus
Ankle
Flexor digitorum longus
Soleus
Gastrocnemius
Knee
Semimembranosus
Semitendinosus
Vastus medialis
Rectus femoris
Adductor magnus
Brachioradialis
Elbow
Brachialis
Triceps brachii
Biceps brachii
Arms
Your brachialis, biceps, and brachioradialis are engaging to pull toward elbow flexion. Your triceps stretch.
Neck
Try to relax your head down and release your neck muscles, allowing your neck to maintain a neutral curve.
Transversus abdominis
Spine
Torso
With as few muscles engaged as possible, your transversus abdominis stabilizes your spine.

» CLOSER LOOK

This stretch can be done with or without a strap, making it accessible for many people. Use your neurophysiology to your advantage to get a more effective stretch with mindfulness tricks.

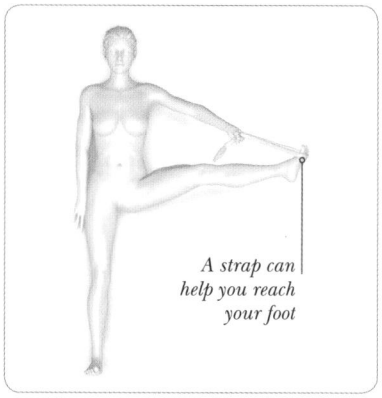

VARIATION

If you are unable to reach your toes without straining, you can hold onto a strap around the sole of your foot. You may also bring your leg out to the side to shift the focus of the stretch to your groin and inner thighs (adductors).

Foot is flexed (in dorsiflexion)

Calf muscles stretch deeply as foot is flexed

Knee is soft, not locked

Elbow is soft, not locked

Reach toward your foot with little to no muscular effort around your shoulders

Allow your ribcage to move with your breath

Eyes can be open or closed

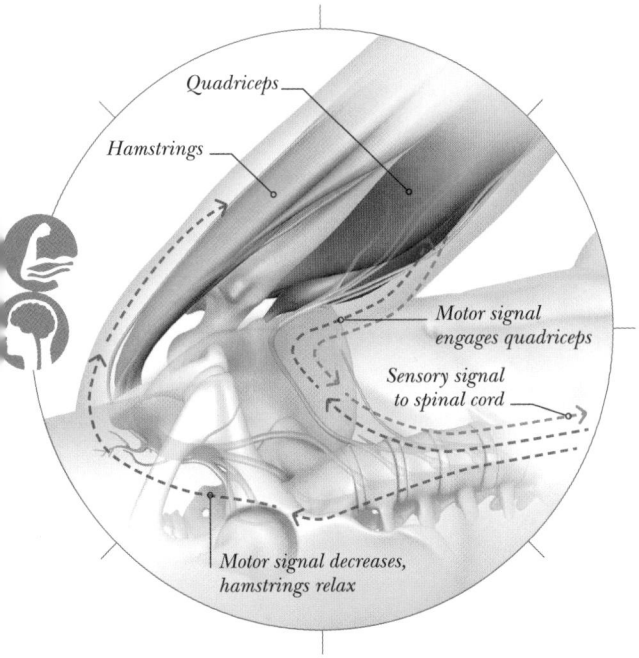

Reciprocal inhibition

Muscles often work in pairs. You can use reciprocal inhibition (RI), a protective physiological phenomenon, to get a deeper stretch safely. To initiate RI, consciously engage your quadriceps for a few breaths. Nerves in your quadriceps send a message to the paired hamstrings, telling them to relax deeper into the stretch.

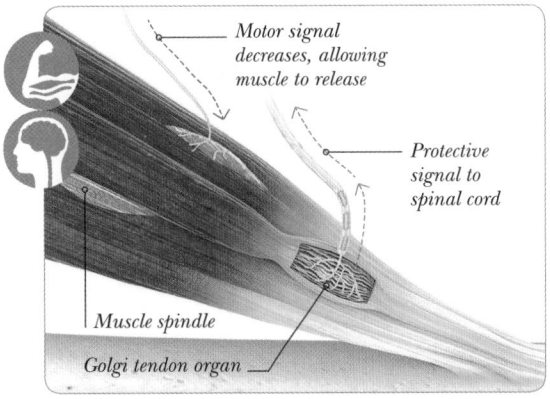

Release

When you first go into a stretching pose, you probably feel a taut pulling in your muscles. After a few breaths, tension peaks, and sensors in tendons called the Golgi tendon organ send a protective signal, inhibiting contraction and resistance in larger muscle fibers. This causes that pleasurable "ahhh" feeling of release.

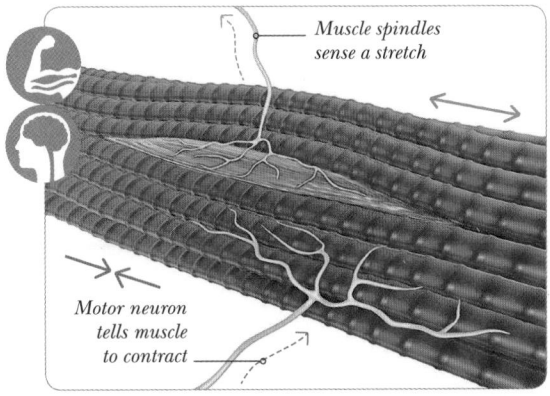

Stretch reflex

Smaller muscle fibers with sensors, called muscle spindles, don't release as quickly, causing the stretch reflex (which involves muscle contraction to protectively resist overstretching). Override this by moving gradually into the pose, allowing muscle fibers to slowly release, to get a deeper stretch without injury.

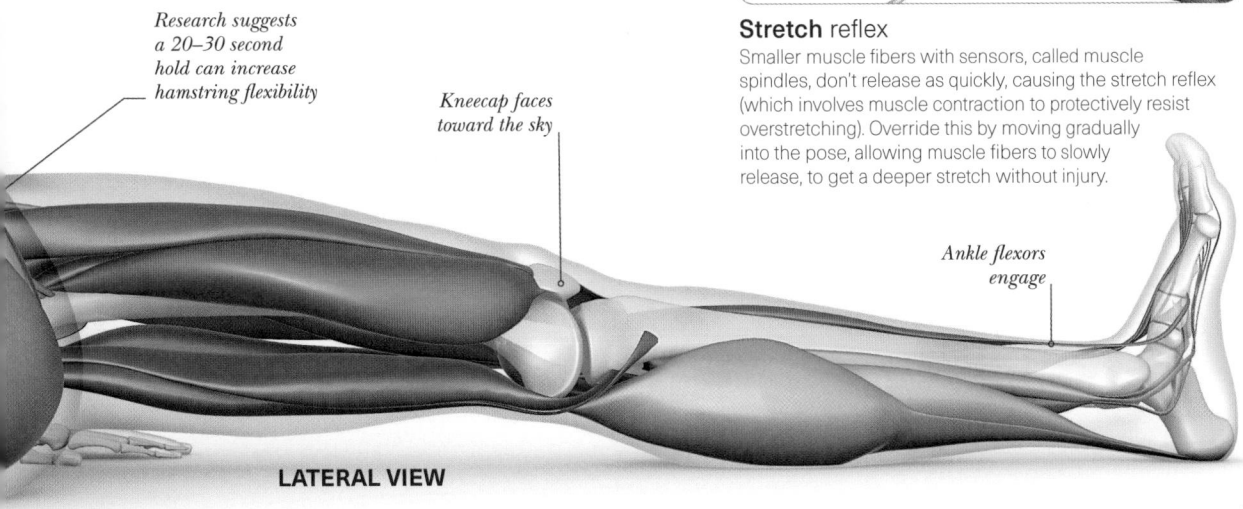

SUPINE TWIST

Supta Matsyendrasana

This relaxing spinal twist is often done at the end of a yoga class to calm your nervous system. Cultivate a sense of groundedness by releasing your body weight down into the floor. Find ease to activate the rejuvenating "rest and digest" part of your nervous system.

THE BIG PICTURE

This pose stretches muscles along your spine, including the small muscles that rotate it. Your shoulders, glutes, and thigh muscles are also stretching, though elsewhere in your body your muscles should be as relaxed as possible.

ALIGNMENT

Release completely to gravity, feeling your bones dropping down. If your shoulders or knee can't completely release, feel free to use a blanket or bolster for support.

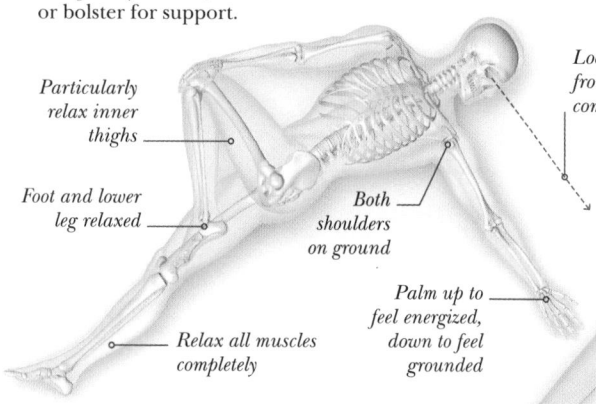

Knee
Vastus lateralis
Iliotibial band
Rectus femoris
Gluteus maximus
Gluteus medius

Thighs

Although you may feel some sensation in your bottom thigh, allow it to be passive. On your top thigh, your **hip abductors** and **quadriceps** stretch. Allow your knee to drop down until you feel a comfortable stretch across your hip and torso into your opposite arm.

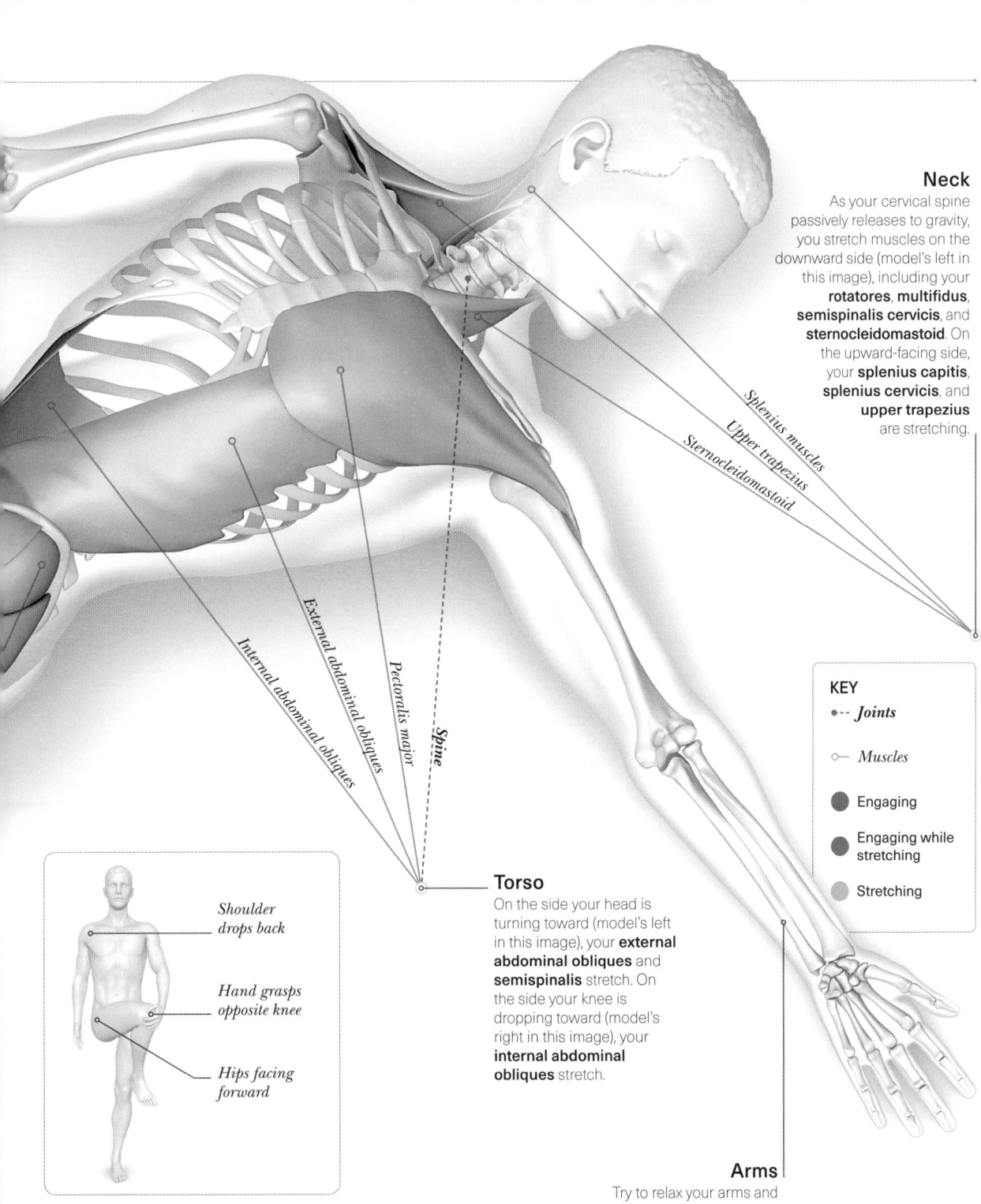

Neck

As your cervical spine passively releases to gravity, you stretch muscles on the downward side (model's left in this image), including your **rotatores**, **multifidus**, **semispinalis cervicis**, and **sternocleidomastoid**. On the upward-facing side, your **splenius capitis**, **splenius cervicis**, and **upper trapezius** are stretching.

Torso

On the side your head is turning toward (model's left in this image), your **external abdominal obliques** and **semispinalis** stretch. On the side your knee is dropping toward (model's right in this image), your **internal abdominal obliques** stretch.

Arms

Try to relax your arms and shoulders completely. You may rest one arm on your top knee; this forearm can rest in supination or pronation, depending on which feels more comfortable.

VARIATION

To challenge your balance and stretch your hip external rotators without lying down, try raising one knee and pulling it gently across your body.

» CLOSER LOOK

For many, Supine Twist is a safe way to do spinal rotation with ease. Wiggle into the pose and use props like a blanket until you find a pain-free position.

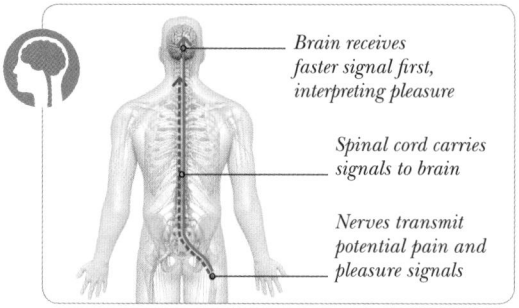

Perceived pain pathway

Imagine two signals like trains simultaneously traveling to your brain: the red train pathway carries a signal that could be perceived as painful (nociceptive), and the green train pathway carries a signal that could be perceived as pleasurable. The green train is faster, reaching your brain first, possibly overriding nociceptive signals. This is called the gate theory of pain.

Spine safety

Supine Twist can be safer than seated or standing twists by changing the orientation of the impact of gravity on your intervertebral disks and spine. Also, spinal flexion often occurs with upright twists and the combination of rotation and flexion increases the risk of spinal issues.

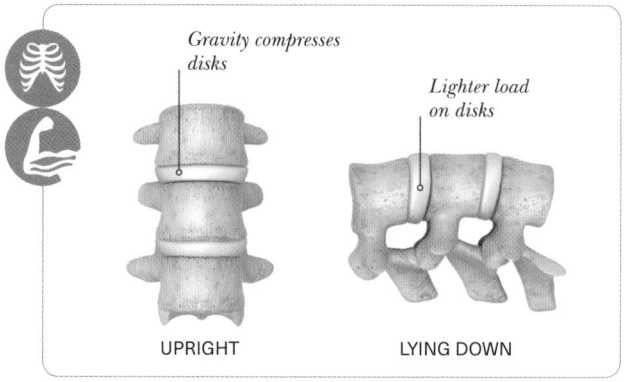

Pectoral muscle stretches

Abdominal obliques stretch

Head is resting and turned to side

Allow gravity to stretch your neck

Arms are passive

Elbow is soft and relaxed

Try using a blanket, pillow, or bolster under this knee for support

SUPERIOR–LATERAL VIEW

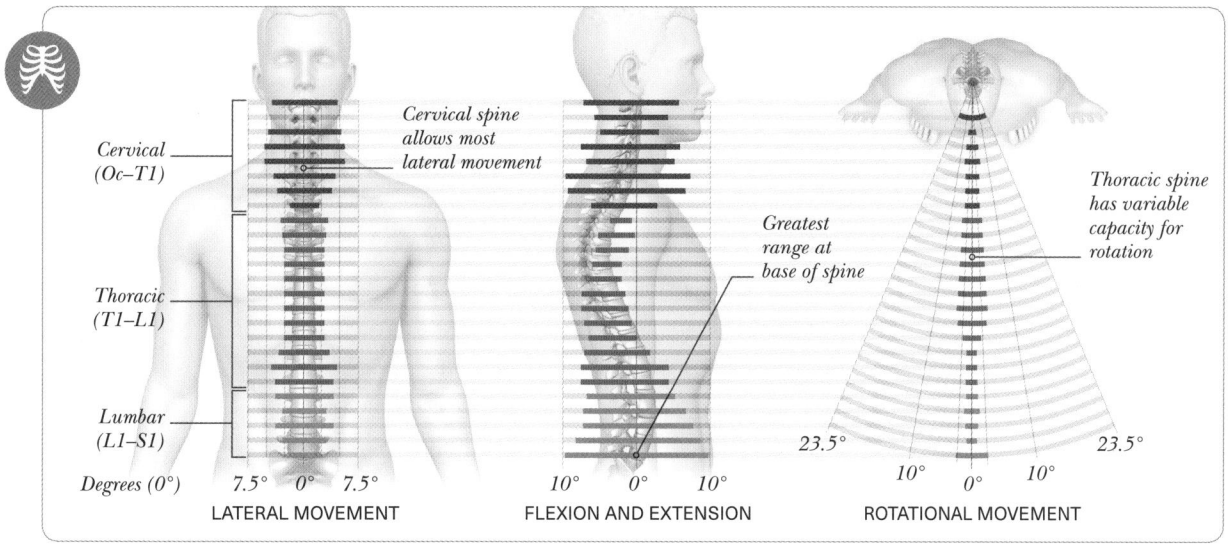

Spinal motion

Notice that your cervical and thoracic spine allow more twisting action than the lumbar. The shape of your vertebrae in each area facilitates or limits the amount of movement. Technically, you won't have a perfectly even twist. That is a visualization to help prevent extreme mobility or pinching in any one area. Different segments of your spine allow varying amounts of other motion.

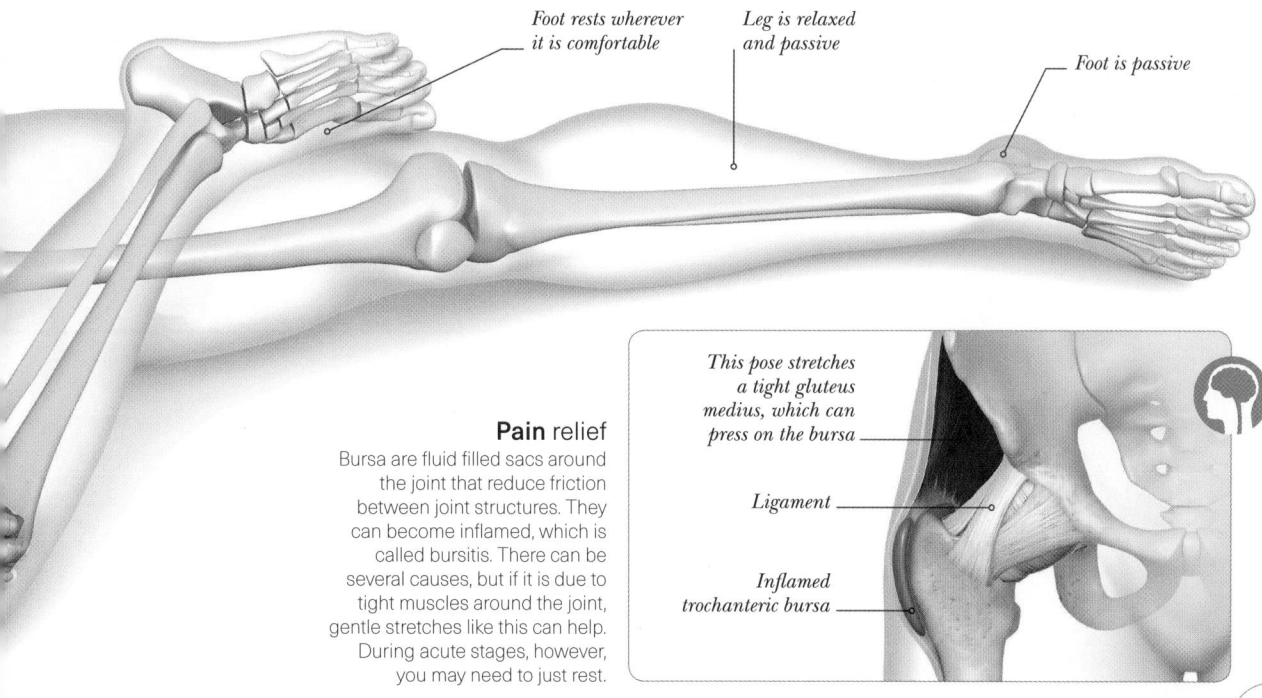

Pain relief

Bursa are fluid filled sacs around the joint that reduce friction between joint structures. They can become inflamed, which is called bursitis. There can be several causes, but if it is due to tight muscles around the joint, gentle stretches like this can help. During acute stages, however, you may need to just rest.

Q&A

QUESTIONS AND ANSWERS

These Q&As are based on common questions I have had from my students over the years. The physical body is addressed first, then mental and more subtle layers of self. It is important to note that although yoga is based in Hindu traditions, its practices and wisdom are adaptable for everyone. Whether you are spiritual, religious, agnostic, or something else, yoga can help you find health and peace.

JOINTS AND FLEXIBILITY

While a certain degree of flexibility is important in accomplishing many asanas and completing daily activities, it's crucial to understand your body and know your limits so you can avoid injury and look after your joints. If you are very flexible, it may be best to focus on strengthening asanas.

> *Yoga has been widely shown to increase* ***flexibility****, so a lack of flexibility only gives you* ***more reason to practice.***

Most of the **360** joints in the body are synovial, or free-moving.

Q CAN I DO YOGA IF I'M NOT FLEXIBLE?

Yes. Yoga has been widely shown to increase flexibility, so a lack of flexibility only gives you more reason to practice. If you have limited range of motion (ROM) in a pose because your muscles are tight or you are recovering from an injury, it can be helpful to visualize your body moving deeper into the pose. Research suggests that this creates a neural map, sending signals to the muscles which leads to increased ROM. Similarly, research has found that visualizing yourself doing a pose and getting stronger can measurably strengthen your muscles, even without moving.

Q WHY DO MY JOINTS "POP"?

Most joints have synovial fluid between the bones, which contains dissolved gas molecules. Creating more space in the joint—for example, by pulling your thumb—pulls gases out of the fluid, similar to how CO_2 bubbles fizz out of carbonated drinks when you open the bottle. The gases redissolve into the fluid, and can be "popped" again after 20–30 minutes. There is no evidence to suggest this causes arthritis, but it may make your joints larger. If your joints pop with no wait, the joint structures may be rubbing against each other. This could slowly damage the joint structures.

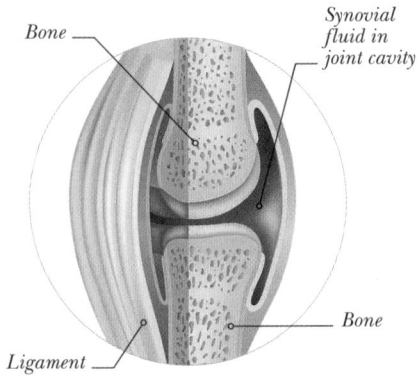

SYNOVIAL JOINT

Q IS IT POSSIBLE TO STRETCH TOO MUCH?

Yes. There is a correlation between hypermobility—the ability to stretch beyond the normal range, or being "double jointed"—and chronic joint pain. When you stretch, you should feel the stretching sensation in the middle of the muscle, not near the joints, and you want to be able to breathe smoothly through the stretch. If you feel sharp or shooting sensations, numbness, pain, or anything that makes you grimace or hold your breath, you are overstretching. Overstretching lengthens your ligaments and/or tendons and, since they don't have much elasticity, they don't recoil well after they have been stretched. In other words, when the stress (load or stretch) on the tissue reaches the yield point it stops being "elastic" and becomes "plastic" (see above right). In clinical terms, this represents a tear. To avoid injury, it's best to strike a balance between using your yoga asana practice to improve your strength and using it to improve your flexibility.

Stress–strain curve

This graph shows how much stress your tissue (muscle, tendon, or ligament) can take before injury. In the elastic region, the tissue can still return to its normal length when the stress is removed, but in the plastic region, it can't recoil. The ultimate fail point is a complete tear. To avoid injury, don't push beyond your limits.

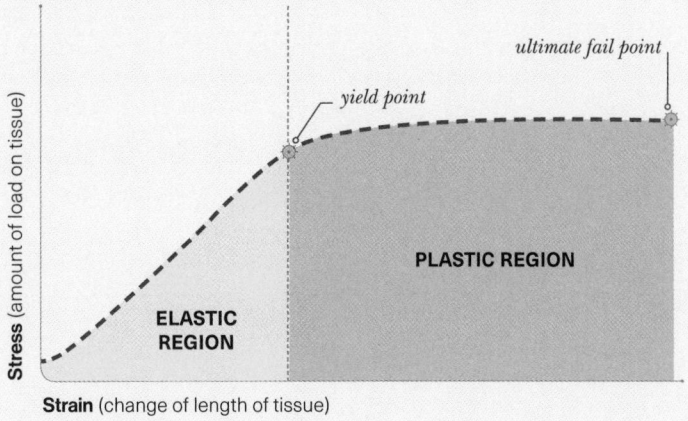

MYTH-BUSTER

Hot yoga makes me more flexible.

It does, but only in the moment; it doesn't necessarily affect how flexible you are afterward. Higher temperatures raise your metabolic rate, warming your tissue quicker so you can stretch deeper. Practicing in hotter conditions makes it easy to stretch beyond your muscles' natural lengthening, which can lead to muscle damage (see above). Move slowly into poses with awareness to prevent injury.

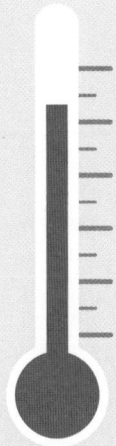

SPINAL CARE

Your spine supports your whole body and protects your spinal cord, so looking after it is crucial for your health and well-being. Yoga helps care for your spine by encouraging good posture and alignment, but you may need to make simple adjustments to prevent or manage specific conditions and diseases.

Leaning forward over a smartphone can increase the load on the neck by **5** times.

Q I SUFFER FROM NECK PAIN FROM TEXTING AND TYPING. CAN YOGA HELP?

Yes. While typing or texting, many of us allow our heads to fall forward. This increases the load on the neck and upper back muscles. With sustained strain, these muscles become inflamed and excessively tight, which can lead to pain. Yoga improves your awareness of how you hold your head throughout the day, which can prevent tech neck. To counteract its effects, you can also strengthen key muscles of proper neck posture by pressing your head back into your hands, a wall, or a car headrest for several breaths.

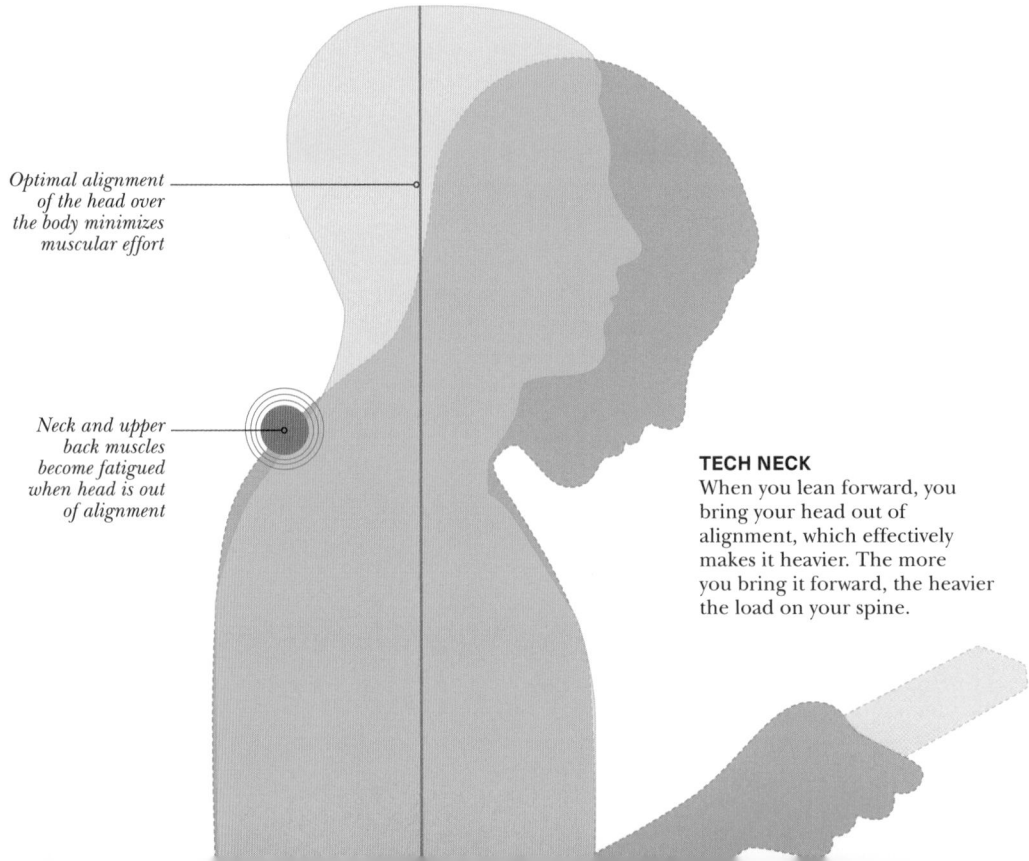

TECH NECK
When you lean forward, you bring your head out of alignment, which effectively makes it heavier. The more you bring it forward, the heavier the load on your spine.

MYTH-BUSTER

I have back problems, so I can't do yoga.

Research suggests that yoga is safe and effective for relieving chronic back pain. However, you may need to make adjustments to certain asanas or avoid some poses completely if you are managing a specific back condition (see pp.202–205). For many people, for example, touching the floor in Standing Forward Fold is not possible or comfortable, particularly for the lower back (the lumbar spine). However, you can still get the main benefits of the pose by bringing the floor closer to you, for example, by resting your hands on a block or on the base of a chair.

Q IS THERE AN ALTERNATIVE TO ROLLING UP FROM A STANDING FORWARD FOLD?

The cue of rolling up from a Standing Forward Fold "vertebra by vertebra" is likely to have come from the dance world. Biomechanically and functionally, this transition has more risks than benefits. For many, it feels good and improves coordination. However, rolling up could lead to or exacerbate a herniated disk or a spinal fracture for those with osteoporosis. This transition also doesn't prepare you properly for real-world activities, such as picking things up safely. To avoid potential injury, and to build the muscle memory of safe movement patterns, try coming out of a Standing Forward Fold in the following way:

1. **Create a wider base** of support with your toes turned out slightly. This reduces the pressure on your knees.
2. **Bring your hands to** your hips or the front of your thighs.
3. **Keeping a neutral spine,** engage your core and push up to standing, as with a hip hinge. This can particularly recruit your transversus abdominis, which may help alleviate lower back pain.

Did you know?

BACK PAIN IS ONE OF THE MOST COMMON **DISABLING AILMENTS** AND IS A LEADING CAUSE OF **LOST PRODUCTIVITY**. RESEARCH SHOWS YOGA NOT ONLY **REDUCES BACK PAIN** BY CLINICALLY SIGNIFICANT LEVELS, BUT ALSO **REDUCES** THE NUMBER OF SICK DAYS TAKEN.

> *Research suggests that yoga is **safe** and **effective** for relieving chronic back pain.*

LIFE STAGES

Not only is it possible and safe to practice yoga during different life stages—from childhood to pregnancy to old age—but research is now building to show that yoga and its accompanying practices, such as meditation, can have additional benefits at these times of life.

*Yoga emphasizes the **whole child,** so it fulfills an important need for **social and emotional** learning.*

There are over **900** "yoga in school" programs in North America.

Did you know?

RESEARCH SUGGESTS THAT YOGA COULD **IMPROVE** CORE SYMPTOMS OF ATTENTION DEFICIT HYPERACTIVITY DISORDER (ADHD) SUCH AS INATTENTION, HYPERACTIVITY, AND IMPULSIVITY, IN **CHILDREN** AND ADOLESCENTS, WHEN PRACTICED AS A **MIND-BODY** THERAPY AND FORM OF EXERCISE.

Q DOES YOGA BENEFIT CHILDREN?

A focus on academic performance can result in children sitting for long periods of time and can lead to other vital life skills being overlooked. As a holistic practice, yoga emphasizes the whole child, so it fulfills an important need for social and emotional learning (SEL). Yoga can affect all components of the social and emotional learning model, which include:

- **Self-awareness:** recognizing and identifying emotions
- **Self-management:** regulating emotions and managing stress
- **Social awareness:** acknowledging the perspectives of others
- **Relationship skills:** creating and maintaining a social network
- **Responsible decisions:** making conscious, positive decisions.

A review of research from Harvard and the Kripalu Center for Yoga and Health, for example, found that using yoga therapeutically was a viable way to improve the physical and mental health of children and adolescents. Meditation programs in schools have also shown strong improvements in resilience to stress and cognitive performance.

Q IS YOGA SAFE DURING PREGNANCY? DOES IT HAVE ANY BENEFITS?

Yes. Prenatal yoga classes are widely available and are often recommended by doctors. Research, including a 2015 study from the Alpert Medical School at Brown University, has suggested that prenatal yoga is not only safe for both the expectant mother and the baby (as measured by fetal heart rate), but it can also be beneficial for the fetus and mother while pregnant, throughout labor and delivery, and postpartum. Small studies have also suggested that, during pregnancy, yoga may have the positive effects shown below.

REDUCES

- pelvic pain and overall pregnancy discomfort
- signs of stress, depression, and anxiety
- postpartum depression.

IMPROVES

- optimism, empowerment, well-being, and social support
- birth weight (by reducing risk of preterm labor).

Q HOW DOES MEDITATION AFFECT MY BRAIN AS I AGE?

Many areas of your brain tend to shrink with age, but Harvard neuroscientist Sara Lazar, PhD, and her team have shown via MRI brain scans that 50-year-old meditators have key brain structures similar to that of 25-year-old nonmeditators. This suggests that meditation may slow or even prevent some of the natural degradation of brain tissue that happens with aging. This is thanks to neuroplasticity (see pp.26–27). While it is likely that other factors are involved, such as lifestyle and diet, it is feasible that meditation and the resulting mindset contribute significantly. Research also suggests that your brain can start to make these changes in eight weeks. A daily 30-minute mindfulness practice (including a body scan, yoga, and seated meditation practice) has been shown to change the brain in ways that result in better memory and improved problem-solving. A mindfulness questionnaire also showed that eight weeks' instruction and practice improved three key qualities that may contribute to a positive mindset as we age: observing internal and external environment; acting with awareness instead of reacting; and the non-judgment of inner experience.

> *Meditation may **slow** or even **prevent** some of the natural degradation of **brain tissue** that happens with aging.*

Continued →

*The yogic concept of **equanimity** teaches us to handle **change** and challenges with **grace**.*

Q HOW DOES YOGA AFFECT HOW WE AGE?

According to experts, yoga has the following benefits that support healthy aging:

- builds muscle strength to counteract the natural skeletal atrophy that happens with aging
- improves flexibility to prevent the loss of range of motion
- improves dynamic and static balance, reducing your risk of falling
- improves mental and physical agility so you can react faster.

Yoga improves strength, flexibility, balance, and agility in both physical and mental realms. Together, all of this may help improve your healthspan—the number of years you live without illness.

Did you know?

IN 2050, **ONE FIFTH** OF THE WORLD'S POPULATION WILL BE **AGE 60** OR OVER. THIS MAKES IT **MORE IMPORTANT** THAN EVER TO PREPARE OUR BODIES FOR **HEALTHY AGING** WITH PRACTICES SUCH AS YOGA.

MYTH-BUSTER

I'm too old to practice yoga.

Studies of yoga and older adults have shown improvements in flexibility, strength, balance, and functional activities, such as getting up from a chair. Yoga is also highly customizable. You can practice simple breathwork and adapt any asana, for example by using a chair, blocks, or blankets.

Q CAN YOGA HELP ME KEEP MY INDEPENDENCE?

Yes. Practicing yoga can help you maintain independence by preserving functional abilities so you can perform daily activities and continue doing what you love. Applying the philosophy of yoga to your life can also help you find purpose and meaning, which contributes to independence and well-being. For example, the yogic concept of equanimity (mental calmness) teaches us to handle change and challenges with grace.

Just **8** weeks of mindfulness practice could slow brain changes associated with aging.

Q HOW DOES YOGA AFFECT MY BONES AS I AGE?

Yoga can feasibly protect you from fractures associated with osteoporosis by preventing falls and strengthening the bone and muscles around common fracture sites, such as T9 and T10 (vertebrae at the base of the upper back), wrists, and the hip, particularly with asanas such as the below. Yoga also helps maintain the ability to safely get up and down from the floor so you can protect your joints and keep active.

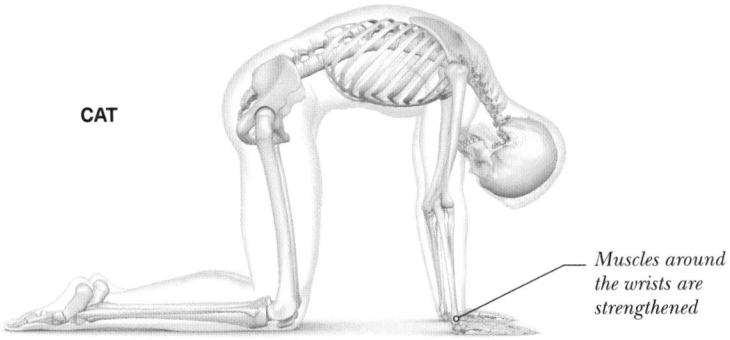

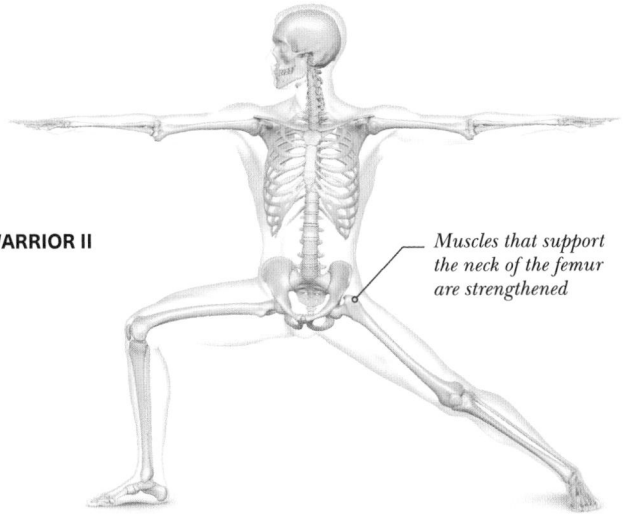

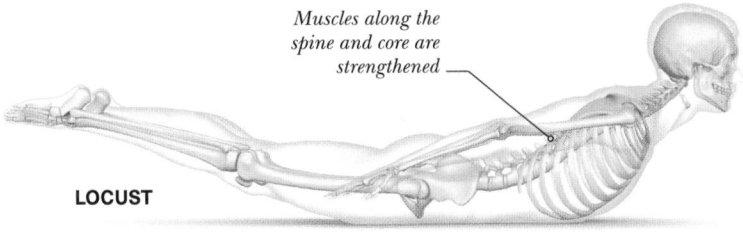

Yoga is becoming more popular with those age **65** and over.

MEDITATION

Yoga was traditionally seen as a way to prepare the body for meditation. Today, many yoga classes include meditative elements, such as mindfulness practices and chanting "om," as ways to relax the body and mind. Science shows that the benefits of these meditative practices also extend into your daily life.

*Simply **observe** your thoughts arising. It's like **watching clouds** pass by while remaining aware of the vast, clear blue sky.*

Q IS MINDFULNESS THE SAME AS MEDITATION? HOW IS IT PRACTICED?

Mindfulness is a simple and popular type of meditation that is often practiced in traditional seated poses. It also refers to a mindset that you can bring into the rest of your life. According to Jon Kabat-Zinn, PhD, founder of the well-researched Mindfulness-Based Stress Reduction (MBSR) program, mindfulness can be defined as deliberately paying attention to the present moment without judgment. It often involves observing breath, thoughts, sounds, or physical sensations, all of which are encouraged in yoga practice.

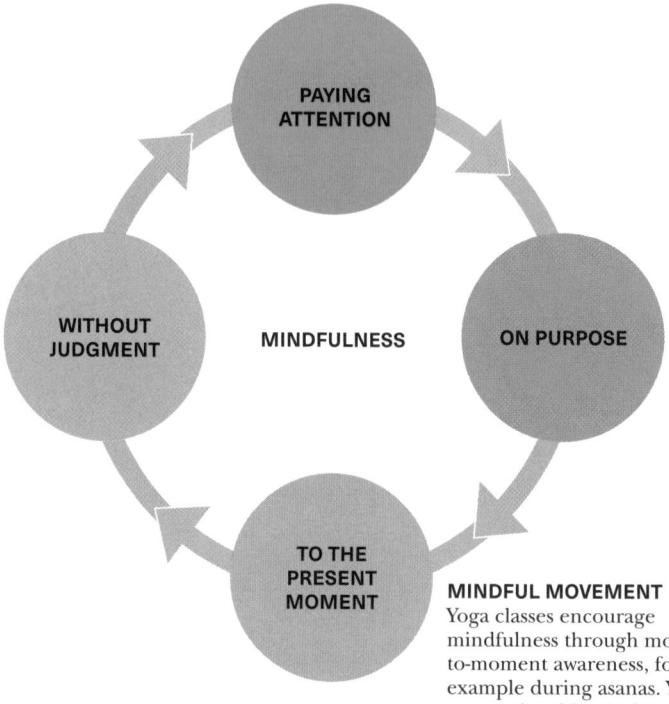

MINDFUL MOVEMENT
Yoga classes encourage mindfulness through moment-to-moment awareness, for example during asanas. You may notice this mindset spilling into your daily life, making even menial tasks, such as washing the dishes, mindful.

Q DOES MINDFULNESS REALLY WORK?

Anatomical MRI scans have shown changes in subjects' brain gray matter concentration after they participated in an eight-week MBSR program, suggesting that MBSR affects areas of the brain involved in learning and memory processes, emotion regulation, self-awareness, and new perspective-taking. Another study showed that even brief training in mindfulness reduced fatigue and anxiety, while longer training seems to particularly improve attention and focus.

Q HOW DO I SIT COMFORTABLY FOR MEDITATION?

Sitting on a cushion, folded blanket, pillow, or bolster helps you to elevate your hips at an angle and tilt your pelvis to neutral, bringing a natural inward (lordotic) curve to your lumbar spine. Another traditional meditation posture is Hero pose (Virasana), or kneeling. If you feel any pain in your knees, you can use blocks or a bolster to elevate your hips. If neither of these positions work for you, you can also sit in a chair to meditate. Try to sit tall and forward in the chair, without leaning back. It may also help to sit on a cushion as this will tilt your pelvis forward slightly. Place your feet directly under your knees or a little ahead of them. If meditating in any of these seated positions is too uncomfortable, meditate in Savasana (see p.186).

Did you know?

EXTREME FOCUS, SUCH AS WHILE PLAYING AN INSTRUMENT, HAS BEEN CONNECTED WITH **MEDITATION**. PSYCHOLOGISTS CALL THIS A "FLOW STATE." IN BOTH MEDITATION AND "FLOW," YOUR **BRAIN WAVES** CHANGE FROM BETA—ASSOCIATED WITH THINKING AND CONVERSING—TO MOSTLY ALPHA AND THETA—ASSOCIATED WITH RELAXATION AND **CREATIVE PROBLEM-SOLVING**.

Q MY MIND IS SO BUSY. DOES THIS MEAN I'M NO GOOD AT MEDITATING?

No. Many people think meditating is about "stopping" thoughts, but it isn't. In the form of meditation most commonly practiced today, you simply observe your thoughts arising. It's like watching clouds pass by while remaining aware of the vast, clear blue sky in which they float. When meditating, your only task is to gently remind yourself to come back to the present in a state of observing.

Q WHY DO WE CHANT "OM?"

An elongated exhale turns on the relaxation response. One small study also found that chanting "om" deactivates parts of the emotional brain related to fear, compared to chanting "sssss," as seen in fMRI brain imaging. This suggests that "om" may have benefits beyond the elongated exhale.

SAVASANA

Also known as Corpse pose, Savasana is the final relaxation pose often practiced for 5–15 minutes at the end of yoga classes. It is also used for meditative practices, such as yoga nidra. While there is still more research to be done, Savasana has been used clinically for its relaxation benefits.

> *Savasana activates the* ***parasympathetic nervous system*** *and all the* ***profound benefits*** *of this relaxation response.*

WHAT IS SAVASANA AND WHAT ARE ITS BENEFITS?

Savasana is practiced lying flat on your back with your legs and arms relaxed symmetrically, palms face up. It can also be used for relaxation and as a meditation posture if sitting is uncomfortable or you are not well. Its many benefits include:

- activating the parasympathetic nervous system (PNS) and all the profound benefits of this relaxation response, including lowering blood pressure and slowing heart rate
- teaching muscles to relax effectively
- increasing heart rate variability, representing resilience.

6
weekly sessions of yoga nidra improved stress, muscle tension, and self-care.

WHAT IS PROGRESSIVE MUSCLE RELAXATION?

Progressive muscle relaxation (PMR) involves squeezing and then releasing your muscles, often sequentially from head to toe, while in Savasana. This encourages neuromuscular connection, giving the body-mind clear examples of tension and release, which helps the body relax physically. Immediately after your muscle fibers contract, they have the capacity to lengthen or relax even more.

WHY IS THERE OFTEN A LONGER, GUIDED SAVASANA AT THE END OF CLASS?

This is a mindfulness practice called yoga nidra. Nidra means sleep, so think of it as a "yoga nap." A general intention of the practice is to remain alert to allow observation of the physiological effects of each stage of sleep. It is usually practiced in Savasana for 15–30 minutes and in small studies has shown promising results for improving sleep, decreasing depression, and managing chronic pain.

Q DOES YOGA NIDRA PROVIDE THE SAME BENEFITS AS SLEEP?

Although it does seem to offer many of the same rejuvenating benefits, yoga nidra does not replace sleep. However, it does produce brain wave patterns similar to those of sleep (see below).

Brain frequency chart

BRAINWAVE	SLEEP STAGE	YOGA NIDRA STAGE	LEVEL OF CONSCIOUSNESS	CHARACTERISTICS
GAMMA	Fully awake	Not nidra	Conscious	High alertness (not well understood)
BETA	Fully awake	Initially when transitioning into the practice	Conscious	Thinking and talking
ALPHA	First stage of sleep	During body scan and relaxation	Conscious—gateway to the subconscious	Relaxation
THETA	Next stage of sleep	May be reached, likely later in the practice	Subconscious	Creative problem solving
DELTA	Deep dreamless sleep	May be reached but there is little to no evidence of this	Unconscious	Rejuvenation and intuition

Q LYING FLAT ON MY BACK IS UNCOMFORTABLE. WHAT CAN I DO?

Many people find Savasana uncomfortable, particularly for their backs. Try using a support under your knees or lying in a constructive rest position—raising your knees and placing the soles of your feet on the floor—to relieve tension in your lower back. This can also help stop you from falling asleep.

MYTH-BUSTER

Savasana prevents lactic acid buildup.

No. Lactic acid, a waste product from muscle engagement, is broken down and removed by your liver within minutes after exertion. To reduce soreness, build the intensity of your asana practice over time. You can also rest the sore muscles by doing a more restorative class or working different muscle groups.

STRESS

Common sense tells us that yoga helps to manage stress by promoting relaxation and holistic well-being. But understanding the science behind the calming power of yoga can empower us to take a more proactive approach to a less stressed life, which enables us to achieve more positive health outcomes.

Q HOW DOES STRESS IMPACT MY HEALTH? AND HOW DOES YOGA HELP?

We tend to think of all stress as bad, but healthy levels of positive stress—eustress—can help us perform at our best. However, too much negative stress is associated with mental health imbalances and chronic pain, along with many of the industrial world's major killers, including heart disease, stroke, and cancer. It's important to recognize that stress doesn't necessarily cause these diseases. Research suggests that the greatest predictor of whether or not you will suffer from these diseases is not how much stress you experience, but how you deal with and think about stress. Those who have more negative emotions amid stress are more likely to experience negative health outcomes. Yoga is an effective tool for managing stress because it helps us regulate our emotional response to stressors by teaching us to become the observer of our thoughts and feelings, and through improving our mind-body connection (see right). As a result, yoga can lead to more positive health outcomes.

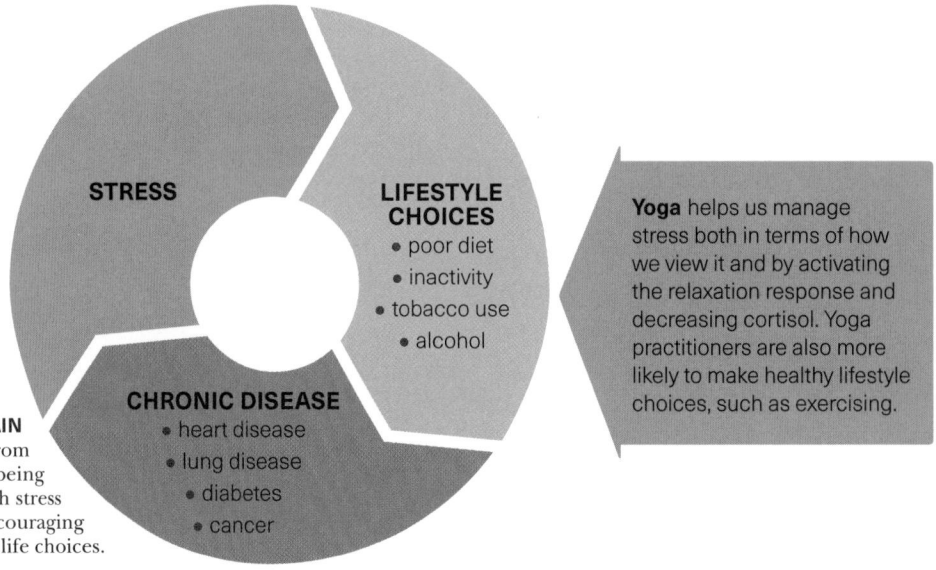

BREAKING THE CHAIN
Yoga can stop stress from affecting our physical well-being by helping us deal with stress more positively and by encouraging healthier life choices.

HOW DOES AN IMPROVED MIND-BODY CONNECTION HELP ME MANAGE STRESS?

Because yoga includes practices that engage both your mind and body, it helps you to regulate your system through both top-down and bottom-up pathways. Enhancing your mind-body and body-mind connections increases your ability to self-regulate and improves your resilience (your ability to bounce back after stress via homeostasis, the body's self-regulation of internal conditions). This all occurs partly due to the complex workings of your vagus nerve (see pp.190–91).

NEUROCOGNITIVE (MIND-BODY) PATHWAY

1 Meditation, mindful movement, and intentional living based on the philosophical teachings of yoga increase your attention

2 Increased attention regulates your nervous system and helps you maintain homeostasis more efficiently

NEUROPHYSIOLOGICAL (BODY-MIND) PATHWAY

1 Yoga practices such as asanas, mudras, and pranayama, give you internal body awareness (interoception)

2 This interoceptive information affects your autonomic nervous system (ANS), which changes your thoughts and neural pathways, building your brain and improving self-regulation

> *Enhancing your mind-body and body-mind* ***connection*** *increases your ability to* ***self-regulate*** *and improves your* ***resilience.***

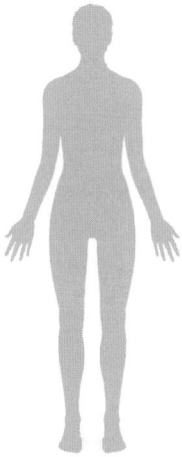

Did you know?

HANS SELYE COINED THE TERM **"STRESS"** IN 1936 TO DESCRIBE THE BODY'S **RESPONSE TO CHANGE**. HE IDENTIFIED TWO TYPES OF STRESS: **EUSTRESS**, WHICH IS BENEFICIAL STRESS, SUCH AS AN ENGAGING WORK PROJECT; AND **DISTRESS**, WHICH IS REAL OR IMAGINED STRESS THAT PUTS MORE PRESSURE ON YOUR SYSTEM.

Continued →

HOW DOES STRESS FIT INTO TRADITIONAL YOGIC PHILOSOPHY?

A 2018 article in *Frontiers in Human Neuroscience* aligns the ancient wisdom of yoga, particularly the gunas, with the role of the vagus nerve in our physiological response to stress and relaxation.

The vagus nerve is the only cranial nerve that leaves the head and neck area. It is mainly responsible for your relaxation response: telling your heart to slow down, improving your digestion, and encouraging social connection. Rather than an "on/off" switch, the stress and relaxation responses work more like a dial or dimmer knob. This allows adjustment to the perfect blend of electrical activity from each branch of your autonomic nervous system (ANS) for the situation (see below).

According to the Polyvagal Theory proposed by American neuroscientist Stephen Porges, PhD, the vagus nerve is split functionally in a way that helps us adjust effectively. Researchers have explained this neural adaptability in terms of the gunas. Gunas means "thread" or quality. The three gunas—sattvic, rajasic, and tamasic—are the three essential aspects of nature that weave together to create what we observe as the reality of the material world (also known as prakriti) with its ever-changing conditions. Each of the gunas is associated with a state of mind and certain characterstics that map against the different functions of the vagus nerve (see below).

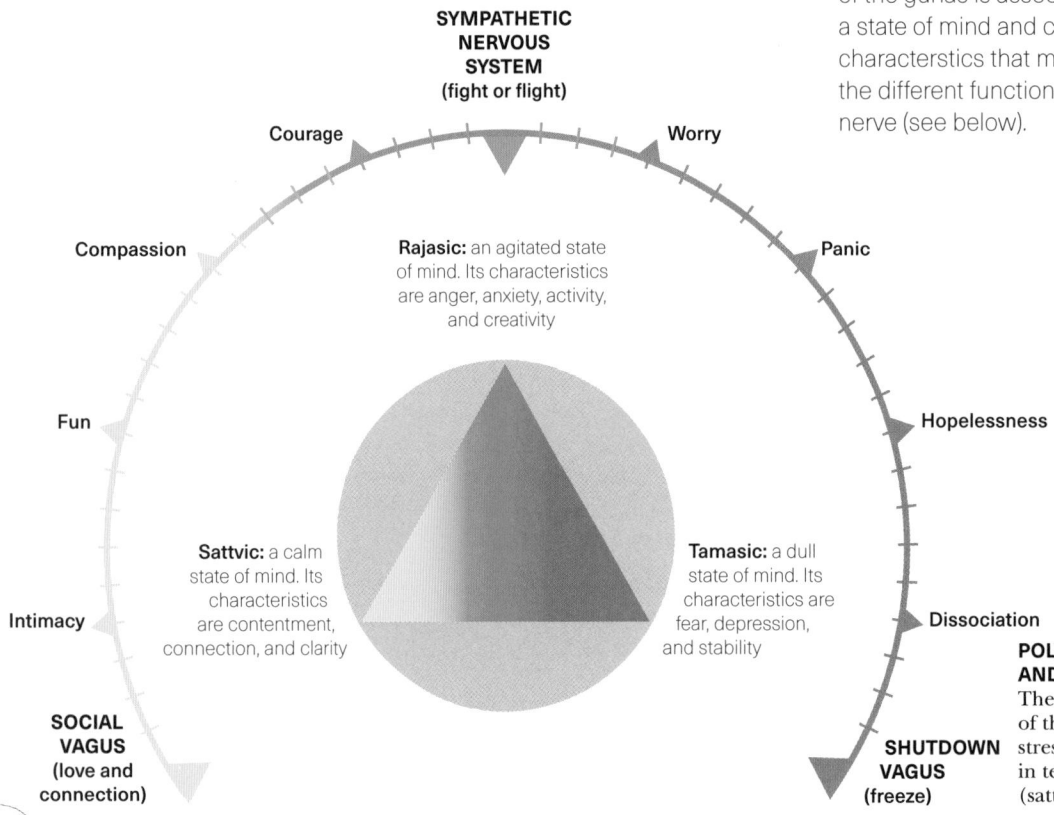

POLYVAGAL THEORY AND THE GUNAS
The adaptable responses of the vagus nerve to stress can be understood in terms of the gunas (sattvic, tamasic, rajasic).

Q SHOULD I BE CALM AND UNDER THE SOCIAL VAGUS OR SATTVIC STATE ALL THE TIME?

No. Yoga does teach our bodies to go into the sattvic state more often and more efficiently. This helps us to find balance in a world dominated by extremes of rajas and tamas. However, there is a misconception that yoga should make you perfectly calm all the time and that if that doesn't happen, you are bad at yoga. Constant calm is not the goal.

Your nervous system dynamically fluctuates, as do the gunas, throughout the day and over the course of your life to help you rise to the challenges your environment presents. Through yoga, you cultivate the capacity to be a non-judgmental observer of the constant changes so they don't control you. The ultimate ideal of this higher state of pure consciousness (also known as Purusha) is self-realization: finding meaning and connection amid the experience of inevitable stressors. Increased consciousness of any level represents increased resilience.

Q HOW CAN I RECOGNIZE AND REBALANCE THE "NEGATIVE" GUNAS?

The first step is to notice the signals of stress and the "negative" gunas in your body. These signals are different for everybody. Does your chest tighten or gut churn in an agitated, rajasic state? Do you tend to slouch or disassociate from sensations in a dull, tamasic state? Once you can recognize, identify, and observe your signals effectively, you can use the tools of yoga—including physical poses, mudras, breathwork, and meditation—to activate the relaxation response. Many yoga practices can be done discreetly throughout the day: no one will know that you are elongating your exhales to calm down, adjusting your posture, or taking fuller breaths for more energy.

Did you know?

80 PERCENT OF THE VAGUS NERVE'S FIBERS SEND **INFORMATION** FROM THE BODY TO THE BRAIN. THIS MAKES IT A KEY PATHWAY OF **INTEROCEPTION** (INTERNAL BODY AWARENESS) FROM YOUR **HEART AND GUT** TO YOUR BRAIN. YOGA CAN IMPROVE YOUR INTEROCEPTION AND **VAGAL FUNCTION**.

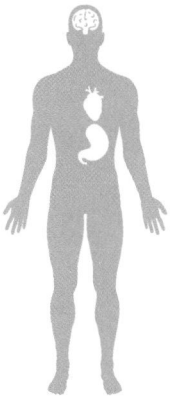

> *Through **yoga,** you cultivate the capacity to be a **nonjudgmental observer** of the constant changes so they don't control you.*

THE BRAIN AND MENTAL WELL-BEING

New research shows us that yoga changes how our brains work, for the better. Due to the neuroplasticity of our brains (see pp.26–27), these changes demonstrate the potential for yoga to become an effective adjunct to our medical and psychological care.

*Yoga gives us the **tools** to break thought and **emotional patterns** that no longer serve us.*

8 weeks of mindfulness meditation can help reduce fear-related activity in the brain.

Q WHAT DOES YOGA DO TO MY BRAIN?

When your brain becomes accustomed to a well-worn neural path, it becomes a habit, such as mindlessly looking at your phone when you're bored. New neural paths can form in the same way, and repeated activation makes these paths bigger and stronger.

By reinforcing positive behaviors, yoga gives us the tools to break thought and emotional patterns that no longer serve us. This allows the choice of healthier patterns when challenges arise, making yoga a powerful practice for our mental health and well-being.

Q HOW CAN YOGA HELP MY MENTAL WELL-BEING?

Sometimes we get stuck in a rajasic (the energy of agitation), reactionary pattern or a tamasic (the energy of resistance) slump. Yoga alone is not enough to manage a serious mental health concern, but it can be an effective supplement to your medical and psychological care because it affects how your brain responds to mental challenges.

In simplified terms, there are three structures within the brain:

- **The instinctual brain** (brain stem), which asks, "Am I safe?"
- **The emotional brain** (limbic system), which asks, "What am I feeling?"
- **The thinking brain** (frontal cortex), which asks, "What does this mean?"

Under trauma, depression, chronic stress, or anxiety, you may have an overactive emotional brain. Signals from your amygdala (the "fear center" of your emotional brain) encourage fight-or-flight responses from your instinctual brain, causing the stress response to override the relaxation response. When this happens often, your thinking brain is less effective at regulating. Yoga—including asanas, pranayama, and meditation—teaches the thinking brain to better regulate mood and emotional states amid stressors in life (see p.188).

Q WHAT EVIDENCE IS THERE TO SHOW THAT YOGA REALLY CHANGES OUR BRAINS?

A number of studies have focused on this. One 2015 review of two decades of research found that specific areas of the brain are commonly affected by the yoga-based practice of mindfulness, as shown in the diagram, right. It showed that key areas of the frontal cortex are strengthened, helping you effectively recognize and regulate emotions. Brain scans reported in a different research article, from 2018, also demonstrated that yoga asanas and meditation both reduced amygdala volume on the right-hand side of the brain, which is more associated with negative emotions and fear. In addition, researchers at Stanford University found that eight weeks of mindfulness meditation enabled people to better reduce fear-related amygdala activity. This seems to work largely as a result of participants being mindful of sensations and emotions instead of pushing them down.

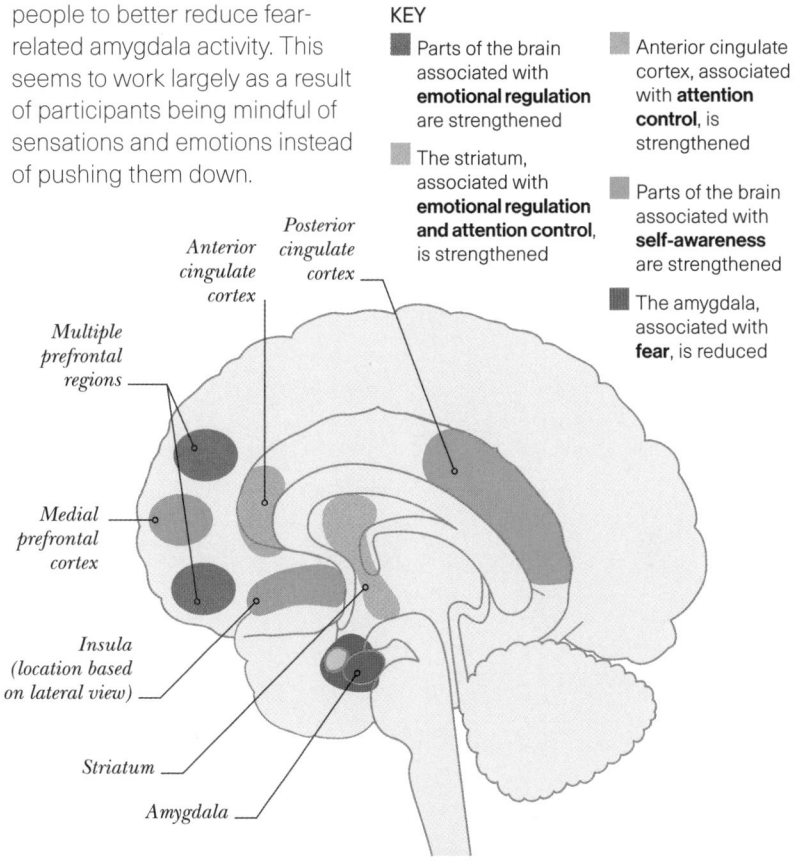

MIDSAGITTAL VIEW OF THE BRAIN

Did you know?

RESEARCHERS BELIEVE THAT **SOMATIC PRACTICES** (OR MOVEMENT PRACTICES THAT EMPHASIZE **PERCEPTION**, SUCH AS YOGA ASANAS) ARE USEFUL FOR HELPING PEOPLE TO **PROCESS TRAUMA** WITHOUT RETRIGGERING BECAUSE THEY HELP US **RELEASE TENSION** HELD IN THE BODY.

CHRONIC PAIN

Acute pain, such as an ankle sprain or a slip-and-fall injury, often needs rest to heal, which may mean avoiding or modifying yoga poses. But when pain becomes chronic, mind-body practices such as yoga have been shown to be well-suited to providing a safe supplement to medical care.

Q CAN YOGA REALLY HELP WHEN PAIN BECOMES CHRONIC?

Yes, there is evidence to show that it can help. Pain becomes chronic when it persists beyond the healing time of about three months. If you suffer from chronic pain, such as many cases of back pain or arthritis, you generally don't need to rest more because there may be little to no physical damage to heal. In fact, you probably need to move more because exercise tends to help relieve chronic pain, along with reducing associated stress.

Yoga practices have been shown to have an analgesic—or pain-relieving—effect. In one study of military veterans with lower back pain, opioid use declined in all subjects after a 12-week, twice weekly yoga program.

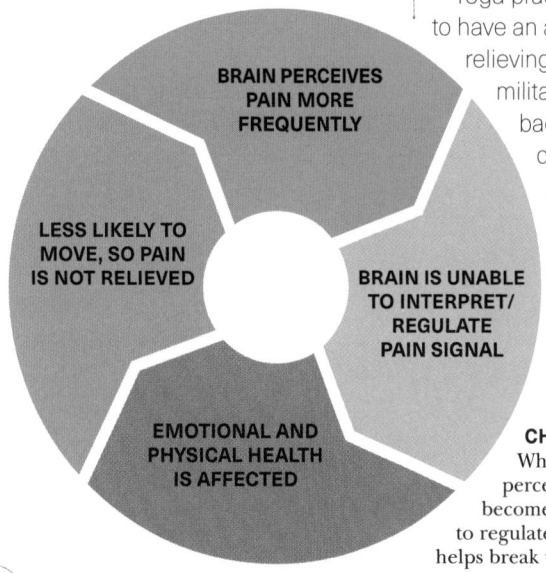

CHRONIC PAIN CYCLE
When the brain frequently perceives signals as pain, it becomes inured and is unable to regulate its response. Yoga helps break the cycle.

MYTH-BUSTER

Meditation relieves pain because of the placebo effect.

Recent research has shown that mindfulness meditation works better than a placebo in reducing pain. Subjects were exposed to a painful heat stimulus before and after receiving treatment: a placebo cream, "fake meditation," and traditional mindfulness meditation. The intensity and unpleasantness of the pain was evaluated psychophysically and by functional neuroimagery. The mindfulness group's pain intensity and unpleasantness reduced most significantly.

Four **20**-minute mindfulness classes can reduce pain's unpleasantness by **57%**

Q WILL ASANA PRACTICE REDUCE MY CHRONIC PAIN?

It depends. Some asanas can help reduce pain by stretching and strengthening the affected area(s). However, biomechanics is just one piece of the puzzle. At its most basic level, what your brain interprets as "pain" starts as a signal received from a receptor (nociceptor) in your body. Research has shown that the amount of pain perceived doesn't depend on the amount of tissue damage as seen in X-ray or MRI scans. This means that without the brain there is no pain; but this doesn't mean pain is imagined. Your brain builds your pain experience just as it constructs your reality and perspective. The level of pain you experience is based on your brain's interpretation of the level of danger those signals represent. So, as with chronic stress, chronic pain is partly a problem with regulation, often related to a faulty alarm system. Research shows that relaxing yoga asanas and practices, such as meditation and pranayama, can help regulate the pain response.

Q HOW MUCH DO I NEED TO MEDITATE TO REDUCE PAIN?

Research has shown that less than 1½ hours of meditation training may help alleviate pain and diminish pain-related brain changes. One study showed that just four 20-minute mindfulness classes reduced the unpleasantness of pain by 57 percent and the intensity of pain by 40 percent. It wasn't just the perception of the pain that changed: the brain's activity also measurably changed. The same study showed, via fMRI scans, that meditation reduced pain-related activation of the primary somatosensory cortex. Instead of a spike of activity in the area of the somatosensory cortex related to the location of the pain, researchers found that, while meditating, participants had more brain activity reflecting sensory awareness of the neck and throat, which represented the participants' mindfulness of their breathing.

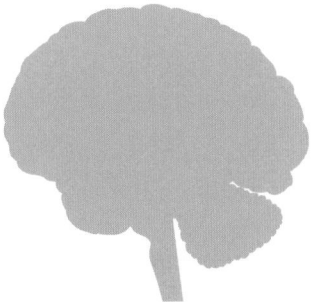

Did you know?

CHRONIC PAIN CAUSES **GRAY MATTER DETERIORATION**, BUT THE AREAS OF THE BRAIN THAT ARE DEGRADED BY CHRONIC PAIN ARE **RESTORED DURING MEDITATION** THROUGH THE INCREASE OF NEURAL CONNECTIONS IN THOSE AREAS.

> *Relaxing **yoga asanas** and practices, such as meditation and pranayama, can help regulate the **pain response***

YOGA THERAPY

Yoga therapy is a growing field in integrative healthcare, based on the mounting research into yoga's therapeutic benefits. With educational standards and a scope of practice beyond those of yoga teaching, yoga therapists use the tools of yoga to empower individuals toward well-being.

> *Lifestyle changes and **mindset** shifts from yoga can help people **move beyond** a disease focus to cultivate **human flourishing**.*

WHAT CAN I EXPECT FROM A YOGA THERAPY SESSION?

Yoga therapy sessions are often one-on-one or in small groups of people with similar conditions or life situations. Yoga therapists will always take your health history into account and, though they don't make medical diagnoses, they provide an individualized assessment of your health using tools including:

- **Observations of posture**, movement, and breath
- **Questions about** mood and lifestyle
- **Observations through the lens** of yogic subtle anatomy, such as the vayus and the five koshas.

The koshas are five layers, or "sheaths," that make up your self, similar to the layers of an onion. The koshas start with your physical well-being and end with bliss (see below). Yoga therapists consider all aspects of your well-being and how they interact in their recommendations. For example, arthritis in your physical body may be affecting your emotions and deeper connection to bliss, while your emotions may be exacerbating the pain. From these observations and considerations, yoga therapists create a personalized plan of care for each client using tools such as poses, breathwork, meditations, and lifestyle suggestions.

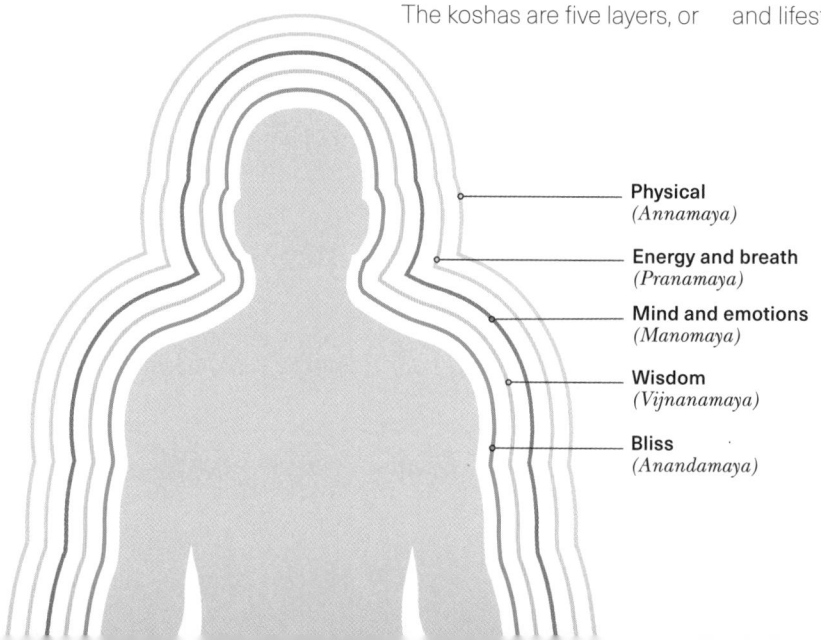

THE FIVE KOSHAS
Each of these five layers or "sheaths" must be looked after if we are to live a healthy, balanced life.

HOW DOES YOGA THERAPY WORK?

Yoga has profound therapeutic potential because it acts on what researchers call a biopsychosocial-spiritual model (see right). Much of yoga research is done through this lens, showing therapeutic yoga's promise for multidimensional conditions such as chronic pain, trauma, and anxiety. Just as with the koshas (see left), the core of yoga therapy is that each aspect of self interacts with the others. To address this, yoga therapy applies a balance of research evidence, client values, and clinician experience.

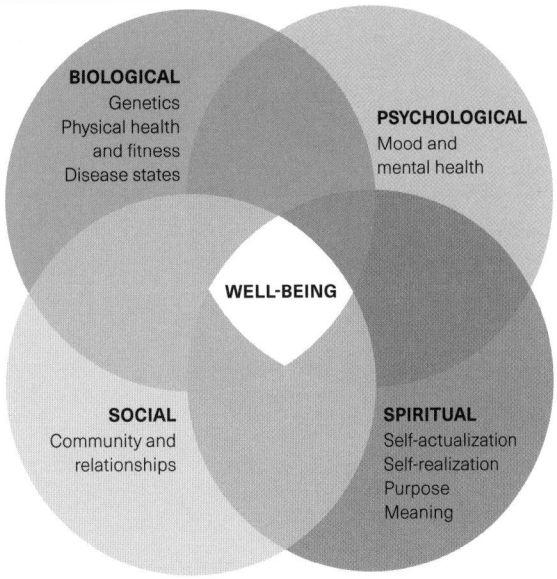

BIOPSYCHOSOCIAL-SPIRITUAL MODEL

DO WE HAVE SCIENTIFIC EVIDENCE TO SUPPORT THE BENEFITS OF YOGA THERAPY?

Yes. The vast majority of scientific research into yoga is focused on understanding its therapeutic benefits, particularly for one of the world's most pressing healthcare issues: lifestyle-based chronic diseases (see pp.178–79, pp.188–91, and pp.194–95). The quality of this research is also improving, although some of the therapeutic benefits of yoga may never be fully understood through Western scientific inquiry. The yoga therapy profession is now growing partly because of the need for highly trained individuals who can work with specialized populations, for example veterans and those in cancer care.

HOW DOES YOGA THERAPY COMPARE TO OTHER HEALTHCARE PRACTICES?

Most healthcare systems work on the level of pathogenesis, which is a disease-based model of healthcare. The primary aim in this model is managing symptoms and "fixing" parts and pieces of the system. Although yoga therapy often is successful at managing symptoms, such as by providing pain relief, it also works on a level of salutogenesis, which is a health-based model. Rather than focusing on the disease to be cured or a problem to be fixed, salutogenesis focuses on creating well-being. Lifestyle changes and mindset shifts from yoga, therefore, can help people move beyond a disease focus to cultivate human flourishing.

TRANSFORMATION

Exercise is the most common reason why people first come to yoga. However, the spiritual side of yoga often becomes more important for those who continue to practice. With advances in technology including neuroimaging, researchers are now exploring yoga's potentially transformative spiritual effects.

> *Neuroscientists are now studying the **brain** during **spiritual** states.*

Q WHAT ARE THE SPIRITUAL STATES THAT ANCIENT YOGIS SPOKE OF?

The "eight limbs" of yoga are outlined in an ancient text called the "Yoga Sutras." The first four limbs concern how we live in the external world, and are intended to prepare your body and mind for the second four, which concern our internal world or consciousness (see below).

Astronauts undergo a similar process to the eight limbs of yoga: from an ethical code to intensive physical exercises to prepare the body and mind. When in space, "Earth gazing" is reportedly so captivating that astronauts spend hours just staring at the planet. This can be seen as similar to yogic concentration (dharana) exercises, such as staring at the flame of a candle to improve concentration and eventually evoke higher states of consciousness.

According to a 2016 paper called "The Overview Effect: Awe and Self-Transcendent Experience in Space Flight," astronauts return to Earth with a new perspective and sense of purpose. The founder of Phoenix Rising Yoga Therapy, Michael Lee, believes we can experience the same transformation on Earth by exploring the last four limbs of yoga.

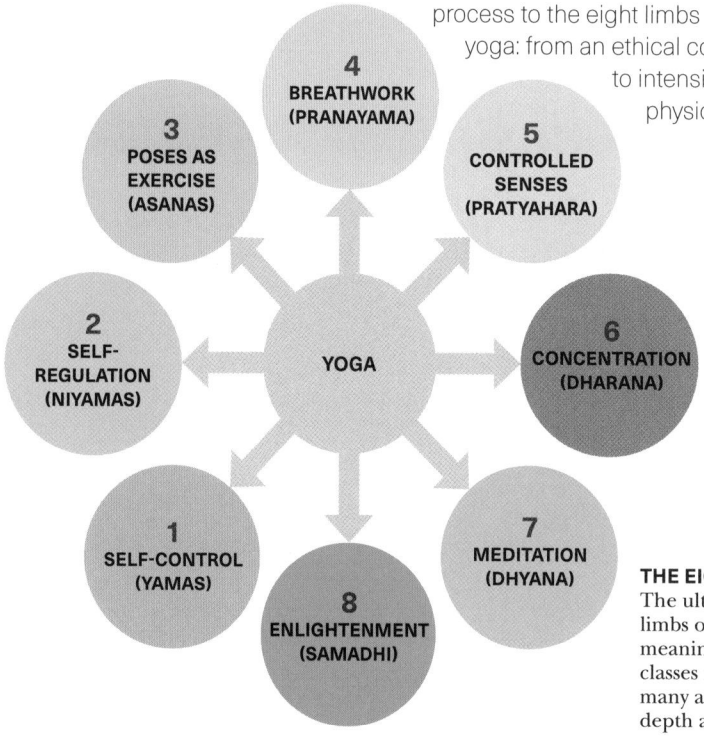

THE EIGHT LIMBS OF YOGA
The ultimate aim of the eight limbs of yoga is to help us live a meaningful life. Not all modern classes incorporate them, but many at least allude to this depth and potential.

HOW CAN WE STUDY THE EFFECTS OF SPIRITUALITY?

Neuroscientists are now studying the brain during spiritual states, with fascinating findings. American neuroscientist Dr. Andrew Newberg from the Marcus Institute of Integrative Health, for example, uses neuroimaging to understand higher spiritual states, including spiritually based meditative states such as deep prayer practices, drug-induced spiritual experiences, and Samadhi (see below).

Four common brain patterns in spiritual experiences

Dr. Newberg has compared the brain at rest and during transcendent spiritual experiences, including Samadhi, to identify specific brain activity patterns associated with spirituality.

AT REST	DURING SPIRITUAL EXPERIENCE		INTERPRETATION
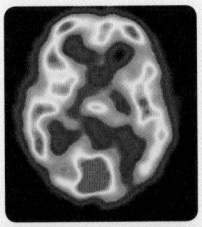	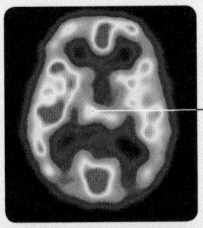	*Increased activity in the limbic system*	**INTENSITY** Increased activity in the limbic system may account for the intense emotional states that people often feel during spiritual experiences. This increase is also likely to make such experiences memorable and life-changing.
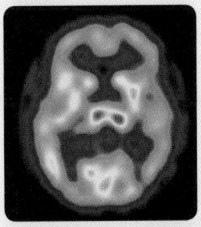	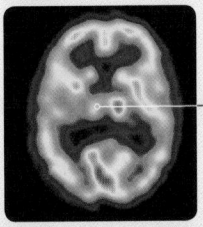	*Decreased activity in the left thalamus*	**CLARITY** The thalamus is a relay center that helps us integrate sensory information to construct our sense of reality. Decreased activity here may result in a sense of increased clarity.
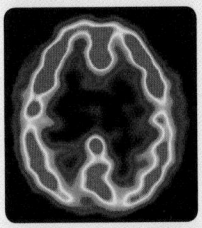	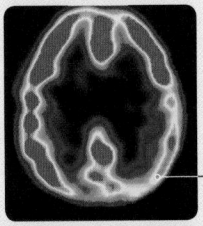	*Decreased activity in the posterior parietal lobe*	**UNITY** The posterior parietal lobe is in charge of spatial orientation. A decrease in activity here may reduce the feeling of being physically separate from what is around us, creating a sense of unity and a lack of boundaries.
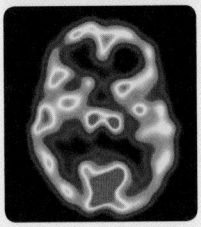	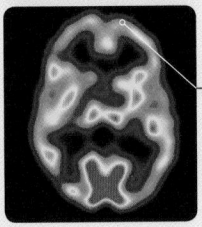	*Decreased activity in the frontal cortex*	**SURRENDERING OF SELF** Though many meditation practices increase activity in the frontal cortex due to the increase in concentration and regulation, spiritual states such as Samadhi may turn off the frontal cortex, the seat of the will, leading to a sense of surrendering to what is.

ON THE FRONTIERS OF SCIENCE

Scientists predict that we only observe and understand 4 percent of the universe in which we live. Similarly, we are only on the frontiers of exploration when it comes to the science of the human brain, mind, and consciousness, which gets to the heart of yoga's capacity for transformation.

> *Bear in mind that* ***extraordinary claims*** *require* ***extraordinary evidence.***

Q HOW DO I KNOW IF A YOGA STUDY IS RELIABLE?

Not all yoga research is created equally, so it is good to approach it critically. Some factors to consider are:

- **What kind of study is it?** The hierarchy of scientific evidence (see below) gives a good idea of how reliable different kinds of studies are. Evidence that is lower on the pyramid is still valuable, but the higher up, the more reliable. There are increasing systematic reviews and meta-analyses on key topics in yoga, including mental health, heart disease, chronic pain, and safety.
- **How large is the sample size?** From case reports of one to randomized controlled trials (RCTs) of 228 people, yoga studies tend to be relatively small, especially compared to pharmaceutical RCTs with up to tens of thousands of participants.
- **Is there a control group? If so, what?** Many yoga studies incorporate a "usual care" control group. A few higher-quality ones have an active control, such as comparing yoga to exercise or talk therapy.
- **What is the conclusion?** Bear in mind that extraordinary claims require extraordinary evidence. This is why many yoga researchers use phrases such as "yoga may improve" or "this suggests that yoga helps." As interest in yoga research increases, scientists will keep questioning results.

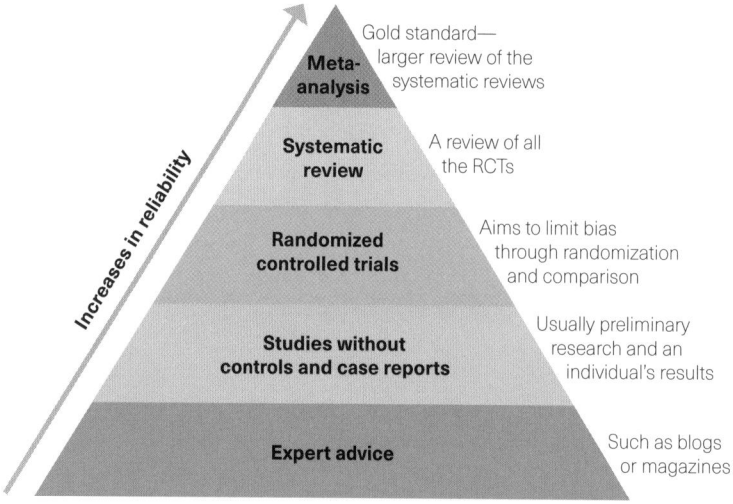

HIERARCHY OF EVIDENCE
You can use this pyramid to gauge the reliability of different types of scientific evidence.

Q IS THERE SCIENTIFIC EVIDENCE TO SUPPORT YOGIC CONCEPTS, SUCH AS PRANA AND THE CHAKRAS?

Yoga research tends to focus on specific health conditions and practical benefits, rather than subtle energetics, as prana and chakras represent a way of knowing that doesn't necessarily translate directly to a straightforward analysis of biology. Some people, for example, claim that the flow of prana is in alignment with the nerves, and the chakras with the glands, but there is no scientific evidence to support this. It may be that before dissection showed us where these structures were, yogis felt them working in their bodies. It is also possible that we are still limited by our current instruments and will one day have the tools to locate and measure prana.

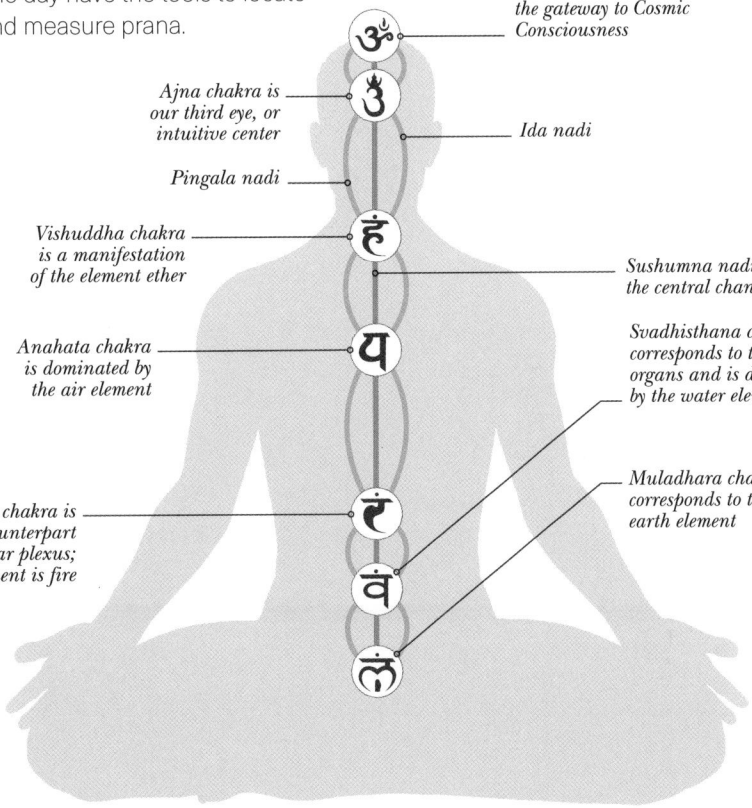

CHAKRAS AND NADIS
Nadis are energy channels along which the seven energy centers, known as chakras, are located.

Q HOW MUCH RESEARCH IS THERE INTO YOGA?

Research into yoga is relatively limited, albeit on the rise, and fast. A bibliometric review of the relevant research from 1967 to 2013, for example, showed an exponential growth in the number of studies conducted from fewer than 25 publications from 1967–1973 to well over 225 publications in 2009–2013, correlating with the rise of popularity of yoga.

The research also identifies that the top four areas of research are:

- mental health disorders
- cardiovascular disease
- respiratory diseases
- musculoskeletal disorders.

CAUTIONS

Just as the Hippocratic Oath states "first do no harm," the first principle of yoga is *ahimsa*, which translates to nonharm. To avoid harm, it is important to know your body and adapt or modify poses and practices based on your needs and health conditions. Everyone is different, so use these pages as a general guide.

Injuries in yoga do happen, as they do in all types of physical activity, from walking down the stairs to lifting weights at the gym. A meta-analysis of randomized controlled trials, however, found that yoga is as safe as other types of recommended exercise. In fact, yoga may be safer than many forms of exercise because it often incorporates slow transitions, present-moment awareness, and an emphasis on nonharm.

That said, if you believe that yoga practices are powerful enough to profoundly benefit you, you must also acknowledge that yoga has the power to harm, and you must treat it with that level of respect. To prevent injury, therefore, practice the first two limbs of yoga—the Yamas and Niyamas—both in yoga class and in life (see p.205). It is also advisable to bear in mind the following guidelines:

- **We all have** differently shaped bones and bodies, so poses will look different when practiced by different people. Some postures may not be accessible to you without modifications
- **Allow recovery** after strains, sprains, tears, breaks/fractures, surgery, or wounds. After surgery, ask your surgeon for guidance
- **The point of yoga** is not to be able to perform an asana perfectly, or to do any particular technique or pose. Enjoy the journey of self-exploration!
- **Avoid anything that** causes pain or increases existing pain
- **Be careful of** sharp sensations inside the body or sharp, shooting sensations down the limbs
- **Avoid anything that causes numbness** in the limbs.

CONDITIONS

The following pages outline any cautions and considerations for specific health conditions that you should bear in mind when practicing yoga, as general guidance. However, you should always ask your professional medical team what is right for you. If in doubt, work with a qualified yoga professional, such as a yoga therapist.

ACID REFLUX/GERD/HEARTBURN

Be careful of or avoid any full or partial inversion where the head goes below the heart, and fast breathing (kapalabhati).

ANKYLOSING SPONDYLITIS

Be careful of spinal flexion and move slowly into gentle spinal extension.

ANXIETY/TENDENCY TOWARD PANIC ATTACKS

Be careful of inversions, backbends, fast breathing (kapalabhati), or holding the breath (kumbhaka) during symptoms.

ARTHRITIS (including osteoarthritis and rheumatoid arthritis, and other conditions that involve joint inflammation)

For osteoarthritis and rheumatoid arthritis, avoid anything that increases joint pain, and focus on modifying poses for comfort, strengthening, and learning to meditate to cope with pain; for rheumatoid arthritis, avoid hot yoga and overheating.

ASTHMA

Be careful when practicing backbends, holding the breath (kumbhaka), and fast breathing (kapalabhati); avoid intense back bending during symptoms.

BURSITIS AND TENDONITIS

Avoid anything that increases pain or swelling; rest the affected area during acute stages.

CARPAL TUNNEL SYNDROME

Be careful of or avoid arm balances or weight bearing while wrists are extended (e.g. Plank or Crow pose), especially if it increases numbness; consider resting your forearms on the floor or blocks, or try using a wedge.

DEGENERATIVE DISK DISEASE
Practice spinal flexion and spinal rotation gently; avoid or be careful during headstands, shoulderstands, or anything that puts pressure on the neck.

DIABETES
For type 1, avoid anything that puts pressure on your insulin pump; for type 1 and 2, eat before class if needed, and rest if lightheaded.

DISK HERNIATION (SLIPPED, BULGING, PROTRUDING)
Be careful of unsupported spinal flexion, such as a Standing or Seated Forward Fold, and spinal rotation; focus on lengthening the spine before gently entering a pose, and consider keeping the spine neutral and bending at the hips into a Forward Fold—Child's or Cat pose may be safer forms of spinal flexion; be careful during headstands, shoulderstands, or anything that puts pressure on the neck.

EAR INFECTION
Be careful with inversions and in balancing poses.

EYE CONDITIONS THAT INCREASE PRESSURE (such as glaucoma, detached retina, diabetic retinopathy, or recent cataract surgery)
Be careful with or avoid any pose in which the head goes below the heart, holding the breath (kumbhaka), and fast breathing (kapalabhati); seek the advice of your ophthalmologist if you are unsure.

FIBROMYALGIA
Consider restorative yoga and yoga nidra; use lots of props and let your teacher know if you prefer not to be touched in a hands-on assist.

FROZEN SHOULDER (ADHESIVE CAPSULITIS)
Move slowly into shoulder stretches and gradually increase the stretch over time.

HEART CONDITIONS
Be careful when performing inversions, holding the breath (kumbhaka), and fast breathing (kapalabhati); you should also seek the advice of your cardiologist.

HIGH BLOOD PRESSURE (HYPERTENSION)
Be careful with any pose where the head goes below the heart, holding the breath (kumbhaka), and fast breathing (kapalabhati); if your blood pressure is not currently regulated, avoid full inversions, intense practice, and hot yoga completely.

HIP REPLACEMENT
Follow these precautions 6–8 weeks after surgery, and with the advice of your doctor. In the anterior approach, be careful of or avoid extension (as in the lifted leg in Warrior III); in the posterior approach, be careful of or avoid hip flexion past 90 degrees, internal rotation, and crossing the midline (crossing legs); after proper healing, you are likely to be able to perform all of these movements, but move slowly into the poses and ask your doctor for advice.

HYPERMOBILITY
Avoid any extreme movement or hyperextension of joints; focus on strengthening.

KNEE LIGAMENT INJURY (ACL, PCL, LCL, MCL)
Be careful with poses that involve rotation (e.g. Triangle pose and Warrior II); for ACL, avoid deep knee flexion and for PCL, be careful of hyperextension/locking your knees; for both, be careful of or avoid jumping into poses.

Continued →

CAUTIONS *continued*

KNEE MENISCUS TEAR/INJURY
Be careful of or avoid deep knee flexion, especially if weight bearing.

KNEE REPLACEMENT
Avoid extreme knee flexion; cushion the knee with blankets or padding when in kneeling poses.

LOW BLOOD PRESSURE (HYPOTENSION)
Move slowly out of any pose where the head goes below the heart; pause a few moments in a restful pose, such as Child's pose, after full inversions to prevent dizziness; move slowly when rising from the floor.

MIGRAINE
Be careful when performing full inversions; try practicing in a room with the lights dimmed.

MULTIPLE SCLEROSIS
Be careful of intense practices that make you feel overheated; avoid hot yoga.

OBESITY
Be careful of unsupported spinal flexion and full inversions, such as headstands, shoulderstands, or anything that puts your weight on your neck.

OSTEOPOROSIS/OSTEOPENIA
For spinal areas, talk to your doctor, as what you can do will depend on the severity of your condition. However, general guidelines are to be careful of unsupported spinal flexion and spinal rotation; move slowly and focus on elongating the spine before coming into twists, and consider flexing at the hips and try keeping the spine neutral in many Forward Folds to avoid the risks of spinal flexion (Child's or Cat pose may be safer forms of spinal flexion); avoid or take extreme caution during headstands, shoulderstands, or anything that puts pressure on the neck; take particular caution to move slowly and gently in movements that combine spinal flexion and rotation such as Triangle pose; take care in transitioning poses and balancing poses to reduce the risk of falling; for nonspinal areas, such as hips or wrists, move slowly into poses and focus on mindfully strengthening muscles around the affected areas.

PARKINSON'S DISEASE
Be careful of inversions and balancing poses; try holding onto the wall or a chair to prevent falls; use props as needed.

PLANTAR FASCIITIS
Avoid or be careful jumping into poses, or any movement that exacerbates symptoms; stretch the feet and legs slowly and mindfully.

PREGNANCY
Be careful of full inversions, especially if you don't already have an inversion practice; be careful of or avoid anything that puts pressure on the abdomen (e.g. Locust pose or extreme abdominal engagement); avoid extreme abdominal stretching (e.g. Wheel pose); don't stay too long lying on your back in later stages of pregnancy if uncomfortable, and consider lying on your side with a pillow between your legs, or propping yourself up to lie at an angle.

ROTATOR CUFF (TEAR, TENDONITIS, INSTABILITY)
Be careful with any shoulder stretches; avoid Low Plank pose (Chaturanga), particularly in acute stages; focus on strengthening over stretching, e.g. consider holding a forearm version of Plank or Downward Dog on the floor or wall.

SACROILIAC (SI) DYSFUNCTION/PAIN
Avoid extreme twists and be careful in wide-legged postures (e.g. Triangle pose). Being in asymmetric poses, such as Warrior poses or Triangle pose, for a prolonged period on one side may be uncomfortable; if so, consider switching sides more often.

SCIATICA
Be careful of anything that increases numbness; if the condition is due to a tight piriformis, consider modified versions of Pigeon pose, e.g. figure 4 on your back (see p.82).

SCOLIOSIS
Avoid anything that causes numbness; consider strengthening your back muscles by practicing Side Plank pose and gently stretching in the opposite direction of the curvature.

SHOULDER DISLOCATION, HISTORY OF
Avoid any extreme shoulder flexion, especially while weight bearing, such as in Wheel pose; consider focusing your practice on strengthening.

SINUSITIS
Be careful of inversions and spinal extensions; you may find the alternate nostril breathing technique difficult.

SPINAL STENOSIS
Be careful of spinal extension.

SPONDYLOLISTHESIS
Ask your doctor what to avoid in your individual case. However, general guidance is: be careful of spinal extension and spinal rotation; avoid deep twisting, moderate or deep backbends, and jumping into poses.

STROKE, HISTORY OR RISK OF
Be wary of inversions and extreme cervical extension; avoid anything that puts pressure on the neck.

VERTIGO/DIZZINESS
See Low Blood Pressure.

Approaching yoga with respect
The Yamas and Niyamas are the ethical guidelines for a yogic lifestyle. Traditionally, a guru would require that a practitioner lives these principles before learning any asana, to prevent ego and injury.

YAMAS (SELF-CONTROL)
- **Ahimsa (nonharm):** don't do anything that hurts or increases current pain
- **Satya (truthfulness):** be truthful with yourself about what your body can do today
- **Asteya (nonstealing/abundance):** focus on the things you can do instead of what you cannot do
- **Brahmacharya (moderation):** practice everything in moderation to regulate your energy
- **Aparigraha (nonpossession):** there is no need to grasp for a body you used to have, or to be jealous of the person practicing next to you.

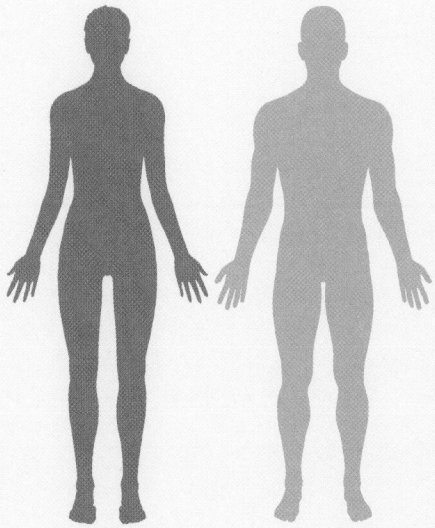

NIYAMAS (SELF-REGULATION)
- **Saucha (cleanliness):** organize your props and practice area to prevent falls or distraction
- **Santosha (contentment):** find contentment with where you are physically and mentally today
- **Tapas (self-discipline):** balance your burning desire to improve with the practice of nonharm
- **Svadhyaya (self-study):** observe your breath and energy today and adjust your practice to respect that
- **Ishvara Pranidhana (surrendering/accepting):** allow a sense of surrendering to what is in the present moment, changing what you can (for example, using a prop for comfort in a pose), but accepting what you cannot change. Just be.

GLOSSARY

Acute When symptoms come on rapidly; acute pain generally lasts for less than 3–6 months.

Alignment In yoga, the way a pose is instructed with the intention of encouraging optimal function and preventing injury; although there are general principles, proper alignment may vary from person to person and day to day, and based on the intention behind the pose.

Anatomy Study of the structure of the body, including the naming of parts.

Antigen Invader that the body's immune system fights with antibodies and white blood cells.

Arthritis Group of joint conditions that involves joint inflammation and/or damage; osteoarthritis is the most common type and involves damage to the cartilage of the joint due to wear and tear.

Asana Yoga pose or posture.

Bile Substance that helps break down fats in digestion.

Cartilage Firm but flexible connective tissue; includes hyaline (glasslike, in synovial joints to reduce friction), fibrocartilage (firm cushioning, in intervertebral disks for cushioning), and elastic (stretchy, in nose and ears for elasticity).

Central nervous system (CNS) The brain and spinal cord; controls the body and perceives the world.

Cerebral cortex Outer shell of the cerebrum.

Cerebrum Largest part of the brain; contains the cerebral cortex and some internal structures such as the hippocampus.

Cervical spine Seven vertebrae of the neck.

Chromosomes Threadlike molecules made of DNA and proteins; humans generally have 23 pairs.

Chronic Long-lasting symptoms, disease, and/or pain; chronic pain generally persists for longer than 3–6 months.

Collagen Key component in many connective tissues; has good tensile strength, allowing it to resist tension or pull.

Concentric contraction Muscle shortening in response to a load, as in lifting a weight in a biceps curl.

Connective tissue Forms connections in your body; subtypes include cartilage, bone, blood, lymph, adipose (fat), and elastic tissue (such as in the ears and nose), along with fibrous connective tissue.

Control group The research group that doesn't receive the intervention being studied; may recieve nothing, or an active control, to act as a comparison.

Deep Further inward from the surface; for instance, your muscles are deep to your skin.

Diaphragm Usually refers to the respiratory diaphragm, which is the primary muscle used in a relaxed breath; there are also the vocal/thoracic outlet diaphragm and urogenital/pelvic floor diaphragm.

DNA Deoxyribonucleic acid; carries hereditary information in genes; within chromosomes.

Eccentric contraction Muscle lengthening in response to a load, as in lowering a weight in a biceps curl.

Engaging When a muscle is contracting; "Engaging while stretching" is used in this book to describe contraction while a muscle is in a neutral or lengthening position, as in an eccentric contraction, but held steady.

Epithelial tissue Forms coverings in your body, such as the superficial layer of skin.

Fascia Fibrous connective tissue that surrounds muscles and other organs.

Fibrous connective tissue Contains either a parallel or irregular pattern of collagen fibers; includes dense regular connective tissue, of tendons and ligaments, and dense irregular connective tissue, of fascia and synovial joint capsules.

fMRI Functional magnetic resonance imaging; machine that measures blood flow in the brain to reflect neural activity.

Gray matter Tissue in the central nervous system that contains mostly cell bodies, dendrites, and synapses (as compared to white matter which contains mostly axons and is white due to myelin).

Heart rate variability (HRV) Measure of the variation between heart beats within a specific increment of time; may be an indicator of cardiorespiratory and stress resilience.

Hip points Colloquial name for the two bony protrusions on the front of the pelvis, called the anterior superior iliac spines.

Homeostasis State of dynamic equilibrium maintained in the human body to support life.

Hot yoga Yoga classes in rooms heated to 92–105°F (33–40.5°C).

Hyperextension Extreme extension of a joint, often past normal range.

Hypermobile Extremely flexible beyond normal limits.

Hypertension High blood pressure.

Inflammation Indication that the body is fighting something locally or systemically; symptoms can include redness, swelling, heat, and pain.

Interoception Sensory body awareness of your internal environment, including of the digestive organs, heart, and muscles.

Intervertebral disk Disks, made mostly of fibrocartilage, which absorb shock between vertebrae and allow movement.

Inversion Poses, like Headstand, where the body is "upside down;" partial inversions include any pose where the head is below the heart.

Isometric contraction Muscle engagement where the muscle stays the same length, such as pushing into a wall or the floor.

Isotonic contraction Muscle engagement where the muscle changes length; can either be eccentric or concentric.
Kinesiology Study of body movement.
Kumbhaka Pranayama practice of breath retention.
Kyphosis Convex curves of the spine, found naturally in the thoracic spine and sacrum; the term can also describe an excessive amount of this convex curve, as in a dowager's hump.
Ligament Connects bone to bone; made of dense regular connective tissue proper, which has parallel collagen fibers.
Lordosis Concave curves of the spine, found naturally in the lumbar and cervical spine; the term can also describe an excessive amount of this concave curve.
Lumbar spine Five vertebrae of the low back.
Lymph Fluid filled with white blood cells to fight invaders; collected from interstitial fluid, it drains back into the heart after being filtered in lymph nodes.
Meditation Concentration or mental focus exercise; includes mindfulness, mantra, loving-kindness, transcendental meditation (TM), and others; Dhyana, in Sanskrit.
Meta-analysis Systematic assessment of previous research in a specific area to derive broad conclusions; the gold standard of review articles.
Mindfulness Paying attention on purpose to the present moment, without judgment (as defined by researcher Jon Kabat Zinn, PhD).
Muscle tissue Contractile tissue; the three types are skeletal, smooth, and cardiac muscle.
Nadis According to Indian medicine and Hindu philosophy, these are channels for prana to flow.
Nerve Bundle of axons of neurons in the peripheral nervous system; conductive tissue that acts like wires through the body, carrying signals to and from the central nervous system. Includes cranial nerves and spinal nerves; a bundle of axons in the central nervous system is called a tract.
Nervous tissue Conductive tissue made of neurons and helper cells.
Neuron Nerve cell; carries electrical signals.
Neuroplasticity Ability of the brain to create neural connections.
Neutral spine Position of optimal load distribution for the spine; maintains the natural curves of the cervical (lordosis), thoracic (kyphosis), and lumbar (lordosis) segments of the spine.
Neutral pelvis Position of the pelvis that best supports the inward curve of the lumbar spine. No excessive anterior or posterior pelvic tilt; hip points are in line with each other; minimized stress on ligaments, muscles, and other tissues.
Osteoporosis Condition where bones become weak and brittle, leaving them at higher risk for fractures.
Parasympathetic nervous system (PSNS) "Rest and digest" branch of the autonomic nervous system; the relaxation response.
Peripheral nervous system (PNS) Includes the cranial and spinal nerves.
Physiology The study of the function of parts and systems in the body; the study of how the body works.
Postural hypotension Also called orthostatic hypotension; a sudden onset of low blood pressure caused by standing up too quickly from the floor or an inversion.
Prana Sanskrit word meaning life-force energy, vital energy, or breath, similar to the Chinese concept of qi; yogis believe you can consciously transform and move your prana.
Pranayama Sanskrit word meaning breath extension or control; breathwork.
Proprioception Type of interoception that focuses on spatial body awareness, particularly while in motion.
Randomized controlled trial (RCT) Randomization of the experimental group and control(s), which can lead to less bias; gold standard of research trials.
Sacroiliac joint Joint between the sacrum and ilium of the pelvis; allows a small amount of movement.
Samskaras According to Indian philosophy, these are imprints or impressions of our past actions.
Sanskrit The ancient Indian language in which many yoga texts were written.
Stretching When muscle fibers lengthen, often beyond resting length.
Sun salutation Series of asanas done in flowing sequence to warm up the body and focus the mind.
Superficial Closer to the surface; for instance, your skin is superficial to your muscles.
Supine Lying on your back, face up.
Sympathetic nervous system (SNS) "Fight or flight" branch of the autonomic nervous system; the stress response.
Synovial joint Most common and most mobile type of joint in the body, such as the shoulders, hips, and knees.
Tendon Connects muscle to bone; made of dense regular connective tissue proper, which has parallel collagen fibers.
Thoracic spine The 12 vertebrae of the mid-back region.
Tissues Collection of cells that come together for a similar function; the four primary tissue types are epithelial, connective, muscle, and nervous.
Vagus nerve Tenth cranial nerve (CN X), important in parasympathetic control of the heart, lungs, and digestive organs.
Vayus According to yoga philosophy, your prana flows in specific patterns called the vayus: Prana (in), Udana (into head), Vyana (into limbs), Samana (around), and Apana vayu (down and out).
Yoga therapy According to the International Association of Yoga Therapists, "Yoga therapy is the process of empowering individuals to progress toward improved health and well-being through the application of the teachings and practices of Yoga;" this developing field has educational standards that exceed those for general yoga instruction, and prepares practitioners to work safely with health conditions.

INDEX

A

abdominal aorta 34, 40
abdominal separation 156
abdominals 18, 66, 68, 76, 152, 153
abduction 16
Accomplished 46–49
acetabulum 15
Achilles tendon 126
acid reflux 202
adduction 16
adductors 18
Adho Mukha Svanasana 124–27
adrenal gland 28, 40
agility 114, 146, 182
aging
  and meditation 181
  and yoga 182–83
agitation 143, 190, 191, 192
agonist muscles 20, 21
ahimsa (non-harm) 202, 205
  diet 39
ajna chakra 201
alignment 178
allostasis 29
alpha waves 25, 185, 187
alternate nostril breathing 33
amygdala 25, 27, 192, 193
anahata chakra 201
analgesic effect 194
anatomy 10–41
Anjaneyasana 98–101
ankle dorsiflexors 18
ankle flexibility 50
ankle plantar flexors 19
ankylosing spondylitis 202
antagonist muscles 20, 21
anterior cingulate cortex 193
anterior superior iliac spine 15
antibodies 37
antigens 37
anus 38
anxiety 185, 202
aorta 34
aparigraha (nonpossession) 205
appendix 38
arches, foot 89
Ardha Matsyendrasana 68–71
Ardha Sarvangasana 132–35
arm strength 128–31, 154–57
arteries 135
artery wall 35
arthritis 92, 194, 196
  osteoarthritis 13, 17, 157, 202
  rheumatoid 17, 37, 202
asanas
  and chronic pain 195
  floor 146–73
  inversion 124–43
  seated 46–83
  standing 86–121
asteya (nonstealing/abundance) 205
asthma 202
astronauts 198
atoms 10
attention 189, 193
Attention Deficit Hyperactivity Disorder (ADHD) 180
autoimmune disease 37
autonomic nervous system (ANS) 22, 23, 189, 190
axial rotation 16
axillary nodes 36
axons 23, 27

B

B-cells 37
back
  care of 178–79
  lying flat on 187
  pain 80, 92, 136, 158, 162, 165, 194
  strength and flexibility 140, 161
backbends 76–83, 136–39, 140–43, 158–61
Baddha Konasana 50–53
Bakasana 146–49
balance 67, 96, 97, 106, 182
  poses 102–5, 106–9, 110–13, 114–17, 146–49, 154–57
  and pressure 97, 104, 130, 148
  three mechanisms of 108–9
Balasana 72–75
ball and socket joints 116
bandhas (energetic locks) 152–53
baroreceptors 32, 134
bee breath (*brahmari*) 32, 33
behavioral changes 26
belly breathing 31
belly-down boat pose 162–65
beta waves 185, 187
Bhujangasana 158–61
biopsychosocial-spiritual model 197
Bitilasana 56–59
bladder 40
bliss (*anandamaya*) 196
blood
  composition of 34
  and nutrients 38
  rush of 138
  and waste 40
blood cells 12, 34
blood flow, venous 135
blood pressure 35, 40, 96, 186, 203, 204
  lowering 132, 134
  regulating 130
blood vessel changes 63
body awareness 110
body-mind connection 189
bone density 120
bone growth 120
bone marrow 12
bone spurs 17
bones 12–13
  strengthening 118
  yoga and aging 183
Bound Angle 50–53
brachial plexus 22
brahmacharya (moderation) 205
brain
  and chronic pain 194, 195
  hemispheres 33
  inside 24–25
  and meditation 181
  and mental well-being 192–93
  nervous system 22, 23
  and perceived pain pathway 172
  plasticity 26
  and regulation of blood flow 138
  and spiritual experience 199
  understanding science of 200
  yoga and 25, 26, 192–93
brain-derived neurotrophic factor (BDNF) 25, 26

brainstem 25, 32, 192
brainwaves 187
breathing 30–33
  and concentration 113
  more efficient 128, 131
  and posture 88
breathwork 32–33, 191, 198
Bridge 136–39, 142
bunions 116
bursa 173
bursitis 173, 202

C

calcaneus 13, 126
calcium 12, 21
calmness 136, 143, 191
Camel 76–79
cancer 188, 197
capillaries 35
carbon dioxide 113
cardiac muscle 18
cardiovascular risk 79
cardiovascular system 34–35
carotid artery 34
carpal tunnel syndrome 202–3
carpals 12
cartilage 12, 13
  arthritis 17, 37
cartilaginous joints 16
Cat 54–55, 183
cataracts 203
caudate nucleus 25
cells 10
central nervous system 22
cerebellum 25
  activation 149
cerebral arteries 35
cerebral cortex 24, 26
cerebral veins 35
cervical extensors 19
cervical hyperextension 79
cervical spine 173
cervical vertebrae 14
cervix 41
Chair 94–97
chakras 201
change, response to 189
chanting 184, 185
children, yoga for 180
Child's pose 17, 72–75
cholesterol 35
chondrocytes 13
chromosomes 10
cingulate gyrus 25
circulation 35
classes, yoga 26
clavicle 12
closed chain movement 139
coordination 106
Cobra 158–61
coccyx 15
collagen 12, 19, 21
compassion 26
concentration 198
  breath and 113
concentric contraction 21
consciousness
  higher states of 191, 198
  levels of 187
  science of 200
constructive rest position 187
contraction, types of muscle 20
control groups 200
core strength 97, 118, 128, 150, 154
Corpse pose 186–87
corpus callosum 24
cortisol 25, 29
  yoga and levels of 27
Cow 56–59
Cow Face 60–63, 142
cramp 117
cranial nerves 22, 190
Crescent Lunge 98–101
Crow 146–49
cysterna chyli 36
cytokines 37

D

Dancer 17, 114–17
deep six external rotators 112
degenerative disk disease 203
dehydration 117
delta waves 187
deltoids 62
dendrite 23
dendritic cells 37
dermis 11
detached retina 203
detoxification 11, 71
dharana (concentration) 198
dhyana (meditation) 198
diabetes
  type 1 203
  type 2 28, 29, 203
diaphragm 30, 31, 41, 153, 156
  compressed 131
  three diaphragms 152
diastasis recti 156
digestion, stimulating 68, 70, 158, 164
digestive system 11, 38–39
disk degeneration 165
disk herniation 92, 203
disk stacking 48
disks 64, 66, 68, 70, 93, 179, 203
disease
  lifestyle-based chronic 197
  stress and chronic 188
  yoga research and 201
distress 189
dizziness 205
DNA 10
dopamine 25
dorsiflexion 16
"double jointed" 177
Downward-facing Dog 124–27
drishti (focal point) 108
dull state 143, 190, 191
dynamic balance 182

E

ear infections 203
Easy pose 46
eccentric contraction 20
edema 134
eight limb structure 25, 198
Eka Pada Rajakapotasana 80–83
elastic region 177
elbows 16
  flexors 18
  locking 157
electrolyte imbalance 117
elimination, stimulating 158
emotional brain 192
emotional states, regulating 192, 193
emulation of movement 26
endocrine system 24, 28–29
endometrium 41
endoneurium 23
energy 143, 150, 191
  flow of dormant 158
energy and breath (*pranamaya*) 196
energy centres (*chakras*) 201
energy channels (*nadis*) 201
enlightenment 198, 199
enteric nervous system (ENS) 39
epidermis 11
epiglottis 30, 38
epineurium 23
equanimity 182
equilibrium 108
erythrocytes 34
esophagus 38
eustress 188, 189
exhalation 31, 32, 59

extension 16, 20
external rotation 16
eye conditions 203

F

fallopian tubes 41
falls 182, 183
fascia 20, 21, 64
fascicles 20
fatigue 185
fear 27, 192, 193
feet 89
femoral artery 34
femoral nerve 22
femoral vein 34
femur 13
  differences 52
fetal position 72, 75
fibromyalgia 203
fibrous joints 16
fibula 13
fibular nerve 22
fight-or-flight response 190, 192
fine motor skills 27
flexibility 90, 114, 146, 173–74, 182
flexion 16, 21
flexion dominance 79
floor asanas 146–73
flow sequences 124
flow state 185
focal point 108
focus 106, 185
food, digestive system 38
foot arches 89
fornix 25
Forward Fold 90–93, 179
fractures 183
friction 173
frontal cortex 192, 193, 199
frontal lobe 24
frozen shoulder 203

G

gallbladder 38
gamma waves 187
Gamma-Aminobutyric acid 25
gate theory of pain 172
genes 10
glands 28–29
glaucoma 203
gluteals 19, 98, 139
glycemic control 29
Golgi tendon organ 169
Gomukhasana 60–63
gravity 86, 118
gray matter
  deterioration 195
  and mindfulness 185
great saphenous vein 34
greater sciatic notch 15
greenhouse gasses 39
groin stretch 50
groundedness 170
gunas 190–91
gut brain 39

H

habits 192
hair 11
Half Shoulderstand 132–35
hamstrings 139
  tight 101, 126
hand-specific sensory areas 27
head, relaxing 204
Headstand 128–31
health conditions, and yoga 202–5
healthcare systems 197
healthspan 182
heart 28, 32, 34, 35
  circulation 35
  pounding 143, 154
heart disease 35, 188, 201, 203
heart openers 138
heart rate variability (HRV) 35, 186
heartburn 202
heel 126
herniated disks 92, 203
Hero pose 185
hip abductors 112
hip extensors 19
hip flexors 18, 47, 51, 98, 101
hip joints 116
hip replacement 203
hip stretches 50–53
hippocampus 25, 27
hips 15
homeostasis 28, 29, 134, 189
hormones 28–29
hot yoga 11, 177
human body 11
humerus 13
hyperextension
  of elbow 157
  of knee 121, 203
  of lumbar region 143
  of neck 57, 79, 160
hyperkyphosis 14, 161
hypermobility 127, 177, 203
hypertension 35, 203
hyperventilation 32
hypotension 204
hypothalamus 24, 28

I

immune system 36
impulsivity 180
incontinence 40
independence, preserving 182
inferior vena cava 34, 40
inflammation 36, 37
inguinal nodes 36
inhalation 31, 32, 59
injuries, yoga 202
inner ear, and balance 106, 108
inner thighs 50
instinctual brain 192
insula 24
insulin 29
integumentary system 11
intercostal muscles 18, 74
internal rotation 16
interoception 189, 191
intervertebral disks 64, 70, 92, 165, 172
inversion asanas 124–43
ischial tuberosity 15
ishvara pranidhana (surrendering/accepting) 205
islets 29
isometric contractions 20

J

Jalandhara bandha 153
joint stabilization 110
joints 12, 16–17
  ball and socket 116
  bursitis 173
  and flexibility 176–77
  inflamed 37
  "popping" 176
jugular vein 34

K

Kabat-Zinn, Jon 184
kapalabhati (breath of fire) 32
karma (action) 26
Khumbhakasana 150–53
kidneys 40
kinetic chains 139
King Pigeon 80–83
knee alignment 105
knee hyperextension 121
knee ligaments 105
  injury 203
knee meniscus tear 204
knee replacement 204
koshas, five 196
kumbhaka (breath holding) 32, 153
kyphosis 14, 161

L

lactic acid 187
large intestine 38
larynx 30
Lazar, Sara 181

learning 185
Lee, Michael 198
left nostril breathing 32
leukocytes 34, 37
life stages 180–83
lifestyle changes 196, 197
lifestyle choices 188
ligaments 12, 13, 116, 177
  knee 105
limbic system 25, 192, 199
liver 11, 38, 187
lobes of the brain 24
Locust 162–65, 183
longevity 94, 96, 149
lordosis 14
low squat 149
lumbar load 92, 96, 97
lumbar plexus 22
lumbar spine 143, 173
lumbar vertebrae 14
lung capacity 88
lungs 30, 31
lymph nodes 36
lymph vessels 36
lymphatic drainage 134
lymphatic system 36–37
lymphocytes 36

M

M-line 20, 21
macrophages 37
Malasana 149
mandible 12
manipura chakra 201
Marjaryasana 54–55
Meatless Monday 39
median nerve 22
meditation 26, 29, 184–85, 189, 198
  and brain activity 192
  and brain aging 181
  for children 180
  and chronic pain 194, 195
  and Savasana 186
memory 27, 185
menstruation 130
mental well-being, brain and 192–93
meta-analysis 200
metabolism 29, 177
metacarpals 12
metatarsals 12
migraine 204
milk ducts 41
mind and emotions (*manomaya*) 196
mind-body connection 58, 118, 188, 189, 194
mind-body therapy 180
mindfulness
  and aging 181, 182
  and brain activity 192, 193
  and chronic pain 194
  and meditation 184, 185
Mindfulness-Based Stress Reduction (MBSR) program 184, 185
mindset shifts 196
mineral density 120
mirror neurons 26
mobility 116
mood, regulating 192
motor neurons 169
motor signals 169
Mountain 86–89
mouth 30, 38
movement
  closed chain 139
  and fascia 21
  mindful 184, 189
  spinal 173
  types of 16
MRI scans 181, 185, 195
mudras 27, 189, 191
mula bandha 40, 153
muladhara chakra 201
multiple sclerosis 204
muscle cramps 117
muscle fibers 20
muscle memory 149
muscle spindles 169
muscles
  building 116
  contraction 21
  damage 177
  progressive muscle relaxation 186
  skeletal 18–19
  sore 187
  structure 20
  working of 20–21
muscular strength 102, 110, 116, 118, 120
muscular system 18–21
myelin 23
myofibrils 20

N

nadis 201
nasal cavity 30
nasal cycle 33
Natarajasana 114–15
neck 178
  problems 160
  relaxing 74
negative emotions 188, 191
nerve signals 23
nerve structure 23
nervous system 22–27, 64, 132
  and breath 30
  and muscle contraction 21
neti pots 30
neural pathways 26, 189, 192, 195
neurocognitive pathway 189
neurogenesis 27
neuroglia 27
neuroimaging 199
neuromuscular fatigue 117
neurons 23, 24, 26, 27
neurophysiological pathway 189
neuroplasticity 26, 181, 192
neutral pelvis 15, 50
neutral spine 14, 46, 50, 166, 179
Newberg, Andrew 199
nidra 186, 187
nipples 41
niyamas 198, 205
nociceptive signals 172
non-judgmentalism 191
norepinephrine 25
nose 31
nostrils 30
numbness 202
nutrients 38, 116

O

obesity 204
occipital lobe 24
olfactory bulb 25
"om" 184, 185
organisms 11
organs 11
osteoblasts 120
osteopenia 92, 204
osteophytes 17
osteoporosis 68, 92, 179, 183, 204
ovaries 28, 41
overstretching 177
oxygen 30, 35

P

pain
  chronic 194–95, 196, 197
  perceived pain pathway 172
  relief 173
pancreas 28, 29
panic attacks 202
parasympathetic nervous system (PSNS) 22, 23, 33, 170, 186
parathyroid gland 28
parietal lobe 24
Parivrtta Janu Sirsasana 64–67
Parkinson's disease 204
patella 123
pathogenesis 197
pectorals 18
pelvic alignment 113

pelvic floor diaphragm 152
pelvic floor muscles 40, 41, 150, 153
pelvis 12, 15
  differences 53
  softening 53
penis 41
perceived pain pathway 172
perception 193
perineurium 23
peripheral nervous system (PNS) 22
peristalsis 39, 70, 71
phalanges 12
pharynx 38
physical well-being (*annamaya*) 196
pineal gland 25, 28
piriformis 82
pituitary gland 28
placebos 194
Plank 150–53
plantar fascia 20
plantar fasciitis 204
plantar rotation 16
plasma 34
plastic region 177
plasticity 26
platelets 34
pleasure signals 172
Polyvagal Theory 190
pons 25
popliteal artery 34
Porges, Stephen 190
positive behaviors, reinforcing 192
positive mindset 181
posterior parietal lobe 199
postural alignment 86
posture 178
  and breathing 88
  counteracting poor 162
prakriti 190
prana 32, 201
pranayama 32, 189, 192, 195, 198
pratyahara (controlled senses) 198
prefrontal cortex 26
pregnancy
  cautions 156, 204
  yoga during 181
present moment 184, 185
pressure, and balance 97, 104, 130, 148
pressure points 67, 75, 89, 109, 120
primary somatosensory cortex 195
problem-solving 181, 185
progressive muscle relaxation 186
proprioception 59, 106, 109
prostate gland 40
psoas major 100
psoas minor 164
pubic symphysis 15
pulmonary artery 35
pulmonary vein 35
Purusha 191
push-ups 150
putamen 25

Q

Q-angle 88
quadratus lumborum 66
quadriceps 18, 96

R

radius 13
rajasic state 190, 191, 192
randomized controlled trials 200
range of motion (ROM) 63, 176, 182
reciprocal inhibition (RI) 169
rectum 38
red blood cells 34
relaxation 72–75, 185, 186, 190
relaxin 53
release 169
reproductive system 41
resilience 189, 191
respiratory diaphragm 152
respiratory muscles 156
respiratory system 30–33
rest and digest response 23, 170
rheumatoid arthritis 37
rhomboids 19, 49
ribcage movement 48
ribs 12, 31
right nostril breathing 32
rotation 16
rotator cuff
  injuries 204
  strength 131
routine, changing 26
running 98

S

sacral plexus 22
sacroiliac dysfunction/pain 204
sacroiliac joint 15, 71
sacrum 13, 15
sahasrara chakra 201
Salabhasana 162–65
salivary glands 38
salt water 30
salutogenesis 197
samadhi (enlightenment) 198, 199
sample size 200
samskaras (impressions) 26
santosha (contentment) 205
sarcomeres 20, 21
sattvic state 190, 191
satya (truthfulness) 205
saucha (cleanliness) 205
Savasana 185, 186–87
scalenes 156
scapula 13
sciatic nerve 22, 83
sciatica 80, 204
scoliosis 156, 157, 205
seated asanas 46–83
Seated Twist 68–71
self-actualization 197
self-awareness 193
self-control 25, 198
self-realization 191, 197
self-regulation 25, 189, 195, 198, 205
Selye, Hans 189
sensory awareness 195
serotonin 25, 39
serratus anterior 160, 161
Setu Bandhasana 136–39
sex hormones 28
sexual reproduction 41
shoulder dislocation 205
shoulder flexion 142
shoulder joints 116, 128, 130
shoulder stretches 60–63, 170
shoulders, tight 142
Shoulderstand, Half 132–35
shutdown vagus 190
Siddhasana 46–49
Side Bend 64–67
Side Plank 154–57
sinuses 30
sinusitis 205
Sirsasana 128–31
sitting, for meditation 185
skeletal muscles 18–19, 116
skeletal system 12–13
skin 11
skull 12, 30
sleep 87
  preparation for 136
small intestine 28, 38
smell 27
smooth muscle 18
social vagus 190, 191
somatic nervous system 22
somatic practices 193
sperm 41
Sphinx pose 160
spinal cord 22, 23, 172, 178
spinal disks 66, 68, 78, 93
spinal extensors 19
spinal flexibility 100, 127, 143
spinal stenosis 205
spinal twists 68–71, 170–73

spine 14
elongating 49, 162
flexion and extension 58, 78, 92, 143, 173
lateral movement 173
motion 173
rotating 68, 70, 170–73
safety 172
scoliosis 156, 157, 205
spinal care 178–79
warming up 54, 56
spirituality 198–99
spleen 36
spondylolisthesis 205
stability 116, 128, 157
standing asanas 86–121
static balance 182
stem cells 27
sternum 12
stimulation, brain 26
stomach 38, 39
strength 146, 150, 182
stress 188–91
and health 188
managing 188, 189
and memory 27
and mindfulness 184
types of 189
and yoga 188–91
stress hormones 25, 29
stress-strain curve 177
stretch reflex 169
striations 18
striatum 193
stroke 79, 188, 205
subclavian artery 34
Sukhasana 46
sulci 24
sun salutations 26, 72, 124
Supine Leg Stretch 166–69
Supine Twist 170–73
suprarenal gland 28
Supta Matsyendrasana 170–73
Supta Padangusthasana 166–69
sushuma nadi 201
svadhisthana chakra 201
svadhyaya (self study) 205
sweat 11, 154
sympathetic nervous system (SNS) 22, 23, 32, 33, 190
synapses 24
synergist muscles 20
synovial fluid 16, 176
synovial joints 16, 176
systems 11

T

T-cells 37
Tadasana 86–89
tamasic state 190
tapas (self-discipline) 205
tarsals 12
teeth 38
telomeres 10
temporal lobe 24
tendonitis 202, 204
tendons 19, 177
tension, release of 193
testes 28, 41
testosterone 41
texting 178
thalamus 24, 199
theta waves 185, 187
thick filament 20, 21
thigh strength 94, 106, 120
thigh stretches 50–53, 76–79, 166–69, 170
thin filament 20, 21
thinking brain 192
thoracic duct 36
thoracic outlet diaphragm 152
thoracic spine 143, 173
thoracic vertebrae 14
thoughts
observing 184, 185
stopping 185
three diaphragms 152
throat 31
thyroid 29
stimulation of 135
tibia 13
tibial nerve 22
tissue 10
toe joints 116
tonsils 36
toxins, wringing out 70, 71
trachea 30
transformation 198–99, 200
transversus thoracis 156
trapezius 100
trauma 193, 197
Tree 110–13
Triangle 16, 20, 118–21
Trikonasana 118–21
Twisted Triangle 120
twists
seated 68–71
spinal 170–73
typing 178

U

Uddiyana bandha 153
ulna 13
ulnar nerve 22
upper back strength 161
Upward-facing Dog 160
Urdhva Dhanurasana 140–43
ureter 40
urethra 40
urinary system 40
Ustrasana 76–79
uterus 41
Utkatasana 94–97
Uttanasana 90–93

V

vagina 41
vagus nerve 32, 189, 190, 191
valves, vein 35, 135
Vasisthasana 154–57
vasoconstriction 33, 63
vasodilation 33, 63
vayus 196
vegetarianism 39
veins 35, 135
venous blood flow 135
vertebrae 14, 22, 173, 179
vertebral column 13
vertigo 205
victorious breath (*ujjayi*) 33
Virabhadasana II 102–5
Virabhadasana III 106–9
vishuddha chakra 201
visual input, and balance 108
vocal cords 33
vritti (thought pattern) 26
Vrksasana 110–13

W

Warrior II 20, 21, 102–5, 183
Warrior III 106–9
waste 40
Wheel 140–43
white blood cells 34, 37
wisdom (*vijnanamaya*) 196
wrists, strengthening 148

Y

yamas (self-control) 198, 205
yoga research 200–201
"Yoga Sutras" 198
yoga therapy 196–97

Z

Z-disks 20, 21

BIBLIOGRAPHY

10–11 R. Chaix et al., "Epigenetic clock analysis in long-term meditators," *Psychoneuroendocrino* 85 (2017); E. Epel et al., "Can Meditation Slow Rate of Cellular Aging? Cognitive Stress, Mindfulness, and Telomeres," *Ann NY Acad Sci* 1172 (2009); D. Ornish et al., "Effect of comprehensive lifestyle changes on telomerase activity and telomere length in men with biopsy-proven low-risk prostate cancer: 5-year follow-up of a descriptive pilot study," *Lancet Oncol* 14 (2013). **12–17** S. H. Moonaz et al., "Yoga in Sedentary Adults with Arthritis: Effects of a Randomized Controlled Pragmatic Trial," *J Rheumatol* 42 (2015); S. Muraki et al., "Quadriceps muscle strength, radiographic knee osteoarthritis and knee pain: the ROAD study," *BMC Musculoskel Dis* 16 (2015); M. Wallden, "The neutral spine principle," *J Bodywork Movement Ther* 13 (2009). **18–21** T. W. Myers, *Anatomy trains* (3rd ed.), Edinburgh, Churchill Livingstone/Elsevier, 2014. **22–27** M. Balasubramaniam et al., "Yoga on our minds: a systematic review of yoga for neuropsychiatric disorders," *Front Psychiat* 3 (2013); B. Rael Cahn et al., "Yoga, Meditation and Mind-Body Health: Increased BDNF, Cortisol Awakening Response, and Altered Inflammatory Marker Expression after a 3 Month Yoga and Meditation Retreat," *Front Hum Neurosci* 11 (2017); R. A. Gotink et al., "Meditation and yoga practice are associated with smaller right amygdala volume: the Rotterdam study," *Brain Imaging Behav* (2018); B. K. Hölzel et al., "Mindfulness practice leads to increases in regional brain gray matter density," *Psychiat Res Neuroim* 191 (2011); D. E. Larson-Meyer, "A Systematic Review of the Energy Cost and Metabolic Intensity of Yoga," *Med Sci Sport Exer* 48 (2016). **28–29** M. Á. D. Danucalov et al., "Cardiorespiratory and Metabolic Changes during Yoga Sessions: The Effects of Respiratory Exercises and Meditation Practices," *Appl Psychophys Biof* 33 (2008); K. E. Innes and T. K. Selfe, "Yoga for Adults with Type 2 Diabetes: A Systematic Review of Controlled Trials," *J Diabetes Res* 2016 (2016); C. C. Streeter et al., "Effects of yoga on the autonomic nervous system, gamma-aminobutyric-acid, and allostasis in epilepsy, depression, and post-traumatic stress disorder," *Med Hypotheses* 78 (2012). **30–33** S. Telles et al., "Blood Pressure and Heart Rate Variability during Yoga-Based Alternate Nostril Breathing Practice and Breath Awareness," *Med Sci Monitor Basic Res* 20 (2014); M. Joshi and S. Telles, "Immediate effects of right and left nostril breathing on verbal and spatial scores," *Indian J Physiol Pharmacol* 52 (2008); R. Kahana-Zweig et al., "Measuring and Characterizing the Human Nasal Cycle," *PLoS ONE* 11 (2016); M. Kuppusamy et al., "Effects of Bhramari Pranayama on health – A systematic review," *J Trad Complem Med* 8 (2018); D. S. Shannahoff-Khalsa et al., "Ultradian rhythms of autonomic, cardiovascular, and neuroendocrine systems are related in humans," *Am J Physiol* 270 (1996); A. N. Sinha et al., "Assessment of the effects of pranayama/alternate nostril breathing on the parasympathetic nervous system in young adults," *J Clin Diag Res* 7 (2013); G. Yadav and P. K. Mutha, "Deep Breathing Practice Facilitates Retention of Newly Learned Motor Skills," *Sci Rep* 6 (2016); F. Yasuma and J. Hayano, "Respiratory Sinus Arrhythmia," CHEST 125 (2004). **34–35** World Health Organization, "Cardiovascular diseases (CVDs)," *World Health Organization* [web article], 17 May 2017, (accessed 20 Aug 2018); H. Cramer et al., "Effects of yoga on cardiovascular disease risk factors: A systematic review and meta-analysis," *Int J Cardiol* 173 (2014); K. E. Innes et al., "Risk Indices Associated with the Insulin Resistance Syndrome, Cardiovascular Disease, and Possible Protection with Yoga: A Systematic Review," *J Am Board Fam Med* 18 (2005); D. Ornish et al., "Can lifestyle changes reverse coronary heart disease? The Lifestyle Heart Trial," *Lancet* 336 (1990). **36–37** Harvard Health Letter, "Inflammation: A unifying theory of disease," *Harvard Health Publishing* [web article], Apr 2006, (accessed 20 Aug 2018); R. I. Falkenberg et al., "Yoga and immune system functioning: a systematic review of randomized controlled trials," *J Behav Med* 41 (2018); T. Oka et al., "Changes in fatigue, autonomic functions, and blood biomarkers due to sitting isometric yoga in patients with chronic fatigue syndrome," *BioPsychoSocial Med* 12 (2018). **38–39** M. Berners-Lee et al., "The relative greenhouse gas impacts of realistic dietary choices," Energy Policy 43 (2012); H. C. J. Godfray et al., "Food Security: The Challenge of Feeding 9 Billion People," *Science* 327 (2010); M. Springmann et al., "Analysis and valuation of the health and climate change cobenefits of dietary change," *P Natl A Sci* 113 (2016); D. Tilman and M. Clark, "Global diets link environmental sustainability and human health," *Nature* 515 (2014). **40–41** S. Prosko, "Optimizing Pelvic Floor Health Through Yoga Therapy," *Yoga Ther Today*, 12 (2016); A. Huang et al., "PD32-01 A Randomized Trial of a Group-Based Therapeutic Yoga Program for Ambulatory Women With Urinary Incontinence," *J Urology* 199 (2018). **46–49** T. W. Myers, *Anatomy Trains* (3rd ed.), Edinburgh, Churchill Livingstone/Elsevier, 2014. **50–53** F. Dehghan et al., "The effect of relaxin on the musculoskeletal system," *Scand J Med Sci Spor* 24 (2013). **60–63** J. M. M. Brown et al., "Muscles within muscles: Coordination of 19 muscle segments within three shoulder muscles during isometric motor tasks," *J Electromyogr Kines* 17 (2007); H. Mason, "Learning to Abide with What Is: The Science of Holding Poses," *Yoga Ther Today* 13 (2017). **76–79** E. J. Benjamin et al., "Heart Disease and Stroke Statistics—2018 Update: A Report From the American Heart Association," *Circulation* 137 (2018); P. Page et al., *Assessment and Treatment of Muscle Imbalance: The Janda Approach*, Champaign (IL), Human Kinetics, 2010; K. W. Park et al., "Vertebral Artery Dissection: Natural History, Clinical Features and Therapeutic Considerations," *J Korean Neurosurg S* 44 (2008). **94–97** L. B. De Brito et al., "Ability to sit and rise from the floor as a predictor of all-cause mortality," *Eur J Prev Cardiol* 21 (2014); A. B. Newman et al., "Strength, but not muscle mass, is associated with mortality in the health, aging and body composition study cohort," *J Gerontol A-Biol* 61 (2006). **102–105** J. L. Oschman et al., "The effects of grounding (earthing) on inflammation, the immune response, wound healing, and prevention and treatment of chronic inflammatory and autoimmune diseases," *J Inflamm Res* 2015 (2015). **118–121** Y. H. Lu et al., "Twelve-Minute Daily Yoga Regimen Reverses Osteoporotic Bone Loss," *Top Geriatr Rehabil* 32 (2016). **128–131** L. M. Fishman et al., "Yoga-Based Maneuver Effectively Treats Rotator Cuff Syndrome," *Top Geriatr Rehabil* 27 (2011); R. Hector and J. L. Jensen, "Sirsasana (headstand) technique alters head/neck loading: Considerations for safety," *J Bodywork Movement Ther* 19 (2015); T. B. McCall, *Yoga as Medicine: The Yogic Prescription for Health and Healing*, New York, Bantam, 2007. **132–135** M. Robin, *A 21st-Century Yogasanalia: Celebrating the Integration of Yoga, Science, and Medicine*, Tucson (AZ), Wheatmark Inc., 2017. **136–139** P. Page et al., *Assessment and Treatment of Muscle Imbalance: The Janda Approach*, Champaign (IL), Human Kinetics, 2010. **146–149** L. B. De Brito et al., "Ability to sit and rise from the floor as a predictor of all-cause mortality," *Eur J Prev Cardiol* 21 (2014); R. T. Proyer, "The well-being of playful adults: Adult playfulness, subjective well-being, physical well-being, and the pursuit of enjoyable activities," *Eur J Humour Res* 1 (2013); United Nations, "Convention on the Rights of the Child," 2 Sep 1990, (accessed 11 Aug 2018). **150–153** D. Frownfelter and E. Dean, *Cardiovascular and Pulmonary Physical Therapy: Evidence to Practice* (4th ed.), St Louis, Elsevier Health Sciences, 2005. **154–157** L. M. Fishman et al., "Serial Case Reporting Yoga for Idiopathic and Degenerative Scoliosis," *Glob Adv Health Med* 3 (2014). **162–165** B. Duthey, "Background Paper 6.24 Low back pain," Priority Medicines for Europe and the World, World Health Organization, 2013; Society of Behavioral Medicine, "Yoga Shown to be Cost-Effective for Chronic Back Pain Management," *PR Web*, [web article], 13 Apr 2018, (accessed 17 Sep 2018). **166–169** H. Mason, "Learning to Abide with What Is: The Science of Holding Poses," *Yoga Ther Today* 13 (2017); W. D. Bandy and J. M. Irion, "The effect of time on static stretch on the flexibility of the hamstring muscles," *Phys Ther* 74 (1994). **170–173** J. Hamill and K. M. Knutzen, *Biomechanical Basis of Human Movement* (2nd ed.), Philadelphia, Wolters Kluwer Health, 2003. **176–177** K. deWeber et al., "Knuckle Cracking and Hand Osteoarthritis," *J Am Board Fam Med* 24 (2011); A. Guillot et al., "Does motor imagery enhance stretching and flexibility?," *J Sport Sci* 28 (2010); A. J. Hakim and R. Grahame, "A simple questionnaire to detect hypermobility: an adjunct to the assessment of patients with diffuse musculoskeletal pain," *Int J Clin Pract* 57 (2003); G. N. Kawchuk et al., "Real-Time Visualization of Joint Cavitation," *PLoS ONE* 10 (2015); V. K. Ranganathan et al., "From mental power to muscle power – gaining strength by using the mind," *Neuropsychologia* 42 (2004); D. Syx et al., "Hypermobility, the Ehlers-Danlos syndromes and chronic pain," *Clin Exp Rheumatol* 35 (2017). **178–179** R. Chaix et al., "Epigenetic clock analysis in long-term meditators," *Psychoneuroendocrinol* 85 (2017); L.-H. Chuang et al., "A Pragmatic Multicentered Randomized Controlled Trial of Yoga for Chronic Low Back Pain: Economic Evaluation," *Spine* 37 (2012); K. K. Hansraj, "Assessment of stresses in the cervical spine caused by posture and position of the head," *Surg Tech Int* 25 (2014); Society of Behavioral Medicine, "Yoga Shown to be Cost-Effective for Chronic Back Pain Management," *PR Web*, [web article], 13 Apr 2018, (accessed 17 Sep 2018). **180–183** B. P. Acevedo et al., "The Neural Mechanisms of Meditative Practices: Novel Approaches for Healthy Aging", *Curr Behav Neurosci Reports* 3 (2016); R. F. Afonso et al., "Greater Cortical Thickness in Elderly Female Yoga Practitioners – A Cross-Sectional Study," *Front Aging Neurosci* 9 (2017); B. Bell and N. Zolotow, *Yoga for Healthy Aging: A Guide to Lifelong Well-Being*, Boulder, CO, Shambhala, 2017; A. J. Cerrillo-Urbina et al., "The effects of physical exercise in children with attention deficit hyperactivity disorder: a systematic review and meta-analysis of randomized control trials," *Child Care Hlth Dev* 41 (2015); B. Chethana et al., "Prenatal Yoga: Effects on Alleviation of Labor Pain and Birth Outcomes," *J Altern Complem Med* (2018); A. Herbert and A. Esparham, "Mind–Body Therapy for Children with Attention-Deficit/Hyperactivity Disorder," *Children* 4 (2017); Q. Jiang et al., "Effects of Yoga Intervention during Pregnancy: A Review for Current Status," *Am J Perinatol* 32 (2015); S. B. S. Khalsa and B. Butzer, "Yoga in school settings: a research review," *Ann NY Acad Sci* 1373 (2016); S. W. Lazar et al., "Meditation experience is associated with increased cortical thickness," *NeuroReport* 16 (2005); P. J. Reis and M. R. Alligood, "Prenatal Yoga in Late Pregnancy and Optimism, Power, and Well-Being," *Nurs Sci Quart* 27 (2014); M. Y. Wang et al., "Physical-Performance Outcomes and Biomechanical Correlates from the 32-Week Yoga Empowers Seniors Study," *Evid-Based Compl Alt* 2016 (2016). **184–185** B. K. Hölzel et al., "Mindfulness practice leads to increases in regional brain gray matter density," *Psychiat Res-Neuroim* 191 (2011); B. G. Kalyani et al., "Neurohemodynamic correlates of 'OM' chanting: A pilot functional magnetic resonance imaging study," *Int J Yoga* 4 (2011); K. Katahira et al., "EEG Correlates of the Flow State: A Combination of Increased Frontal Theta and Moderate Frontocentral Alpha Rhythm in the Mental Arithmetic Task," *Front Psychol* 9 (2018); F. Zeidan et al., "Mindfulness meditation improves cognition: Evidence of brief mental training," *Conscious Cogn* 19 (2010). **186–187** R. Anderson et al., "Using Yoga Nidra to Improve Stress in Psychiatric Nurses in a Pilot Study", *J Altern Complem Med* 23 (2017); H. Eastman-Mueller et al., "iRest yoga-nidra on the college campus: changes in stress, depression, worry, and mindfulness," *Int J Yoga Ther* 23 (2013); S. A. Gutman et al., "Comparative Effectiveness of Three Occupational Therapy Sleep Interventions: A Randomized Controlled Study," *OTJR-Occup Part Heal* 37 (2016); M. M. Hall et al., "Lactate: Friend or Foe," *Am Acad Phys Med Rehabil* 8 (2016); M. S. McCallie et al., "Progressive Muscle Relaxation," *J Hum Behav Soc Envir* 13 (2008); T. H. Nassif et al., "Mindfulness meditation and chronic pain management in Iraq and Afghanistan veterans with traumatic brain injury: A pilot study," *Milit Behav Heal* 4 (2016). **188–189** A. Ross et al., "National survey of yoga practitioners: Mental and physical health benefits", *Complement Ther Med* 21 (2013); M. B. Sullivan et al., "Yoga Therapy and Polyvagal Theory: The Convergence of Traditional Wisdom and Contemporary Neuroscience for Self-Regulation and Resilience," *Front Hum Neurosci* 12 (2018); S. Szabo et al., "'Stress' is 80 Years Old: From Hans Selye Original Paper in 1936 to Recent Advances in GI Ulceration," *Curr Pharm Des* 23 (2017); R. M. Yerkes and J. D. Dodson, "The relation of strength of stimulus to rapidity of habit-formation," *J Comp Neurol Psychol* 18 (1908). **192–193** R. A. Gotink et al., "Meditation and yoga practice are associated with smaller right amygdala volume: the Rotterdam study," *Brain Imaging Behav* (2018); P. A. Levine, *In an Unspoken Voice: How the Body Releases Trauma and Restores Goodness*, Berkeley (CA), North Atlantic Books, 2010; K. Nila et al., "Mindfulness-based stress reduction (MBSR) enhances distress tolerance and resilience through changes in mindfulness," *Ment Health Prev* 4 (2016); P. Payne et al., "Somatic experiencing: using interoception and proprioception as core elements of trauma therapy," *Front Psychol* 6 (2015); Y.-Y. Tang et al., "The neuroscience of mindfulness meditation," *Nat Rev Neurosci* 16 (2015). **194–195** M. C. Bushnell et al., "Cognitive and emotional control of pain and its disruption in chronic pain," *Nat Rev Neurosci* 14 (2015); E. J. Groessl et al., "Yoga for Military Veterans with Chronic Low Back Pain: A Randomized Clinical Trial," *Am J Prev Med* 53 (2017); G. L. Moseley and D. S. Butler, "Fifteen Years of Explaining Pain: The Past, Present, and Future", *J Pain* 16 (2015); N. Vallath, "Perspectives on Yoga inputs in the management of chronic pain," *Indian J Palliative Care* 16 (2010); F. Zeidan et al., "Mindfulness Meditation-Based Pain Relief Employs Different Neural Mechanisms Than Placebo and Sham Mindfulness Meditation-Induced Analgesia," *J Neurosci* 35 (2015); F. Zeidan et al., "The Effects of Brief Mindfulness Meditation Training on Experimentally Induced Pain," *J Pain* 11 (2010); F. Zeidan et al., "Brain Mechanisms Supporting Modulation of Pain by Mindfulness Meditation," *J Neurosci* 31 (2011). **196–197** International Association of Yoga Therapists, "Educational Standards for the Training of Yoga Therapists," *IAYT*, [web article], 2012, (accessed 10 Sep 2018); W. B. Jonas et al., "Salutogenesis: The Defining Concept for a New Healthcare System," *Global Adv Health Med* 3 (2014); International Association of Yoga Therapists, "Introduction to the IAYT Scope of Practice," *IAYT*, [web article], 2016, (accessed 10 Sep 2018); M. J. Taylor and T. McCall, "Implementation of Yoga Therapy into U.S. Healthcare Systems," *Int J Yoga Ther* 27 (2017). **198–199** C. L. Park et al., "Why practice yoga? Practitioners' motivations for adopting and maintaining yoga practice," *J Health Psychol* 21 (2014); M. T. Quilty et al., "Yoga in the Real World: Perceptions, Motivators, Barriers, and Patterns of Use," *Global Adv Health Med* 2 (2013); D. B. Yaden et al., "The overview effect: Awe and self-transcendent experience in space flight," *Psychol Consciousness* 3 (2016); A. B. Newberg, "The neuroscientific study of spiritual practices," *Front Psychol* 5 (2014). **200–201** M. Hagins and S. B. Khalsa, "Bridging yoga therapy and scientific research," *Int J Yoga Ther* 22 (2012); P. E. Jeter et al., "Yoga as a Therapeutic Intervention: A Bibliometric Analysis of Published Research Studies from 1967 to 2013," *J Altern Complem Med* 21 (2015). **202–203** H. Cramer et al., "The Safety of Yoga: A Systematic Review and Meta-Analysis of Randomized Controlled Trials," *Am J Epidemiol* 182 (2015).

Research on yoga is constantly evolving. For updated resources go to: **www.scienceof.yoga**

ABOUT THE AUTHOR

Ann Swanson, MS, C-IAYT, LMT, E-RYT500, is a mind-body science educator. She holds a Master of Science in yoga therapy from Maryland University of Integrative Health, where she went on to become an adjunct faculty member. With years of experience tutoring and teaching anatomy and physiology in colleges, massage therapy schools, and yoga teacher training programs, she has refined the ability to make complex scientific concepts simple to understand. Ann uniquely applies cutting-edge research practically to yoga while maintaining the heart of the tradition. In her private practice, she makes yoga therapy, qigong, and mindfulness meditation accessible and convenient in an online format, helping people worldwide manage pain and stress effectively.

For more about Ann, head to **www.AnnSwansonWellness.com**

ACKNOWLEDGMENTS

Author acknowledgments

Gratitude to my mentors and colleagues: Yogi Sivadas and Alice from Kailash Tribal School of Yoga in India, Yang Yan and Mahendra from Yogi Yogi in China, John Pace, Steffany Moonaz, Marlysa Sullivan, Laurie Hyland Robertson, and Michel Slover. Gratitude to my family: Mom, Dad, Joe, Aunt Sandy, and Pop. Gratitude to the brilliant DK team: Ruth, Clare, Arran, and everyone else.

Publisher acknowledgments

DK would like to thank Rebecca Fallowfield and Luca Frassinetti for production, Alison Gardner and India Wilson for design assistance, John Friend for proofreading, and Helen Peters for the index.

Picture credits

The publisher would like to thank the following for their kind permission to reproduce their photographs:

(Key: a-above; b-below/bottom; c-center; f-far; l-left; r-right; t-top)

13 Science Photo Library: Biophoto Associates (cla). **18 Science Photo Library:** Professors P.M. Motta, P.M. Andrews, K.R. Porter & J. Vial (clb). **27 Science Photo Library:** Thomas Deerinck, Ncmir (cl). **32 Science Photo Library:** Zephyr (bl). **33 Science Photo Library:** Zephyr (cla). **37 Science Photo Library:** (clb)

All other images **© Dorling Kindersley**
For further information see: **www.dkimages.com**